MICROBIAL ECOLOGY OF FOODS

VOLUME I

Factors Affecting Life and Death of Microorganisms

90 0587581 1

4 Reduced Water Activity

5 pH and Acidity

6 Oxidation–Reduction Potential

7 Organic Acids

8 Curing Salts and Related Materials

9 Antibiotics in Foods

10 Gases as Preservatives

11 Packaging

12 Injury and Its Effect on Survival and Recovery

13 Interacting Factors Affecting Mixed Populations

14 Cleaning, Disinfection, and Hygiene

C. Definitions

Four major physiological groups of bacteria may be distinguished by their temperature ranges of growth: thermophiles, mesophiles, psychrophiles, and psychrotrophs (Table 1.1). *Mesophiles,* many of animal or human origin—including all pathogens and many food spoilage types, prefer moderate temperatures with an optimum generally between 30° and 45°C and a minimum growth temperature ranging from about 5° to 10°C. The generation time for many mesophilic types in a favorable medium and at optimum temperature may be 0.5 hr or less. With *thermophiles* the whole curve relating growth rate to temperature is shifted to a higher range. Optimum temperature for growth usually is between 55° and 65°C with the maximum for some as high as 75°–90°C and the minimum about 35°C.

There is still some confusion over the term psychrophile. In keeping with the definition of thermophilic and mesophilic organisms it is logical to define psychrophiles in terms of the optimal growth temperature, which should be clearly different from that of mesophiles. Many authors have classed as psychrophiles all organisms capable of growth at 0°C, regardless of their optimal temperature. Others separate cold-tolerant organisms into obligate psychrophiles with optimal temperatures below 20°C and facultative psychrophiles with optimal temperatures above 20°C (Ingraham and Stokes, 1959). Morita (1975) called for consistency in definition on the basis of optimal temperatures, with *psychrophiles* being defined as organisms having an optimum temperature for growth at about 15°C or lower, a maximum temperature at about 20°C or lower, and a minimum temperature at 0°C or lower. This usage is followed in this book. True psychrophiles are more common in cold environments than was at first realized, particularly where the temperature remains consistently low. Actually, the dominant environment of the biosphere is cold because polar regions and the oceans

TABLE 1.1

Cardinal Temperatures for Prokaryotic Microorganisms

	Temperature (°C)		
Group	Minimum	Optimum	Maximum
Thermophiles	40–45	55–75	60–90
Mesophiles	5–15	30–45	35–47
Psychrophiles	−5–+5	12–15	15–20
Psychrotrophs	−5–+5	25–30	30–35

MICROBIAL ECOLOGY OF FOODS

VOLUME I

Factors Affecting Life and Death of Microorganisms

by

The International Commission on Microbiological Specifications for Foods

Editorial Committee

J. H. Silliker (Chairman), R. P. Elliott (Editorial Coordinator)
A. C. Baird-Parker, F. L. Bryan, J. H. B. Christian, D. S. Clark
J. C. Olson, Jr., T. A. Roberts

ACADEMIC PRESS, INC.
Harcourt Brace Jovanovich, Publishers
San Diego New York Boston
London Sydney Tokyo Toronto

ACADEMIC PRESS, INC.
San Diego, California 92101

United Kingdom Edition published by
ACADEMIC PRESS LIMITED
24-28 Oval Road, London NW1 7DX

Library of Congress Cataloging in Publication Data

International Commission on Microbiological Specifications for Foods.
Factors affecting growth and death in microorganisms.

(Microbial ecology of foods ; v. 1)
Includes index.
1. Food--Microbiology. 2. Micro-organisms--Physiology. 3. Food--Preservation. I. Title. II. Title: Microorganisms in foods 3. III. Series.
[DNLM: 1. Food microbiology. 2. Ecology. QW85 I61mb]
QR115.M45 vol. 1 576'.163s [664'.028] 80-10521
ISBN 0-12-363501-2 (cloth)
ISBN 0-12-363521-7 (paper)

PRINTED IN THE UNITED STATES OF AMERICA

91 92 93 9 8 7 6

Dedicated to

MAURICE INGRAM
1912–1977

Founder of the ICMSF

DISTINGUISHED SCIENTIST, WARM FRIEND

Contents

4 Reduced Water Activity

5 pH and Acidity

6 Oxidation–Reduction Potential

7 Organic Acids

8 Curing Salts and Related Materials

9 Antibiotics in Foods

Preface

Microbial Ecology of Foods was written by a group of over 30 scientists chosen from 22 countries for their expert knowledge in special fields of food microbiology. The work is not, however, a collection of separately authored chapters, but each chapter fits into an overall scheme to provide complete coverage of all important aspects of food microbiology. The book is intended as a source of information for those who must interpret the results of microbiological analyses conducted on foods.

Volume I, "Factors Affecting Growth and Death of Microorganisms," encompasses the environmental factors that affect microorganisms in foods, namely, temperature, irradiation, water activity, pH, E_h, organic acids, curing salts, antibiotics, gases, packaging, and cleaning systems. Special chapters treat the effects of cell injury on survival and recovery of microorganisms in food and the metabolic aspects of mixed populations. The treatment accorded each of these factors includes methods of measurement, effects on spoilage organisms and pathogens, control measures, and interrelationships with the other factors.

Volume I can stand alone for those who want a basic theoretical background in food microbiology. In this respect it is particularly suitable as an undergraduate or postgraduate textbook for students who have had at least one course in general microbiology. Volume I also gives valuable background information in the theoretical aspects, for those whose interest is primarily the applied microbiology of Volume II.

Volume II, "Food Commodities," is a comprehensive treatise on the microbiology of specific commodity groups. Each chapter covers (1) the important properties of the food commodity that affect the microbial content, (2) the initial microbial flora on flesh foods at slaughter or on vegetable foods at harvest, (3) the effects of harvest, transport, processing, and storage on the microbial content, and (4) the means of controlling the process and the microbial content. Each chapter is an up-to-date review of applied microbiology, compiled by leading authorities selected solely for their expert knowledge.

Volume II is meant to be used by those interested primarily in applied aspects of food microbiology, namely, food processors, food microbiologists, food technologists, veterinarians, public health workers, and regulatory officials. Although some will use Volume II alone, most will want Volume I as well, to establish a background of understanding of the theoretical aspects of foods as substrates for microbial development and destruction.

The subject of each chapter in this book could justify a separate volume, and accordingly this text does not include all the material that has been written on a given subject. In each chapter, the reader has been directed to appropriate key publications for further study. The ICMSF is concerned specifically with foods which move in international commerce. Thus discussion of products of local or regional importance is necessarily minimal.

This book is a logical sequel to the two earlier works published by the Commission. See Appendix I.

Acknowledgments

Microbial Ecology of Foods is the result of four years of study by ICMSF members and consultants, involving workshops held in Alexandria (1976) and Cairo (1977, 1978), Egypt. The content was planned by Dr. J. H. Silliker and Dr. J. H. B. Christian and debated and approved by the Commission in plenary session. Chapters were assigned to subcommittees, whose chairmen, listed in the Table of Contents, were responsible for writing the texts. The Editorial Committee was under the Chairmanship of Dr. J. H. Silliker. R. Paul Elliott coordinated and edited the various parts of the book and proof-read copy.

The Commission is most grateful for the generous financial sponsorship from the U.S. Department of Health, Education, and Welfare, Public Health Service, Center for Disease Control (CDC); the Ministry of Health, Arab Republic of Egypt; the World Health Organization; the Ministry of Health of Kuwait; and the various companies within the food industry (see Appendix II). This assistance, of course, does not constitute endorsement of the findings and views expressed herein. Special thanks are given to Dr. D. J. Sencer, Dr. A. E. Najjar, Dr. F. L. Bryan, and others at the Center for Disease Control, U.S. Department of Health, Education, and Welfare, and to Dr. Fouad Mohy El-Din, Dr. Ibrahim Badran, Dr. Ahmed El-Akkad, Dr. Hekmat El-Sayed Aly, Dr. Mohammed Fahmi Saddick Ahmed, and others at the Ministry of Health, Arab Republic of Egypt, for developing and coordinating the project that supported workshops at which much of the work on this book was done. Appreciation is extended to Dr. J. C. Olson, Jr., and Dr. A. C. Baird-Parker for organizing meetings of the Editorial Committee. Finally, thanks are expressed to the respective national governments, universities, and private companies for supporting the participation of their staff and the work of the Commission, of which the present text is but one result.

Contents of Volume II

FOOD COMMODITIES

1

Temperature

I. INTRODUCTION

A. Importance

Throughout the ages, man has learned by trial and error to exploit temperature extremes to preserve his food supply. He learned that chilling foods retards spoilage and maintaining foods in the frozen state effectively preserves them for long periods. He also found that heating eliminated the cause of microbial spoilage and that, if recontamination was prevented by suitable packaging, heated foods would keep even when stored at ambient temperatures.

Also, through trial and error, man learned that some foods, when held at ambient temperature, changed in organoleptic character but remained edible, and became considerably more stable. Thus, a wide variety of fermented foods and beverages evolved, many of ethnic origin, and characteristic of the foods locally available. Hundreds of such fermented foods are commonplace in today's societies, and, although many have benefited by application of recently discovered scientific principles, they remain essentially identical to those available generations ago. Lactic and alcoholic fermentations, or a combination of the two, are the major fermentations employed. If the food contained a fermentable sugar and was lightly salted, a lactic fermentation was likely to occur; if tart to the taste (e.g., fruit juices), alcoholic fermentation was expected. In either instance, temperature control, generally a cool environment, was essential to yield the desired characteristics of the particular fermented food.

Over the thousands of years of human history only the past two or three generations have been able to capitalize on recently discovered scientific principles in food preservation, fermentation, and food safety. Temperature manipulation and control rank among the more critical factors needed to achieve a safe and wholesome food supply. The scientific principles related to exploitation of temperature to control microorganisms likely to be in man's foods will be discussed in this chapter.

B. Growth Temperature Range

Temperature is probably the most important environmental factor affecting the growth and viability of microorganisms. Although microbial growth can occur at temperatures from about -8° to +90°C, the range of temperature that permits growth of any specific organism seldom exceeds 35°C. Within this range, temperature affects the duration of the lag phase, the rate of growth, the final cell numbers, the nutritional requirements, and the enzymatic and chemical composition of the cells. Temperatures above the maximal for growth of a particular organism are lethal for that organism; the higher the temperature the more rapid the loss of viability. However, the lethality of any exposure to any given temperature above that which is maximal for growth depends on the inherent heat resistance of the organism of interest. Survival is possible at temperatures below the minimum for growth. The lethal effect of chilling or freezing depends on the organism, the microenvironment and the time/temperature conditions of storage. Some microorganisms remain viable for long periods when kept frozen at sufficiently low temperatures.

In a favorable environment growth of bacteria normally proceeds as shown in Fig. 1.1. After a short period of adjustment to the environment (the *lag phase*),

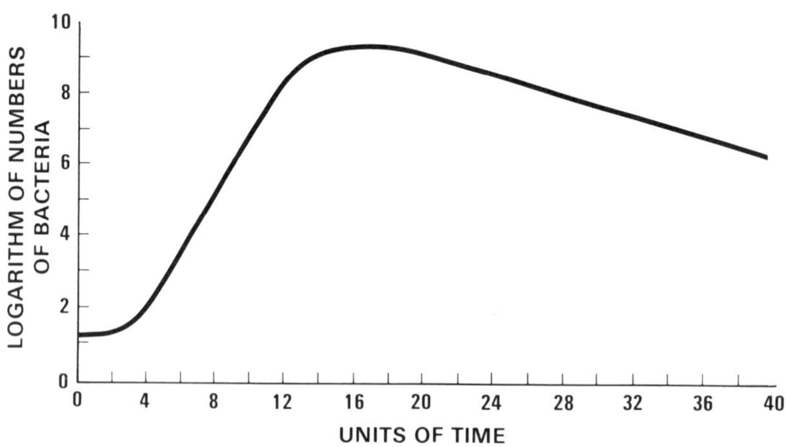

Fig.1.1. A hypothetical bacterial growth curve.

growth begins and accelerates to a phase of rapid, constant, exponential growth, termed the *logarithmic growth phase*. Depletion of nutrients and accumulation of toxic metabolic products eventually slow the growth rate to the point where all cell division and cell death are in balance and the population remains stationary for some time, the *maximum stationary phase*. This is followed by a period wherein the population gradually decreases due to death of the cells. For more detail of the normal growth curve, see any recent textbook on general microbiology.

The influence of temperature on the activity of microorganisms is greatest in moist foods, i.e., at water activity (a_w) levels above 0.85. Most bacteria do not grow below this a_w (see Chapter 4). Figure 1.2 shows the relationship between generation times and temperature for a typical psychrotroph and a typical mesophile. Three effects are readily discernible. As the temperature increases, growth accelerates, i.e., generation times decrease rapidly. At the higher temperatures there is a range in which the rate of growth (generation time) is relatively stable (optimal growth), i.e., from about 20°–30°C for the psychrotroph and about 35°–45°C for the mesophile. Inhibition of growth occurs at a temperature only slightly above the maximum for optimal growth. Thermophilic bacteria respond similarly, but their growth response curve would be shifted substantially to the right. The nature of response to any given temperature is profoundly affected by the length of exposure at that temperature. Even at optimum temperature, growth in a confined environment will eventually cease due to the gradual depletion of nutrients.

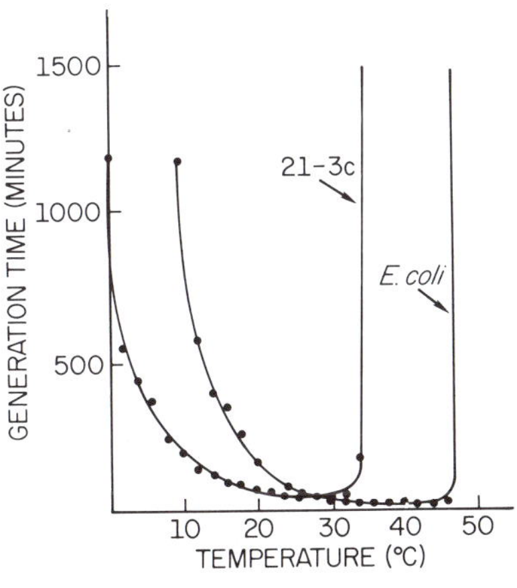

Fig. 1.2. Effect of temperature on the generation time of a typical mesophile (*Escherichia coli*) and a psychrophile (21.3c). (From Ingraham, 1958.) Note: According to current terminology culture 21.3c would be termed a psychrotroph.

C. Definitions

Four major physiological groups of bacteria may be distinguished by their temperature ranges of growth: thermophiles, mesophiles, psychrophiles, and psychrotrophs (Table 1.1). *Mesophiles,* many of animal or human origin—including all pathogens and many food spoilage types, prefer moderate temperatures with an optimum generally between 30° and 45°C and a minimum growth temperature ranging from about 5° to 10°C. The generation time for many mesophilic types in a favorable medium and at optimum temperature may be 0.5 hr or less. With *thermophiles* the whole curve relating growth rate to temperature is shifted to a higher range. Optimum temperature for growth usually is between 55° and 65°C with the maximum for some as high as 75°–90°C and the minimum about 35°C.

There is still some confusion over the term psychrophile. In keeping with the definition of thermophilic and mesophilic organisms it is logical to define psychrophiles in terms of the optimal growth temperature, which should be clearly different from that of mesophiles. Many authors have classed as psychrophiles all organisms capable of growth at 0°C, regardless of their optimal temperature. Others separate cold-tolerant organisms into obligate psychrophiles with optimal temperatures below 20°C and facultative psychrophiles with optimal temperatures above 20°C (Ingraham and Stokes, 1959). Morita (1975) called for consistency in definition on the basis of optimal temperatures, with *psychrophiles* being defined as organisms having an optimum temperature for growth at about 15°C or lower, a maximum temperature at about 20°C or lower, and a minimum temperature at 0°C or lower. This usage is followed in this book. True psychrophiles are more common in cold environments than was at first realized, particularly where the temperature remains consistently low. Actually, the dominant environment of the biosphere is cold because polar regions and the oceans

TABLE 1.1

Cardinal Temperatures for Prokaryotic Microorganisms

	Temperature (°C)		
Group	Minimum	Optimum	Maximum
Thermophiles	40–45	55–75	60–90
Mesophiles	5–15	30–45	35–47
Psychrophiles	−5–+5	12–15	15–20
Psychrotrophs	−5–+5	25–30	30–35

(95% by volume below 5°C) are 14 and 71% of the earth's surface, respectively (Morita, 1975). Special care must be taken when isolating psychrophiles to avoid exposure to temperatures above 10°C. Neglect of this precaution has been responsible for past failures to detect large numbers of psychrophiles. Because of their sensitivity to higher temperatures, psychrophiles are less important than psychrotrophs (see below) in food processing. True psychrophiles are mainly of marine origin and importance and include only a few genera. They are of little consequence in foods. The psychrophilic bacteria are discussed in a review by Morita (1975).

Organisms capable of growth at about 0°C, but not meeting the optimal and maximal temperature requirements for psychrophiles are called *psychrotrophs* (Eddy, 1960). Since these organisms grow best at moderate temperatures, they may be considered a subgroup of the mesophiles capable of growth below the minimum temperature common to most mesophiles. Psychrotrophs include gram-positive and gram-negative bacteria; aerobes, anaerobes, and facultative anaerobes; motile and nonmotile organisms; spore formers and nonspore formers. Species among at least 27 genera are represented (see tabulation).

Genera that include psychrotrophic bacteria

Acinetobacter	*Clostridium*	*Lactobacillus*	*Serratia*
Aeromonas	*Corynebacterium*	*Leuconostoc*	*Streptomyces*
Alcaligenes	*Enterobacter*	*Microbacterium*	*Streptococcus*
Arthrobacter	*Erwinia*	*Micrococcus*	*Vibrio*
Bacillus	*Escherichia*	*Moraxella*	*Yersinia*
Chromobacterium	*Flavobacterium*	*Proteus*	*Listeria*
Citrobacter	*Klebsiella*	*Pseudomonas*	

Among the yeasts, psychrotrophic strains are found in the genera *Candida, Torulopsis, Cryptococcus,* and *Rhodotorula*. Psychrotrophic molds exist in such important genera as *Penicillium, Cladosporium, Trichothecium,* and *Aspergillus*.

II. EFFECT OF CHILLING (COOLING, REFRIGERATION)

A. General Effects

Chilling temperatures are those close to but above the freezing point of fresh foods, usually −1° to +7°C. The effect of chilling on the microflora in a particular food will depend on the temperature characteristics of the organisms as well as the temperature and time of storage. As temperature is lowered from the optimum, growth slows and eventually stops. In the range near the minimal growth temperature, the lag phase increases rapidly with decreasing temperature and approaches infinity. For *Cladosporium herbarum,* an extremely cold-tolerant mold, the lag phase ranges from 1 day at room temperature to about 18 days at −5°C. Lag periods as long as 414 days have been reported, but accurate control of temperature for this length of time is open to question (Michener and Elliott, 1964). In the temperature range where both typical mesophiles and psychrotrophs can grow, the latter have a much shorter lag phase (see Fig. 1.3).

The rate of growth is sensitive to changes in temperature; below the optimum, growth becomes slower and in the lower ranges (below 0°C) the generation time may exceed 100 hr.

In a mixed flora of psychrotrophs and mesophiles, low temperatures have an important selective action and may affect the composition of the initial contamination on particular foods or lead to changes in the flora developing during processing or storage. For example, chilled beef prepared in a semitropical climate had a longer storage life under chilled conditions than did similar beef from cooler areas (Empey and Scott, 1939). This difference was due to the effect

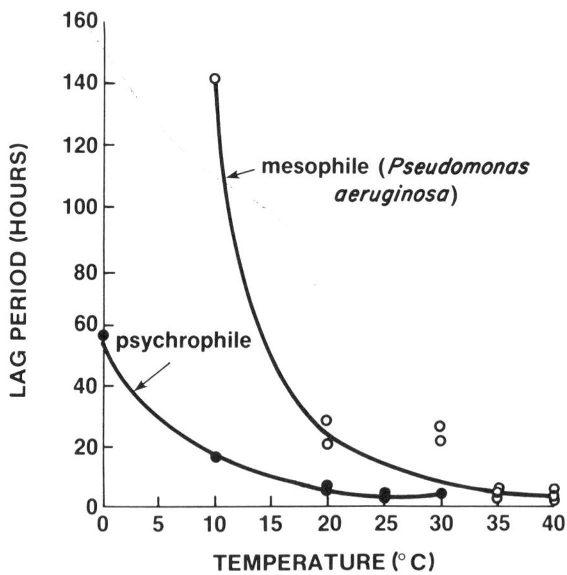

Fig. 1.3. Effect of temperature on duration of lag phase of a mesophile and a psychrophile. (From Elliott and Michener, 1965.) Note: According to current terminology the latter organism would be termed a psychrotroph.

of temperature on the proportion of cold-tolerant organisms in the initial contamination derived from the soil and the hide of the animal; beef from the cooler areas had a larger proportion of psychrotrophs. The storage temperature has a marked influence on the flora of milk. Raw milk kept at about 10°C develops a dominant flora of lactic acid streptococci, whereas that held near 0°C develops primarily psychrotrophic gram-negative bacteria (Mocquot and Ducluzeau, 1968).

B. Cold Shock

The rapid cooling of mesophilic bacteria from the normal growth temperature to about 0°C may result in the death or injury of a proportion of the culture (see Chapter 12). Gram-negative bacteria, including *Escherichia coli, Pseudomonas aeruginosa, P. fluorescens, Salmonella* spp. and *Enterobacter aerogenes* appear to be more susceptible to cold than are gram-positive organisms, although cold shock of *Bacillus subtilis* and *Clostridium perfringens* has been observed (Sato and Takahashi, 1969; Traci and Duncan, 1974). *Staphylococcus aureus* is resistant to cold shock, but a culture of this organism in trypticase-soy broth incubated at 5°C showed evidence of injury manifested by increased sensitivity to mannitol-salt agar (Jackson, 1974). Psychrotrophic organisms appear to be less susceptible to cold (Farrell and Rose, 1968). For further discussion of injury induced by cold and other agents see Chapter 12.

C. Physiological Adaptation to Low Temperatures

Growth of microorganisms at temperatures below the optimum for growth can cause a number of morphological and physiological changes. Examples of morphological changes include increases in the cell size of the yeast *Candida utilis* (Rose, 1968) and *Esherichia coli* (Shehata and Marr, 1975); the formation of filaments in *E. coli* (Shaw, 1968) and mesosome deterioration and double cell wall formation in *B. subtilis* (Neale and Chapman, 1970).

Differential retardation of the activities of enzymes at low temperatures can change metabolic pathways and end products. For example, organisms that synthesize phenazine and carotenoid pigments tend to produce greater amounts of pigment at lower temperatures of growth. This could be of concern in the spoilage of foods, e.g., occurrence of color defects at low temperatures (Witter *et al.*, 1966). Production of extracellular dextrans of *Leuconostoc* and pediococci is favored at temperatures below the optimal for growth of these bacteria (Rose, 1968). Lipase and proteinase production by *Pseudomonas* and certain other genera occurs preferentially at low temperatures (Jezeski and Olsen, 1962; Alford *et al.*, 1971; Juffs, 1976). Some of these enzymes are thermoresistant and may still be present after heat treatment of the food. Many of the regulatory processes of cell metabolism are sensitive to temperatures below the optimum so that exposure to low temperatures can lead to metabolic imbalance and cessation of growth (Rose, 1968).

Incubation at low temperatures can also change the lipid composition of microbial cells (Marr and Ingraham, 1962). The amount of lipid in bacteria is independent of the growth temperature (Cronan and Vagelos, 1972), but in yeasts there is a small increase in total lipids at lower temperatures (Kates and Baxter, 1962). Both bacteria and yeasts have been reported to contain an increasing proportion of unsaturated fatty acids as the growth temperature decreases, and in bacteria less cyclopropane fatty acids are formed (Gill, 1975). Increases in unsaturation of alcohols from waxes produced by a mesophilic strain of *Acinetobacter* grown at low temperatures have also been reported (Gallagher, 1971).

The increasing proportion of unsaturated fatty acids with decreasing temperature is believed to be essential for membrane function at low temperatures, since this change would lower the temperature at which the membrane lipids "freeze." The proportion of unsaturated fatty acids may determine the ability of the cell to withstand cold shock, but in fatty acid auxotrophs of *E. coli,* provided minimum levels of saturated and unsaturated fatty acids are present, the fatty acid composition of the membrane lipids can vary between wide limits without apparently affecting the ability of the organism to grow at low temperatures (Cronan and Gelmann, 1975).

D. Effect on Spoilage Organisms

Foods may be spoiled by representatives of any of the four microbial groups characterized by their responses to temperatures, i.e., psychrophiles, thermophiles, mesophiles, and psychrotrophs. However, in food chilled and held at proper chill temperatures (below 7°C) only psychrotrophs will cause spoilage. Many of the psychrotrophic genera previously listed (Section I,C) are important food spoilage types.

Although generation times of psychrotrophs given in Table 1.2 seem very long, the long periods of refrigerated storage common for many foods permit psychrotrophic populations to reach many millions per gram of product in only a few days—often resulting in objectionable changes in odor, taste, and texture. Because growth rate increases rapidly as temperature increases, fluctuations in storage temperature should be avoided. This will be discussed further in Section V,A,2.

E. Effect on Pathogens and Indicator Organisms

Most pathogens are mesophiles, and with few exceptions growth of these organisms in chilled foods is not a problem. Salmonellae do not grow at tempera-

TABLE 1.2

Examples of Generation Time for Psychrotrophic Bacteria[a]

Organism	Generation time	Reference
Pseudomonas fluorescens[b]	6.68 hr at 0.5°C	Hess (1934)
	30.21 hr at 0°C	
Pseudomonas fluorescens[c]	10.65 hr at 5°C	Hess (1934)
	26.41 hr at 0°C	
Pseudomonas sp.	2.66 hr at 10°C	Ingraham (1958)
	10.33 hr at 0°C	
Strain 82	21.58 hr at 5°C	Upadhyay and Stokes (1962)
	21.23 hr at 0°C	
Pseudomonas fluorescens[d]	4.17–8.20 hr at 4°C	Olsen and Jezeski (1963)
Bacillus psychrophilus	6.30 hr at −5° to −7°C	Larkin and Stokes (1968)
Micrococcus cryophilus	28.33 hr at 0°C	Tai and Jackson (1969)
Pseudomonas strain 92	11.1 hr at 2°C	Frank *et al.* (1972)

[a] From Morita (1975).
[b] Inoculum prepared at 20°C.
[c] Inoculum prepared at 5°C.
[d] Generation time depends on type of substrates, stationary or aerated cultures.

tures below about 6°C (Matches and Liston, 1968), and in a rich broth medium at 10°C the lag phase for *S. typhimurium* is approximately 12 hr and the generation time about 8 hr. In foods the growth rate may be slower. For instance, *Salmonella senftenberg, S. enteritidis,* and *S. manhattan* failed to grow at 10°C in ham salad or custard, although growth occurred in chicken à la king at 7°C (Angelotti *et al.,* 1961). In raw ground beef, an inoculum of five *Salmonella* serotypes did not grow at 7°C and increased by only 300-fold in 5 days at 12.5°C (Goepfert and Kim, 1975).

Clostridium perfringens can grow at temperatures between 12° and 50°C, but below 15°C growth is slow (Michener and Elliott, 1964; Roberts and Hobbs, 1968; Hobbs, 1969). Vegetative cells of *C. perfringens* are sensitive to low temperatures, and prolonged refrigerated storage of foods containing this organism is likely to result in a slow destruction of vegetative cells. Spores are not affected to the same extent by storage at low temperature (Canada *et al.,* 1964). The growth rate is related to pH and medium as well as temperature, so that growth of *C. perfringens* is slower in media at pH 5.8 than at pH 7.2 (Table 1.3; Barnes *et al.,* 1963). Germination of clostridial spores has been observed at 5°C, i.e., below the minimal growth temperature (Roberts and Hobbs, 1968).

Staphylococcus aureus can withstand low temperatures and grow at temperatures down to about 7°C (Angelotti *et al.,* 1961), but the lower limit for toxin production is somewhat higher. For example, enterotoxin has been detected in

TABLE 1.3

The Relationship of pH and Temperature to Growth Rate of *Clostridium perfringens* (*welchii*) F2985/50[a]

Incubation temperature (°C)	Hours to visible turbidity in RCM broth at pH	
	5.8	7.2
15	>700	>700
20	74	48
25	30	24
30	24	8
37	5	5

[a] From Barnes *et al.* (1963).

foods held at 10°C (Genigeorgis *et al.*, 1969; Tatini, 1973), but toxin production at temperatures below about 20°C is slow. One strain producing enterotoxin B required 158 hr at 13°C to produce detectable levels of toxin (1 μg/ml). Other strains grew at 13°C but failed to produce toxin below 19°C (Scheusner *et al.*, 1973). In a rich medium at pH 7 the time taken for the production of measurable toxin ranged from 78–98 hr at 19°C to 14–16 hr at 26°C. Under less favorable conditions toxin production was slower (Scheusner *et al.*, 1973). Despite attainment of high *S. aureus* populations, enterotoxin production may be inhibited by the independent and interactive effects of temperature, pH, oxygen tension, water activity, and competitive growth of other organisms (Tatini, 1973).

Vibrio parahaemolyticus is sensitive to low temperatures and, in Japan, food poisoning caused by this organism generally occurs in the warmer months of the year. On the surface of sea fish, growth appears to cease below about 5–8°C, although the organism may survive for long periods at these temperatures (Sakazaki, 1973). *Vibrio parahaemolyticus* can multiply to dangerous levels when oysters are stored at 10°C for more than a week (Thomson and Thacker, 1973). The lowest temperature reported for growth of *V. parahaemolyticus* in laboratory media is 5°C (Beuchat, 1973); however, growth at this temperature was markedly affected by pH.

Bacillus cereus grows at temperatures from 7° to 45°C (Chung *et al.*, 1976) and *B. subtilis* from 12° to 55°C (Riemann, 1969). No evidence has been found that enteropathogenic *E. coli* grows at temperatures lower than other *E. coli*. Numbers of enteropathogenic *E. coli* recoverable from soft cheese increased during storage at 4°C, but it was not clear whether this was due to growth or recovery from cold shock (Fantasia *et al.*, 1975). During a 7-week ripening period at 10°C for Camembert cheese, four of six strains of enteropathogenic *E. coli* decreased markedly in number; two of the cultures increased slightly during the first week and decreased rapidly in the following 6 weeks (Park *et al.*, 1973). *Yersinia enterocolitica*-like organisms have been isolated from vacuum-packaged meat stored at 1°–3°C (Hanna *et al.*, 1976). Aflatoxin, ochratoxin A., and tremortins A and B have been produced in various foods (the latter in eight types of cereal grains) at 4°–5°C (Diener and Davis, 1967; Farhat and Koburger, 1975; Trenk *et al.*, 1971; Hou *et al.*, 1971).

Clostridium botulinum type E and nonproteolytic strains of *C. botulinum* types B and F are capable of growth and toxin production at temperatures as low as 3.5°–5°C (Schmidt *et al.*, 1961; Eklund *et al.*, 1967; Roberts and Hobbs, 1968). However, the optimum growth temperature for *C. botulinum* type E is about 35°C. The temperature range for growth and toxin production by *C. botulinum* type A and other strains of type B extends from 10° to 50°C. The amount of toxin ultimately produced in laboratory media shows little dependence on temperature in the range 20°–25°C (Riemann, 1969).

The effect of chilling on the organisms that are regarded as indicators of fecal pollution is also of significance. Growth of these organisms during storage of a food could lead to the food's rejection or condemnation even though originally the organisms had not exceeded an acceptable level. *Escherichia coli* and *Streptococcus faecalis* have similar minimum growth temperatures, about 8°–10°C. Other Enterobacteriaceae which give a positive presumptive coliform test at 30°C as specified by the International Standards Organization (ISO, 1977) can grow at temperatures down to about 0°C. These organisms include species of *Klebsiella, Enterobacter (Aerobacter)* and *Hafnia*. Among psychrotrophic cultures responsible for spoilage of dairy products, Schultze and Olson (1960) found about 10% were coliforms.

III. EFFECT OF FREEZING

A. Initial Effects

Growth will cease for many organisms at temperatures above the freezing point of the growth medium. For others, growth continues until the medium freezes. The factors influencing the effect of freezing on microorganisms have been reviewed extensively (Mazur, 1966; Ray and Speck, 1973b; Meryman, 1974) as has inactivation of microorganisms by cold (Ingram and Mackey, 1977).

Microorganisms differ in their responses to freezing: some survive virtually unharmed; some resist freezing but are susceptible to damage during frozen storage or thawing; others are sensitive to freezing, storage, and thawing under only some conditions; others are inactivated by freezing under nearly all conditions. Most spores and some vegetative cells survive virtually unchanged (Table 1.4). Most other non-spore forming organisms are sensitive to one or more steps of the freezing process (Fennema *et al.*, 1973).

Higher organisms are more sensitive to low temperatures than are bacteria. Thus, freezing and frozen storage may be used to destroy parasitic protozoa, cestodes, and nematodes in various foods (ICMSF, 1978).

TABLE 1.4

Effect of Freezing at −30°C on the Proportion of Bacterial Genera in the Aerobic Plate Count on Raw Beef and Raw Minced Meat[a]

	Aerobic plate count (average) (bacteria per gram)			
	Raw beef		Raw minced meat	
	Before freezing	After freezing	Before freezing	After freezing
	385,000	77,000	400,000	47,000
Bacterial group	**Percent of Aerobic Plate Count**			
Gram-positive bacteria	15	70	22	75
CMB-group (*Corynebacterium*, *Brevibacterium* and *Microbacterium*)	5	45	2	20
Aerococcus or *Peptococcus* (*Gaffkya*)	0	21	6	15
Micrococcus	5	2	6	4
Leuconostoc	0	0	3	32
Lactobacillus	5	0	5	4
Bacillus	0	2	0	0
Gram-negative bacteria	85	30	78	23
Pseudomonas	75	22	69	23
Vibrio + *Aeromonas*	6	7	2	0
Acinetobacter/*Moraxella* (*Achromobacter*)	4	1	7	0
Yeasts	0	0	0	2

[a] From Partmann (1975). Reprinted with permission from Water Relations in Food. Copyright by Academic Press Inc. (London) Ltd.

Relatively high freezing temperatures are generally more lethal than lower temperatures. More microorganisms are killed or injured in the range from −2° to −10°C than at −15°C, while at −30°C the lethal effect is even less pronounced (Christophersen, 1968). In minced chicken breast salmonellae survived best at −20°C where 60–83% of the organisms survived after 126 days, and worst at −2° and −5°C where only 1.3–5.8% survived after 5 days (Foster and Mead, 1976). Such a temperature effect does not apply in salt solutions where the freezing point is depressed below -10°C. Unfortunately, high freezing temperatures that cause maximum damage to microorganisms also tend to cause the maximum harm to many foods, thereby limiting their use.

B. Nonlethal Injury

The freezing of microorganisms may result in death or nonlethal injury. From a public health point of view, the occurrence of cryoinjury can be very important. Injured bacteria, which may be difficult to detect in a frozen product, can later, after thawing, recover the ability to grow or to produce toxin at the normal rate. Bacteria differ in their response to the harmful effects of freezing. This selective effect could influence the distribution of particular types in frozen foods. The number of viable organisms in frozen foods may markedly exceed that determined by the usual routine methods. Hence, estimates of numbers of viable bacterial cells in frozen foods are likely to be underestimated. Sublethal injury may greatly affect the interpretation of data gathered preparatory in issuance of microbiological criteria, i.e., specifications, standards, or guidelines (FAO/WHO, 1977) or of data obtained in regulatory surveillance and industry quality control programs. The consequences of sublethal effects of various agents on microorganisms are discussed in Chapter 12.

C. Survival after Freezing

During frozen storage, further destruction of microorganisms may take place (Table 1.5). Although the rate of storage death may be high initially, it generally decreases with time, and below −60°C destruction is slow. Storage in the presence of oxygen is more deleterious than storage in nitrogen (Ray and Speck, 1973b).

The microbial flora of frozen food comprises the most resistant elements of the initial contamination. Spores are highly resistant to freezing and are likely to survive without significant decrease in numbers. The toxins of *Clostridium botulinum* and *Staphylococcus aureus* are also unaffected by freezing; thus it is possible for poisoning to occur from the ingestion of a frozen product containing pre-formed toxin.

TABLE 1.5

Survival of Bacteria in Peas during Frozen Storage[a]

Microorganisms	APC/gm[b] after freezing	Proportion of "after freezing" count surviving storage (%)							
		24 hours		1 month		12 months		24 months	
		−18°C	−29°C	−18°C	−29°C	−18°C	−29°C	−18°C	−29°C
Leuconostoc mesenteroides	17,000	107	147	98	104	113	106	18	68
Escherichia coli	1,900	116	101	49	88	14	36	15	20
Staphylococcus aureus	4,400	99	92	101	94	30	65	8	11
Streptococcus faecalis	2,500	53	112	37	57	41	53	48	24
S. faecium	4,800	167	143	133	104	139	131	145	104
S. lactis	23,700	92	104	85	86	28	37	4	98
Micrococcus conglomeratus[c]	7,300	87	127	105	118	58	59	68	65
Sarcina sp.	7,000	95	81	86	81	62	79	11	24
Pseudomonas sp. I.	3,000	67	90	78	111	33	56	0	43
Flavobacterium suaveolens	8,000	83	54	75	63	35	40	13	27
Total mesophiles	79,600	97	110	90	92	60	66	24	31
Rhodotorula sp.	1,000	200	170	300	267	93	87	10	20
Flavobacterium sp.	2,300	100	86	75	157	87	133	43	1
Pseudomonas sp. II	2,900	80	92	39	109	0	3	0	0.3
Brevibacterium sp. } *Arthrobacter* sp. }	20,000	84	106	77	92	64	79	13	59
Total psychrotrophs	26,400	84	107	81	108	60	77	14	47
Total population	106,000	95	109	88	95	60	68	22	40

[a] From Hall (1969).

[b] APC = Aerobic plate count.

[c] Not recognized or considered as *incertae sedis* in the 8th Edition of Bergey's Manual (Buchanan and Gibbons, 1974).

Most gram-positive organisms, including *Bacillus, Clostridium, Lactobacillus, Staphylococcus, Micrococcus,* and *Streptococcus* are relatively resistant to freezing and cold storage, and freezing will not eliminate staphylococci or other gram-positive organisms from foods. On the other hand, vegetative cells of *C. perfringens* are very sensitive to frozen storage (Canada *et al.,* 1964; Strong and Canada, 1964; Trakulchang and Kraft, 1977). After a holding time of 48 hr in various media, the survival of *C. perfringens* vegetative cells ranged from 0 to 2.4% (Canada *et al.,* 1964).

Gram-negative organisms, such as *Escherichia, Pseudomonas, Alcaligenes, Vibrio,* and *Salmonella,* are more sensitive to freezing than are gram-positive organisms. *Vibrio parahaemolyticus* in oysters rapidly loses its viability during storage at −20°C (Thomson and Thacker, 1973). However, in certain foods, enteric pathogens and other gram-negative organisms may be protected to some extent against the harmful effects of freezing, and if they are present in sufficient numbers before freezing, may survive even long periods of storage. For example, 20% of *Salmonella typhimurium* inoculated into chow mein were recovered after storage at −25°C for 9 months and *S. enteritidis* and *S. typhimurium* survived in ice cream held at −23°C for 7 years (Georgala and Hurst, 1963). The frequently reported isolation of *Salmonella* from many frozen products, including eggs, fish, meat, dairy products, and precooked meals, is ample evidence that freezing is no safeguard against the transmission of foodborne salmonellosis.

Storage at low temperature affords a measure of protection against certain pathogenic protozoa, cestodes, and nematodes. For example, storage of meat at −15°C for 20 days will kill the larvae of *Trichinella spiralis;* cysts of *Taenia saginata* (*Cysticercus bovis*) were destroyed after 6 days at −9.5°C (Dolman, 1957). Storage for 24 hr at −10°C appears to be lethal for *Taenia ovis* (Whitten, 1971). *Toxoplasma gondii* generally does not survive for more than a few days at -20°C, but there appear to be strain differences (Sommer *et al.,* 1965; Dubey, 1974). The larvae of anisakine nematodes, parasitic to fish and sometimes pathogenic for man, have been killed after freezing and holding at −20°C for 24 hr; however, recent studies indicate that the larvae of some types may survive for as long as 52 hr at −20°C (Bier, 1976).

Although microbial growth does not occur in frozen foods held below −10°C, many enzymes remain active at the usual storage temperatures for frozen products. These enzymes, which include some of microbial origin as well as others naturally present in the food, limit the shelf-life during frozen storage. Lipases continue to produce free fatty acids from fats at temperatures far below freezing (Christophersen, 1968). Hence, lipase-producing organisms in foods before freezing can markedly affect the storage life of the frozen product. Large numbers of proteolytic organisms may also contribute to a high level of enzyme activity and consequent deterioration in quality during frozen storage (Peterson and Gunderson, 1960). Freezing will slow but not arrest the development of enzymatic spoilage after it has begun.

D. Growth after Thawing

The growth of microorganisms on foods after thawing depends on the numbers and types that survive freezing and thawing as well as on the properties of the food itself. The composition of the microbial flora after thawing will be related to the organisms present in the food before freezing, but it will be modified by the selective action of the freezing process which eliminates a greater proportion of the more sensitive gram-negative cells. The freezing conditions and the time and temperature of storage will also affect the number of surviving organisms and the proportion of injured cells that require a period of recovery before regaining full activity.

Uncontrolled thawing can result in significant increases in bacterial population. The conditions during thawing and the time/temperature of holding after thawing are most important. The surfaces of large blocks of frozen product may reach the temperature of the thawing medium several hours before the center is completely thawed. If the thawing temperature is sufficiently high, bacterial growth at the surface may be rapid. For example, thawing pork legs in air at 10°C with an air velocity of 0.25 m/sec had very little effect on the mean bacterial count. Increasing the air velocity at 10°C to 5.5 m/sec caused a relatively large increase in bacterial numbers (>1000 times) but in air at 20°C and 0.25 m/sec increase was small (<tenfold). With water and vacuum thawing also, 10°C was the most satisfactory temperature (Bailey *et al.*, 1974). The use of higher temperatures to give more rapid thawing involves a risk that food poisoning organisms as well as spoilage microbes may grow.

Freezing may affect the structural integrity of the food, thus making it more susceptible to microbial attack. Condensation on the food during thawing and the resulting accumulation of moisture in packaged products can provide a favorable growth environment.

There is a mistaken belief among food processors that frozen-thawed foods are more highly susceptible to rapid spoilage than are foods that have never been frozen. However, practical tests have demonstrated no difference (Elliott and Straka, 1964).

IV. EFFECT OF HIGH TEMPERATURES

A. General Effects

As the temperature increases above that at which growth of microorganisms ceases, injury and, inevitably, death occurs. Sublethal or mild exposures (temperature/time) induce stress which may cause injury. Injured cells may remain viable but unable to reproduce until the damage is repaired (see Chapter 12). At higher temperatures, a homogeneous population of a pure culture will begin to decrease in an orderly manner with time due to death of individual cells. While exceptions have been noted, it is well established that the order of death is essentially logarithmic (Pflug and Schmidt, 1968; Brown and Melling, 1971; Stumbo, 1973), which makes it possible to develop time temperature heat processes that will assure with precision the destructive effect against microorganisms. The principles governing the behavior of microorganisms at high temperature in foods will be discussed below, and applications of these principles will be considered further in chapters dealing with various foods or food commodity groups.

B. Rate of Destruction

A typical bacterial "survivor curve" showing the logarithmic (or exponential) rate of destruction is illustrated in Fig. 1.4. By plotting the logarithm of the number of survivors against time of exposure at a given temperature a "straight line" is obtained. Under given conditions the rate of death is constant at any given temperature and independent of the initial number of cells. From such a curve the decimal reduction time or *D* value may be determined, being the time

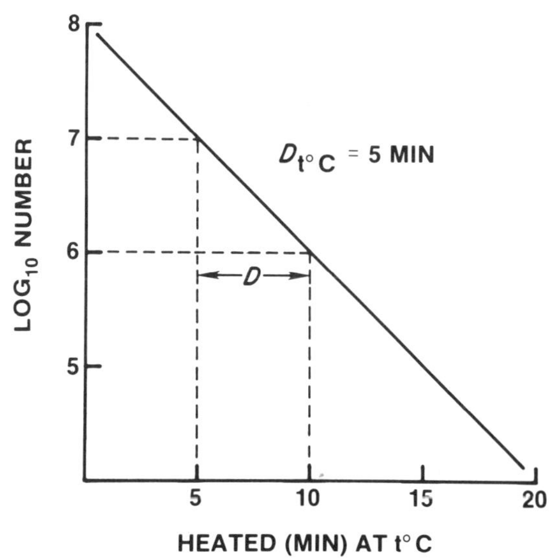

Fig. 1.4. Graphical illustration of decimal reduction time *D*.

in minutes required to destroy 90% of the population. D values should state the temperature (in degrees Celsius) in a subscript. For example, from Fig. 1.4, assuming a temperature of exposure of 65°C, the D value would be expressed as D_{65} = 5 min. The straight line of Fig. 1.4 theoretically extends below the baseline shown, eventually into the area of negative logarithms—for example, $\log_{10}{}^{-2}$ = 1 cell in 100 gm; $\log_{10}{}^{-3}$ = 1 cell in 1000 gm, etc. It is evident, then, that the population would never be reduced to zero. It would follow, then, that if a known population in each of a number of unit volumes or containers were subjected to a lethal temperature for any given time, there would always be a probability of survival in any one container. This fact makes it possible to predetermine a heat process that will result in a desired level of destruction. The most extensive application of this principle is in the determination of heat processes for canned foods.

C. Thermal Death Time

While the D value reflects a measure of heat resistance of bacteria at a given temperature, points on a thermal death time curve indicate the relative resistance at different temperatures, thus permitting calculation of effects of differing temperatures. Thermal death time curves can be constructed in several ways. A

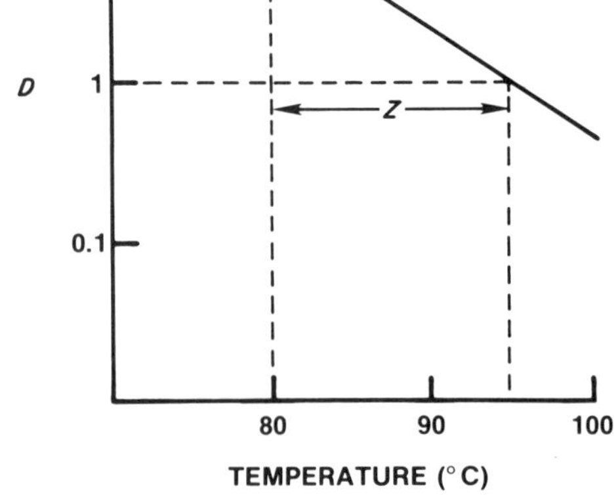

Fig. 1.5. Graphical illustration of thermal death time curve. D = decimal reduction time; z = °C required for thermal death time curve to traverse one $\log_{10}$ cycle.

convenient method is illustrated in Fig. 1.5. Here the logarithms of D values are plotted against exposure temperatures. The slope of the curve is expressed by the term z, which is the degrees Celsius (originally degrees Fahrenheit) required for the thermal death time curve to traverse one log cycle. D and z values are used in calculating heat processes. Nickerson and Sinskey (1972), Shapton *et al.* (1971), and Stumbo (1973) have provided detailed guidance in calculating "sterilizing" and "pasteurizing" heat processes for foods. The sterilizing value of a heat process often is expressed as the F_0 value, which is the number of minutes required to destroy a specified number of spores at 121.1°C (250°F) when $z = 10$ (or 18 in terms of °F). For example, heat processing of low acid canned foods (>pH 4.6) must be sufficient to destroy the spores of *Clostridium botulinum* (see Sections IV,D,8 and V,C,2). Such a heat process has been arbitrarily established as that which will reduce any population of the most resistant *Clostridium botulinum* spores to 10^{-12} of their original numbers or, in other words, by a factor of 12 decimal values, the $12D$ concept. From the studies of Esty and Meyer (1922) and Townsend *et al.* (1938) it was determined that the time required at 121.1°C (250°F) to reduce the survivor population in phosphate buffer by a factor of 12 decimal values was 2.45 min, an F_0 of 2.45 in phosphate buffer. The F_0 for *C. botulinum* in any other heating menstruum, e.g., a canned vegetable, would require determination by inoculated-pack studies.

D. Factors Affecting Heat Resistance

1. General Considerations

Numerous studies have been made to determine the influence of various factors on the heat resistance of microorganisms. Those who have reviewed and attempted to interpret the results of reported studies are unanimous in concluding, as perhaps best expressed by Schmidt (1957), that "The many variables introduced by the type of organism studied, the techniques used, and the unrecognized involvement of the many factors previously discussed make a sound interpretation of previously reported data very difficult." Schmidt further emphasized the need for a critical evaluation of the conclusions reached in past studies and to identify, in light of new knowledge, work that is not invalidated or rendered questionable. While such an evaluation of past studies has not been fully undertaken even to this date, several excellent reviews have been made of the various factors which influence heat resistance of microorganisms (Schmidt, 1957; Hansen and Riemann, 1963; Pflug and Schmidt, 1968; Roberts and Hitchins, 1969; Brown and Melling, 1971). Factors that can effect heat resistance are of three general types:

1. Inherent resistance, e.g., the differences among species, strains within the same species, and spores versus vegetative cells.

2. Environmental influences active during growth and formation of cells or spores (e.g., age of cells, growth temperature, growth medium).

3. Environmental influences acting during the time of heating cells or spores, e.g., pH, a_w, suspension menstruum (type of foods), salts, and other organic and inorganic compounds likely to be present.

In light of present knowledge, the only practical proof of death of microorganisms is their failure to reproduce when provided with suitable conditions for growth. In practice, evidence of death may be more apparent than real. For example, the growth response of injured cells is different from that of normal cells (see Chapter 12). Failure to recognize this in selecting postheat growth conditions may result in serious misinterpretation of the heat resistance of the organism in question. Also, it is likely that in most foods microbial destruction during heat processing will be influenced by interactions among the various factors mentioned above.

2. *Type of Organisms*

Bacterial spores are highly resistant to extremes of temperature, some being capable of surviving for minutes at 120°C or for several hours at 100°C. Vegetative cells of spore formers, like the bacterial nonspore-forming types and yeasts and molds, are no more resistant than vegetative bacteria. Like yeasts and molds, most are killed after a few minutes at 70°–80°C, and in moist foods none survive even momentary exposure at 100°C. Thus, treatments short of complete destruction of the original microflora leave a residual population that determines the microbiological character of the food.

3. *Number of Cells*

The higher the initial population the longer it takes to reduce the population to any predetermined number, e.g., in fig. 1.4, from an initial population of 10^8, 10^5 would remain after 15 min, whereas, from 10^7 initially present, 10^4 survivors would remain after the same heating period. In practice many heat processes are designed to accommodate an anticipated bacterial "load" in the raw product. Faulty handling practices that allow excessive microbial buildup in the product can seriously affect the destructive efficiency of the heat process applied.

4. *Age of Cells*

The effect of age on heat resistance is not clear. Some have reported that young spores are more resistant than old spores (Esty and Myer, 1922), others have observed the opposite effect (Magoon, 1926; Sommer, 1930; Curran, 1935), and no correlation between age and resistance was found among several species of spore formers (Williams, 1936). With *Escherichia coli*, young cells were the most sensitive (Watkins and Winslow, 1932). Quite likely, failure to differentiate between the factors of age per se and stage of growth may account in some measure for the lack of clarity with respect to the effect of age on heat resistance.

5. *Stage of Growth*

Schulz and Ritz (1910) found heat resistance of ''coli-bacillen'' (*Escherichia coli*) greatest in the early lag, late logarithmic, and early maximum stationary phases. Extremely low resistance occurred early in the logarithmic growth phase. Others have made similar observations (Reichenback, 1911; Sherman and Albus, 1923; Stark and Stark, 1929; Elliker and Frazier, 1938). These and more recent studies (Beuchat and Lechowich, 1968; Kenis and Morita, 1968; Griffiths and Haight, 1973; Hurst *et al.*, 1974b) appear to justify the conclusion that cells in the logarithmic phase of growth are more sensitive to heat than at any other stage of growth.

6. *Growth Temperature*

The temperature at which bacteria are grown markedly affects their heat resistance. Weil (1899) was probably the first to show that heat resistance of spores increased as the temperature at which sporulation occurred was increased. This general effect has been observed in subsequent studies (Williams, 1929; El-Bisi and Ordal, 1956; Lechowich and Ordal, 1962). Similarly, the cells of non-spore formers grown at the upper end of their growth temperature range are more heat resistant than those grown at lower temperatures (Dorner and Thoni, 1936; Sherman *et al.*, 1929; Elliker and Frazier, 1938; Olson *et al.*, 1952; Tomlin and Ordal, 1976). Sherman *et al.* (1929), and later Theophilus and Hammer (1938), called attention to a practical application of the effect of growth temperature on bacteria. They emphasized the importance of maintaining raw milk at low temperatures during storage and prior to pasteurization, for, in addition to other benefits, refrigeration would preclude the emergence of the more heat-resistant cells that might develop at higher temperatures. It would also prevent the growth of the more heat-resistant mesophilic types initially present in the raw milk. This same principle has application to the heat processing of other foods.

7. *Growth Medium*

Nutrient conditions may increase or decrease heat resistance (Williams, 1929; Curran, 1935; El-Bisi and Ordal, 1956), but the variable effects of different components of a growth medium, singly or in combinations, make it impossible to generalize.

8. *Medium or Menstruum in Which Microorganisms Are Heated.*

The physiochemical nature of the menstruum profoundly influences the heat resistance of microorganisms. The literature relative to these influences, e.g., pH, a_w, salts, carbohydrates, proteins, and fats has been reviewed by Stumbo (1949), Pflug and Schmidt (1968) and others. Some generalizations can be made. Perhaps the most important factor affecting heat resistance is pH or acidity. In addition to affecting heat resistance directly, the pH of the growth medium will determine which of the survivors of a heat treatment will actually grow. For example, the division of heat-treated foods into "acid" (pH 4.6 and below) and "low acid" (pH $>$ 4.6) categories is based on the inability of *Clostridium botulinum* to grow at or below pH 4.6. Thermal resistance generally is greatest between about pH 6.0 and 8.0. Differences within this range are usually small, while above 8.0 and below 6.0 resistance decreases. There are exceptions: the heat resistance of several serotypes of *Salmonella* in liquid whole egg was found to decrease with an increase in pH from about 5.5 to 8.5 (Anellis *et al.*, 1954).

Reduced water activity as in concentrated and dried foods greatly increases heat resistance of vegetative cells and spores. This is reflected in the higher pasteurization temperature time requirements for more concentrated foods, e.g., ice cream mix (about 15% sugar) vs. that for milk. Heat resistance of salmonellae in heart infusion broth containing NaCl, glycerol, or sucrose could not be related directly to the a_w of the heating menstruum, but depended more on the solutes present (Baird-Parker *et al.*, 1970). Resistance of heat resistant strains generally decreased in presence of NaCl or glycerol, whereas that of heat-sensitive strains was increased. However, all strains showed increasing heat resistance with increasing concentrations of sucrose. The z value of *S. senftenberg* 775W was also affected by the presence of different solutes in the heating menstruum. The z values ranged from 13 in the presence of NaCl (a_w 0.90) to 7.2 and 8.9 in the presence of glycerol and sucrose, respectively (each also at a_w 0.90). Accordingly, it was concluded that it is impossible to predict the heat resistance of salmonellae in a food of low a_w; rather, D and z values must actually be determined in the particular food of concern in order to find the proper heat treatment. In extremely dry products (dried egg white, chocolate) one must depend largely on sampling and testing to determine the level of destruction attained after heat treatments have been applied (see Section V,C,4).

Foods high in fat and low in moisture (e.g., milk chocolate) require severe heat treatments for destruction even of vegetative cells. D values are given in Table 1.6 for two serotypes of *Salmonella* in milk chocolate (less than 2% moisture and high in cocoa fat). For *S. senftenberg* 775W the D_{70} value ranged from about 6 to 8 hr in three trials and for *S. typhimurium* from about 12 to 18 hr. The relative resistance of these two serotypes was the reverse of that observed in a moist food such as liquid eggs (Ng, 1966). Addition of a small amount of water to milk chocolate had a marked effect in reducing the heat resistance of *S. anatum* (Barrile and Cone, 1970). When 2% moisture was added the D_{72} value decreased from 20 hr in the normal product to 4 hr. Milk chocolate has been the vehicle involved in a few outbreaks of salmonellosis. Recent studies (Senhaji and

TABLE 1.6

***D* Values of *Salmonella senftenberg 775W* and *Salmonella typhimurium* in Milk Chocolate (the Results of Three Trials at Each Temperature)[a]**

Temperature (°C)	*D* values (min)	
	S. senftenberg 775W	*S. typhimurium*
70	360	720
	480	678
	480	1050
80	144	222
	96	222
	108	222
90	36	78
	30	72
	42	

[a] From Goepfert and Biggie (1968).

Loncin, 1977; Senhaji, 1977) show conclusively that the protective effect of fat is due to increased solubility of water in fat as heating occurs and the consequent decrease in a_w level in the fat which greatly increases the heat resistance of organisms present therein. However, if water present exceeds that necessary to saturate the fat, any organisms present in the aqueous phase would not be protected to the same extent. The latter situation probably explains the observations of Barrile and Cone (1970).

E. Relative Heat Resistance among Various Groups of Microorganisms

1. General Spectrum of Heat Resistance

The two parameters *D* and *z* (see Sections IV,B and C above) may be used to compare heat resistance among different organisms. Both *D* and *z* values may be considerably altered by conditions under which heat resistance is determined, e.g., the suspension medium (type of food), and strain differences. At best the difference can only be approximate. Several useful summaries of published *D* and *z* values have been provided. Those by Ingram (1969, 1971) shown in Tables 1.7 and 1.8 and of Stumbo (1973) in Table 1.9 will serve to illustrate the differences.

From Table 1.7 if one assumes an exposure time of about 10 min, treatment at 35°C would destroy true psychrophiles; at 55°C the psychrotrophs and the more

TABLE 1.7

Approximate Resistance of Some Nonsporing Bacteria to Heat[a]

	D value (min) at			
Species	65°C	55°C	45°C	35°C
Thermophiles	100	—	—	—
Mesophiles				
Fecal streptococci	5–30	—	—	—
Salmonella senftenberg	1–30	—	—	—
Streptococcus lactis	(0.01)[b]	2	—	—
Pseudomonas aeruginosa	—	2	—	—
Escherichia coli	0.1	5	—	—
Psychrotrophs				
Pseudomonas fluorescens	(0.01)[b]	1–2	—	—
Pseudomonas fluorescens	—	3–4	—	—
Strict psychrophiles				
Vibrio fischeri (*marinus*)	—	—	(0.1)[b]	3
Serratia sp.	—	—	—	2

[a] From Ingram (1971).
[b] Extrapolated values in brackets assuming $z = 5°C$.

heat sensitive mesophiles would begin to be eliminated; while all but the more resistant mesophilic nonspore formers would be destroyed at 65°C; the latter (certain of the enterococci and lactobacilli) would require about 70°C for their destruction. As shown in Table 1.8, except for the few cold-tolerant spore formers (e.g., *Clostridium botulinum* type E), there is then a gap in the spectrum of resistance up to organisms that resist exposures of about 1 min at 100°C. Beyond this the spectrum of resistance is more or less continuous up to about 1 hr at 100°C, which is necessary to destroy spores of *C. botulinum* type A. Beyond this there are a few highly resistant types. One of the most resistant recorded is a rare strain of *Clostridium thermosaccharolyticum* (Xezones *et al.*, 1965).

2. *Heat Resistance of Pathogens*

Table 1.9 gives a comparison of the heat resistance of a number of foodborne pathogens (see also Tables 1.7 and 1.8) and indicates that, depending on the characteristics of the food, heating moist foods (a_w above 0.93) at 65°C for about 10 min is likely to eliminate nonspore-forming pathogens present in modest numbers as well as the vegetative cells of spore formers. The spores of *C. botulinum* type E are somewhat more resistant, although by comparison to those of *C. botulinum* types A and B they are relatively sensitive. As shown in Table 1.8, D_{100} values for *C. perfringens* range from about 0.3 to 20 and the range of D_{80} values for *C. botulinum* type E are from about 0.3 to 3. Generally, *C.*

TABLE 1.8

Approximate Resistance[a] of Some Bacterial Spores to Heat or to γ-Radiation[b]

Bacterial species	Food class[c]	Heat resistance: D value (min) at 120°C	100°C	80°C	z value (°C)	Radiation resistance D_γ (mrad)
Thermophilic anaerobes						
Clostridium thermosaccharolyticum	LA	3-4	—	—	12-18	—
Desulfotomaculum (Clostridium) nigrificans	LA	2-3	—	—	—	—
Thermophilic aerobes						
Bacillus stearothermophilus	LA	4-5	3000	—	7	0.22
Bacillus coagulans (thermoacidurans)	A	0.1	—	—	—	—
Mesophilic anaerobes						
Clostridium sporogenes (including PA 3679)	LA	0.1-1.5	—	—	9-13	0.15-0.22
C. botulinum types A and B	LA	0.1-0.2	50	—	10	0.1-0.35
C. perfringens	LA	—	0.3-20	—	10-30	0.1-0.35
C. caloritolerans[d]	LA	—	3	—	—	0.15
C. histolyticum	LA	—	1	115	10	0.18-0.22
C. butyricum (or *C. pasteurianum*)	A	—	0.1-0.5	—	—	—
C. sordellii	LA	—	—	40	—	0.15
C. subterminale	LA	—	—	30	—	0.16
Mesophilic aerobes						
Bacillus cereus	LA	—	5	—	10	0.16
B. licheniformis	LA	—	13	—	6	0.22
B. megaterium	LA	—	1	—	9	0.12
B. subtilis	LA	—	11	—	7	0.22
B. macerans	A	—	0.1-0.5	—	—	—
B. polymyxa	A	—	0.1-0.5	—	—	—
Vegetative bacteria	LA	—	—	—	1	most
	A	—	—	—	1	<0.1

[a] Approximate values for broad comparisons.

[b] From Ingram (1969). Reprinted with permission from The Bacterial Spore. Copyright by Academic Press, Inc. (London) Ltd.

[c] LA = Low-acid food (pH > 4.6); A = acid food (pH ≤ 4.6).

[d] Not recognized in the 8th Edition of Bergey's Manual (Buchanan and Gibbons, 1974).

TABLE 1.9

Comparative Heat Resistance of Bacteria Important in Pasteurized Foods[a]

Bacterial group	Approximate range of heat resistance: D (°F)	z (°F)
Pathogenic and toxin-producing microorganisms	D_{150}	
Mycobacterium tuberculosis	0.20–0.30	8–10
Brucella sp.	0.10–0.20	8–10
Coxiella burnetii	0.50–0.60	8–10
Salmonella sp. (other than *senftenberg* 775W)	0.02–0.25	8–10
Salmonella senftenberg 775W	0.80–1.00	8–12
Staphylococcus aureus	0.2–2.00	8–12
Streptococcus pyogenes	0.2–2.00	8–12
	D_{180}	
Clostridium botulinum type E (spores)	0.10–3.00	9–16
Spoilage microorganisms	D_{150}	
Non-spore-bearing bacteria, yeasts, and molds	0.50–3.00	8–12

[a] From Stumbo (1973). Reprinted with permission from Thermobacteriology in Food Processing.

botulinum (types A and B) is thought to be the most heat-resistant pathogen. D_{120} values of 0.1–0.2 in phosphate buffer indicate its heat resistance (Tables 1.8 and 1.9). Recently, however, a strain of *Bacillus cereus* of unusual heat resistance (D_{121} = 2.35) was reported (Bradshaw *et al.*, 1975).

3. *Heat Resistance of Spoilage Organisms*

The spectrum of heat resistance of the food spoilage microorganisms generally parallels that of the pathogens; however, many spore-forming and nonspore-forming spoilage types are considerably more heat resistant than their pathogenic counterparts. For example, in a recent study of the heat resistance of psychrotrophic types causing flavor defects during extended refrigerated storage of pasteurized milk, species of the genera *Arthrobacter, Microbacterium, Streptococcus, Corynebacterium,* and *Bacillus* were represented (Washam *et al.*, 1977). *D*-values were not determined, but most survived heat treatments commonly used for pasteurization of milk which, of course, are adequate to destroy pathogens likely to be present.

Similarly, among the spore-forming anaerobes several of the thermophilic spoilage types are much more resistant than the pathogenic *Clostridium botulinum* types A and B. Even among the aerobic spore formers the thermophilic *Bacillus stearothermophilus* is far more resistant than *C. botulinum* (Table 1.8). When both pathogens and spoilage types are present in foods the relatively higher resistance of the spoilage types often is the determining factor in establishing the heat processes that are applied.

V. USE OF TEMPERATURE TO CONTROL MICROORGANISMS IN FOODS

A. Low Temperatures above Freezing (Chilling)

1. Benefits Derived

The chilling and subsequent maintenance of foods at temperatures between about 5°C and freezing for the purpose of extending shelf-life will also provide protection against the growth of pathogens. For example, in ham kept at temperatures below 6°C there would be no risk of enterotoxin formation by *Staphylococcus aureus*. Similarly, growth of *S. aureus* in milk or condensed milk is prevented by refrigeration (George and Olson, 1960). Even at slightly higher temperatures that may support staphylococcal growth, competition from psychrotrophic saprophytes, if present, may cause foods to become organoleptically unacceptable and therefore likely to be discarded before toxin production reaches a hazardous level. However, the absence of spoilage is not always a reliable guide to the safety of a food. Foods which are normally free from spoilage organisms, e.g., certain cooked or canned products, are particularly hazardous if they become contaminated with pathogens after processing. Storage of vacuum-packaged smoked fish at chilled temperatures may delay the onset of spoilage but permit toxin production by *C. botulinum* type E because this organism may be present and can grow at temperatures down to about 3°C. On the other hand, prolonged refrigeration of vacuum-packed red meat does not pose a type E botulism hazard because this organism rarely, if ever, occurs in red meats; and equally important, it probably would not compete well with the natural flora of red meats. Nevertheless, prolonged refrigeration of meats may pose a recently identified hazard (Hanna *et al.*, 1976) due to selective growth of psychrotrophic *Yersinia enterocolitica*-like organisms.*

With the exception of *C. botulinum* type E and possibly *Yersinia enterocolitica,* growth of foodborne pathogens is effectively prevented by rapid cooling and storage of foods below 5°C. However, some saprophytic bacteria are capable of growing on frozen foods at temperatures down to −5°C, e.g., *Pseudomonas, Alcaligenes,* and *Lactobacillus* spp. Some yeasts and several molds including *Cladosporium herbarum, Sporotrichum carnis, Penicillium* and

*The pathogenicity of all biotypes included in this taxon is not certain at present.

Thamnidium spp. may grow at temperatures below -5°C (Michener and Elliott, 1964). Reports of microbial growth as low as −10°C are not so common and may be open to question because of technical difficulties of maintaining constant temperatures for the length of time required for visible growth. However, there seems to be no doubt that *Cladosporium herbarum* can develop at temperatures as low at −8°C although it may take many weeks for evidence of growth to appear.

The lower temperature limit for growth is probably partly determined by the water activity (a_w) of the substrate. In a food, the a_w before freezing is below that of pure water, and as ice is formed during freezing, the concentration of solutes in the frozen phase increases, lowering the a_w still further. At a temperature of −10°C, the a_w of a solution in equilibrium with ice would be 0.907 (Scott, 1962; see also Chapter 4).

2. *Control Measures to Enhance Effectiveness of Chilling.*

Foods capable of supporting microbial growth must be cooled rapidly before storage. This is extremely important, especially for cooked or heated foods that are not consumed shortly thereafter. Since the temperature range between 50° and 5°C embraces the optimum temperature for growth of all pathogens and most food spoilage bacteria, failure to cool foods rapidly through this range may allow large microbial populations to accumulate in a relatively short time.

Control measures must aim to reduce the microbial population before chilling, assure an effective chilling process, prevent subsequent contamination, and avoid fluctuations in storage temperatures.

Prevention of contamination before and after chilling requires a systematic study of processing techniques in relation to possible sources of contamination and ways in which this contamination spreads to the food. Hygiene measures designed to prevent spread of contamination must be developed. This is particularly important for perishable foods held under time/temperature conditions that retard rather than prevent microbial growth. For these foods the level of contamination is often the most important factor determining shelf-life and acceptability.

It is not possible to produce one specification suitable for the chilling of all foods. Even with a single product this is difficult. For example, the storage life of chilled meat is determined by the interaction of a number of factors including time, temperature, relative humidity, air velocity, and packaging as well as the initial level of contamination (Elliott and Michener, 1965; Shaw, 1972). In addition, the undesirable toughening effect of low temperatures on pre-rigor muscle must be taken into account (Locker *et al.*, 1975). Thus, processors may operate to a number of different specifications designed to meet the requirements for a particular product. Provided this does not adversely affect the safety and quality of the product, there should be no objection.

Small differences in temperature can have a significant effect on the growth

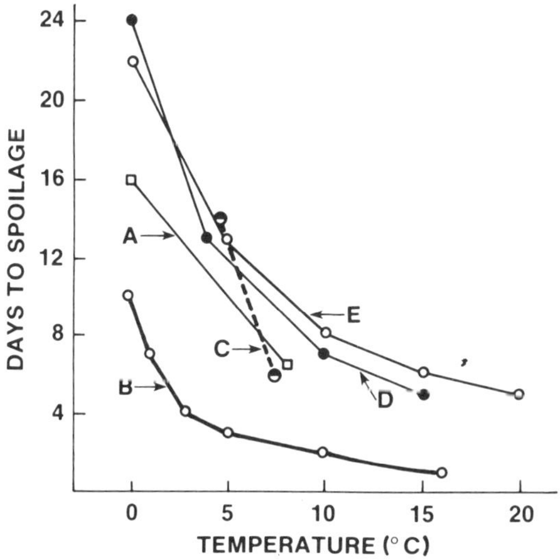

Fig. 1.6. Effect of storage temperature on shelf-life of beef: A, minced; B, joints; C, sliced; D, sliced; E, sliced. (From Elliott and Michener, 1965.)

and metabolic activity of bacteria and thereby on the keeping quality of chilled foods. A reduction in temperature from 2.8° to −0.3°C doubled the keeping time of haddock and cod fillets (Castell and MacCallum, 1950); also holding milk at 1°C, instead of 4°–5°C as is usually recommended, markedly delayed microbial growth and associated changes in the product (Solberg, 1970). The effect of temperature on the storage life of beef in various forms is shown in Fig. 1.6.

Frozen Food Storage

1. The Freezing Process

The freezing process can be divided into three stages:

1. Cooling down from the initial temperature of the product to the temperature at which freezing begins.
2. Chilling at the latent heat "plateau" during which there is a change of state of water into ice.
3. Further cooling to the ultimate temperature for storage.

The effects of freezing on the microbial population of a food are related to these three stages. The initial cooling may kill or injure a proportion of the flora through cold shock. The decrease in temperature will reduce the growth rate of those cold-tolerant organisms still able to grow. The conversion of water into ice may mechanically injure or kill a proportion of the microorganisms. The third stage of cooling to the storage temperature further inhibits growth of the most cold-tolerant microorganisms until it ceases at about −8°C.

2. *Factors Determining the Effectiveness of the Freezing Process*

The effectiveness of the freezing process as a means of preservation depends on many factors including the initial microbial contamination, the dimensions of the material being frozen, the packaging material, the freezing rate, the time and temperature of storage, the constancy of the storage temperature, and the time/temperature conditions of thawing. Freezing must be fast enough to minimize the development of microbial and enzymatic changes in the product. A freezing process taking several days is likely to permit microbial spoilage. Packaging can be used to protect the product from contamination or moisture loss, but its effect on the freezing and thawing rates must be taken into account. Because of the complexity of the freezing process, it is essential to operate to a properly designed specification with appropriate checks on the physical parameters as well as on the hygienic aspects of the process.

3. *Control Measures to Assure Adequate Freezing and Storage.*

a. Temperature Measurement and Monitoring. To ensure compliance with a specification for freezing and storage, the temperature of the product as well as that of the freezing and storage facility must be measured. This is not simple since it involves obtaining an accurate temperature at the point of measurement and selecting sufficient representative locations of measurement to obtain not only a reasonable average but also the variation in temperature, especially during freezing. During freezing the product temperature should be measured at both its surface and thermal center. Temperature variation within and among cartons (packs) is frequently much greater than hitherto supposed, as is temperature fluctuation during storage. Probe-type instruments, with small temperature-sensing elements connected to the measuring device in such a way as to minimize conduction of heat from outside, are the most satisfactory for measuring temperatures (International Institute of Refrigeration, 1971). Instruments used for checking the temperature of frozen products should be tested at least monthly to ensure their accuracy. Fluctuating temperatures can have a marked effect on the storage life of a frozen product, thus storage temperatures should be monitored continuously as a check on the performance of the temperature-control equipment.

For highest quality, the cold chain should be maintained right through from freezing to food preparation. A number of chemical and electronic time/temperature integrating devices have been developed in attempting to provide a simple but reliable means of monitoring temperature changes during the movement of frozen products from the producers cold store to the retail cabinet (Olley, 1976), but none is entirely satisfactory

b. Product Inventory Control. Control systems in freezing and cold storage facitities should include appropriate methods of identifying products and recording the location, age, and other details of all foods in the cold store. This system should have a foolproof method to indicate when any stock of product is nearing the limit of its designated storage life. Each product should be clearly identifiable from its packaging as to both its nature and age to facilitate stock control.

c. Analytical Methods. As a further check on the effectiveness of the control systems, various chemical and physical tests can be applied to frozen foods during storage including changes in ascorbic acid content, reflectance, amount of chlorophyll converted to pheophytin, extent of denaturation of protein, and release of free fatty acids. Although the correlation between chemical changes and growth of microorganisms is often poor, conditions preventing chemical changes tend to inhibit microorganisms. Taste panels, particularly when results are analyzed statistically, are also useful in evaluating the changes in organoleptic criteria such as color, texture, and flavor during cold storage.

Microbiological methods can be used to detect lapses in hygiene during the preparation of food for freezing and to determine the microbial load of raw material and of food contact surfaces. Methods that give a rapid result are particularly useful since they can provide management with information on the state of products actually under preparation. The number of organisms that have been subjected to nonlethal cold injury may not always be reflected in bacterial counts of frozen foods. For example, a rapid loss of viability of vegetative cells of *Clostridium perfringens* occurs when food samples associated with food poisoning outbreaks are frozen for storage or shipment to the laboratory. Therefore, cultural methods used for counting this organism are often inadequate for the accurate determination of population levels in suspected foods. It has been suggested that determination of α-toxin would provide a more satisfactory means of estimating population levels that may have occurred in such foods (Harmon and Kautter, 1976). *Clostridium perfringens* enterotoxin is not usually produced in foods; thus, it is not practicable to make a direct check for the presence of toxin in the food.

d. Specifications for Freezing and Storage. The International Institute of Refrigeration (1971) has published recommendations for the processing and handling of frozen foods in a table giving the estimated storage life at $-18°$, $-25°$, and $-30°C$ of various frozen food products. In the absence of more precise or demanding specifications for particular products, these recommendations should be followed.

4. Thawing after Storage.

The thawing of bulk products for retail sale or further processing requires just as strict control as the original freezing. Since heat diffuses more slowly through a thawed product than through a frozen product, thawing may take longer than freezing. Thawed materials spoil in much the same way as their unfrozen equivalents and must be kept chilled. Consequently, the use of high air temperatures for thawing large blocks of frozen product should be avoided (Buttiaux, 1973; Vanichseni *et al.*, 1972; Bailey *et al.*, 1974).

C. Heat Processes

1. General Considerations

Heat is used to control microorganisms in foods either by applying a temperature adequate to kill them or by maintaining a temperature somewhat above that which permits growth, as in the holding of cooked foods just prior to serving (see Volume II, Chapter 29). Heat treatments applied to foods primarily for other purposes may also have antibacterial effects. Blanching (scalding) which is widely applied in vegetable processing to inactivate enzymes, fix color, reduce bulk, etc., destroys most vegetative cells of bacteria as well as molds and yeasts (Splittstoesser *et al.*, 1961). Similarly, various so-called precook or cooking systems such as those applied to crustaceans to facilitate shell removal or to tuna to remove oil prior to canning also cause bacterial death. Since lethal effects of heat are additive, mild heat treatments, such as blanching and precooking, may increase the effectiveness of a full (lethal) process that follows immediately, by eliminating some heat-sensitive bacteria and sensitizing the heat-resistant types.

Heat designed to kill microorganisms in foods may be applied in different ways according to whether the food is liquid or solid. Very short treatments (seconds) are possible with liquids, but hours may be necessary with some solids because heat penetration is slow. Heat treatments are commonly described as "pasteurization" or "sterilization." One or more of the factors that influence the heat resistance of microorganisms (See Section IV,D) may be applicable when either of these heat processes is applied to foods. This is because foods differ in composition, moisture content, acidity, etc., thus presenting different environments to the microflora involved. For this reason any stated level of heat resistance of an organism is meaningless without an accompanying description of the conditions that prevailed at the time the heat-resistance determinations were made. Nevertheless, present knowledge does permit some generalization.

2. *Commerical Sterilization*

Severe sterilizing heat treatments are most often applied to low acid foods (pH > 4.6) packaged in hermetically sealed containers (i.e., canned foods). Temperature of treatment ranges from about 115° to 150°C.

Tables 1.8 and 1.10 summarize the heat resistance of spoilage bacteria and pathogens of importance to the canning industry.

For the low acid canned foods commercial sterility is necessary, i.e., foods must be *free of pathogens, as well as of microorganisms capable of growing in the food under normal nonrefrigerated conditions of storage*. The spores of *Clostridium botulinum* must be destroyed; if they are not, they may germinate and grow and form the deadly toxin of botulism. Fortunately, the *D* values of important spore formers causing spoilage of low-acid canned foods are greater than those for the more resistant types of *C. botulinum*. The D_{120} values for *C. botulinum* range from about 0.1 to 0.2; those of *C. sporogenes* and related types of putrefactive anaerobes including PA 3679 are 0.1–1.5; while the D_{120} values for the important highly resistant thermophilic types [e.g., *Bacillus stearothermophilus, Clostridium thermosaccharolyticum* and *Desulfotomaculum (Clostridium) nigrificans*] range from 2.0 to 5.0. To avoid serious economic loss due to spoilage, the severity of heat processes applied generally to low acid canned foods is greater than that necessary to assure destruction of *C. botulinum,* i.e., the generally acknowledged 12*D* process (Section IV,C). These severe heat processes are commonly applied where the hazard of thermophilic bacterial growth is expected to be high, as in foods destined for markets in tropical climates.

The presence of spores of *C. botulinum* in acid foods (pH 4.6 and below) is of little health significance, since growth of this organism below pH 4.7 has not been demonstrated. This fact and the relatively low heat resistance of the principal spoilage types for these foods (e.g., *Bacillus coagulans, B. polymyxa, B. macerans, Leuconostoc* sp., and certain lactobacilli, yeasts, and molds) makes it feasible to use heat processes for canned acid foods that are considerably less severe (even extending into the pasteurization range) than those commonly used for low acid canned foods.

3. *Pasteurization*

Pasteur's application of mild heat to wines to delay or prevent spoilage led to the designation of the process as "pasteurization." The term has a broad meaning in the modern technology of food processing. In this sense it means a heat treatment of food that is intended to destroy vegetative forms of pathogens or that which will destroy all or most vegetative forms of organisms that cause spoilage or that interfere with desirable fermentations.

TABLE 1.10

Comparative Heat Resistance of Bacteria Important to Canned Foods[a]

Bacterial groups	Approximate range of heat resistance[b] D	z
	D_{250}	
Low acid and semi-acid foods (pH above 4.5)		
Thermophiles (spores)		
Flat-sour group (*Bacillus stearothermophilus*)	4.0–5.0	14–22
Gaseous-spoilage group (*Clostridium thermosaccharolyticum*)	3.0–4.0	16–22
Sulfide stinkers (*Desulfotomaculum* [*Clostridium*] *nigrificans*)	2.0–3.0	16–22
Mesophiles (spores)		
Putrefactive anaerobes		
C. botulinum types A and B	0.10–0.20	14–18
C. sporogenes group (including P. A. 3679)	0.10–1.5	14–18
Acid foods (pH 4.0–4.5)		
Thermophiles (spores)		
B. coagulans (facultatively mesophilic)	0.01–0.07	14–18
	D_{212}	
Mesophiles (spores)		
B. polymyxa and *B. macerans*	0.10–0.50	12–16
Butyric anaerobes (*C. pasteurianum*)	0.10–0.50	12–16
	D_{150}	
High acid foods (pH 4.00 and below)		
Mesophilic non-spore-bearing bacteria		
Lactobacillus sp., *Leuconostoc* sp., and yeasts and molds.	0.50–1.00	8–10

[a] From Stumbo (1973). Reprinted with permission from Thermobacteriology in Food Processing. Copyright by Academic Press, Inc.

[b] D and z values in terms of °F.

[c] Recently an organism of this group was described by Xezones *et al.* (1965) as having a D_{250} value in excess of 50 min. At present this must be considered as a very rare exception.

Since pasteurization processes do not inactivate all microorganisms present, some additional controlling factors may be necessary, such as low temperature storage, acidity, reduced a_w (concentration or drying or combinations of these), and curing salts.

Generally, temperatures below 100°C are used in pasteurization processes. The severity of the heat treatments vary greatly depending on the nature of the food and various purposes to be served. For some foods, to be consumed without further treatment, e.g., milk and liquid eggs, destruction of pathogens to assure public health safety is foremost in importance. For other foods the principal objective is to reduce the level of spoilage organisms so that the products may have adequate shelf-life (the spoilage-free life of the product). Pasteurization often is applied prior to a fermentation process to destroy organisms that might cause abnormal fermentations or otherwise interfere with the normal fermentation. Pasteurization of wine, for example, is designed almost exclusively to prevent spoilage; the alcoholic content provides safety against pathogens.

A pasteurization process often accomplishes all three purposes. For example, in several countries laws or regulations specify the minimum pasteurization temperature/time requirement for milk to be made into cheese if that cheese is to be labeled "made from pasteurized milk." In this case, pasteurization destroys vegetative pathogens as well as many spoilage types and others that could seriously interfere with desirable fermentation by added "starter" cultures.

Viable spores of *Clostridium botulinum* will survive pasteurization, but combinations of other factors, such as low a_w, refrigeration, preservatives, frozen storage, acidity, are used to prevent their outgrowth.

The range of heat resistance for several pathogens of concern in pasteurized foods, as given by Stumbo (1973), is shown in Table 1.9. Also shown is the range of D_{65} values for the spoilage types of non-spore-forming bacteria, yeasts and molds.

4. *Control Measures to Assure Adequate Heat Processing.*

Prescribed heat treatments must be applied invariably and properly to protect against both hazards to health and premature spoilage. Modern food processing involves heat treatment of a wide range of foods from large volumes in highly complex equipment, often continuously over many hours of operation, to small batches which may not require large complicated equipment. In both cases there is a need for "process assurance." Sampling and testing of finished products is inadequate as the sole measure of product safety, since adequate analytical methods are not available to detect the presence of all pathogens. Those methods that are available to detect certain pathogens relatively simply and in a reasonable period of time are subject to serious limitations in their application. Study of a table of probabilities of acceptance of food lots will illustrate the importance of interpretation of negative results (see ICMSF, 1974). For example, if 10 sample units are tested for the presence of a defect (i.e., salmonellae) from a lot of food having 5% defective (contaminated) packages or units, all 10 samples would yield negative results 60% of the time, i.e., the sampling and testing plan would fail to detect/reveal a 5% contamination level (defective unit level) 60% of the time—hardly an adequate mechanism for giving process assurance!

One should not conclude that sampling and testing has no place in assuring product safety and the attainment of other parameters of product quality. If the knowledge of the microbiology of a product is adequate, especially as it reflects changes throughout various stages of processing, it is possible to apply certain selected analytical methods. which will provide useful and reasonably reliable information about the sanitary practices used in handling the food and the safety of the final product. Sometimes sampling and testing is the only means at hand to obtain information about the acceptability of a food. Also, sampling and testing may be necessary as a supplement to some other criterion of judgment. For example, unpasteurized egg white (albumen) is dried to produce flakes, crystal albumen, or powder. If salmonellae are present they will survive the drying process. It is necessary to hold the packaged dried product in a ''hot room'' at 52°C for a week or more to destroy the organisms. In recognition of the difficulty of predicting the heat treatment necessary to destroy the organisms in a dry environment (see Section IV,D,8), it is common practice to test the egg white for salmonellae following this hot storage, and release it for distribution only if the test is negative; if positive, storage is continued until the test is negative. Some regulatory agencies require a minimum holding period plus sampling and testing. The combination of the minimum hot storage period (with adequate monitoring of temperature) with sampling and testing, provides a reasonable mechanism to assure safety (see also Chapter 19).

The increasing complexity of modern food processing and the limitations of sampling and testing have led to other approaches to process assurance. Various ''fail safe'' mechanisms have become integral components of heating equipment. Perhaps the best example is in the continuous pasteurization of liquid egg products and milk where a flow diversion valve will automatically divert product from forward flow whenever a sensing device detects the temperature of the product below that prescribed. Additionally, a pump that controls flow rate (holding time) is set so that the rate of flow through a holding tube is never exceeded. Thus, these two components of the heating system assure that the product is always held for the required time at a prescribed temperature, provided, of course, that the devices are properly installed and monitored. Numerous other failsafe mechanisms are in use. Others are being developed as new technologies evolve or existing ones are modified. It is not the purpose here to delineate these, rather, only to emphasize the importance of adequate controls over processes, especially those designed to assure food safety. Their installation and use require specialized knowledge in engineering design, and many manufacturers of heat processing equipment maintain technical service units staffed with personnel fully trained and available to assist in installation and proper operation of their equipment. For further discussion and details the following references will be helpful: Harper (1976); Heldman (1975); Farrall (1976); ICMSF (1974); U.S. Department of Health, Education and Welfare (1978); U.S. Department of Agriculture (1975).

VI. SUMMARY

The response of microorganisms to temperature varies with the type of organisms involved. Some (the thermophiles) are capable of rapid growth at temperatures far above common ambient temperatures. These organisms are important in the spoilage of canned foods during storage and distribution where higher than normal ambient temperatures may be encountered (e.g., warehouses in the summer and in tropical environments and in holds of ships during lengthy voyages). Mesophiles which grow well at moderate temperatures include nearly all the pathogens and most food spoilage organisms. Others, the psychrotrophs, grow well at refrigeration temperatures or until freezing causes temperature and a_w to become the growth-limiting factors. At temperatures only a few degrees above the maximum growth temperature for any particular type, injury and eventually death will occur. Death occurs logarithmically, which makes it possible to develop heat processes that will assure their destructive effects with precision. Thus, the process in terms of time and temperature depends largely on the heat resistance of the contaminating organisms and the numbers of them that must be destroyed. Numerous applications of the principles involved in the relationship of microorganisms to temperature, for better or for worse, will be found in Volume II, (Chaps. 15–29) which deal with various food commodity groups.

2

Ultraviolet Irradiation

I. INTRODUCTION

Among the spectra of radiation potentially available for application to foods (Proctor and Goldblith, 1951), it is useful to distinguish two different categories. The first is radiation of relatively low frequency and wavelengths longer than those of visible light, roughly from 10^7 to 10^{10} Hz*, i.e., roughly from shortwave radio (10 megacycles) past radar wavelengths, and into the infrared —roughly 10^{12}–10^{14} Hz. These radiations have low quantum energy and affect molecules by general thermal agitation. Despite occasional reports of specific effects, the balance of evidence strongly indicates that their lethal effects on microorganisms are solely due to this heating, hence they are not considered here.

In the second category are radiations shorter than visible light, with frequencies of about 10^{15} Hz or more, which have sufficiently high quantum energies to excite or destroy organic molecules and hence have a specific lethal action. Those of the lower frequency and energy, in the ultraviolet (uv) part of the spectrum, suffice only to excite molecules and are the subject of this chapter. Those of higher frequency, from about 10^{18} Hz and upward, are sufficiently energetic to break individual molecules into oppositely charged parts called ions, and these radiations are called ''ionizing'' (see Chapter 3.)

A. Type and Sources of Radiation

The uv region of the spectrum extends below wavelengths of ca. 450 nm. The most effective wavelength for the destruction of microorganisms is near 260 nm, of quantum energy about 4.9 electron volts (eV). Wavelengths below 200 nm are

*Hz = Hertz. 1 Hz = 1 cycle per second; frequency (in Hz) × wavelength (in meters) = 3×10^8.

ineffective because they are too readily absorbed by atmospheric oxygen. Wavelengths from 360 to 450 nm are commonly called "long-wave ultraviolet," or "black light," or "near ultraviolet." They are frequently used to excite fluorescence, for example, in ultraviolet candlers to demonstrate the presence of pigments of *Pseudomonas* in eggs. These wavelengths penetrate glass readily and have limited effects against microorganisms (Thomas, 1977).

Ultraviolet is present in light from any high-temperature source (e.g., sunlight or arc lamps), and from some fluorescent tubes. In practice, the usual source of uv is a low-pressure mercury vapor discharge lamp, emitting at several specific wavelengths, some visible, but with about 80% of the total uv emission at 254 nm, which has 85% of the biological activity of 260 nm. The lamp should have special glass to absorb emission below 200 nm which, if absorbed by atmospheric oxygen, produces ozone which is undesirable (see Section II,C). High-pressure mercury lamps used for lighting have little germicidal effect. Laser-activated lamps with much greater outputs are now becoming available.

B. General Properties

The intensity of irradiation* is measured by the energy absorbed per unit area, usually expressed as ergs/sec or $\mu W/cm^2$ (10^7 ergs/sec = 1 watt). A 50-watt low-pressure mercury vapor lamp gives an effective intensity of about 100 $\mu W/cm^2$ at a distance of 1 m. The quantity of "dose" of irradiation absorbed is measured by the total energy (i.e., with a given intensity, the product of time by intensity as above defined) expressed as ergs or μW sec/cm^2. Doses used against microorganisms, up to about 10^5 μW sec represent quantities of energy too small to produce a significant rise in temperature (4.2×10^7 ergs = 1 calorie).

Because of its low quantum energy, uv irradiation is completely utilized by those molecules which absorb it. The molecule is excited by the energy absorbed and becomes potentially capable of unusual reactions which may lead to its destruction, with possible lethal effects on microbes.

There is strong preferential absorption by particular substances, e.g., nucleic acids. The most lethally effective wavelength, near 260 nm, is that maximally absorbed by nucleic acid bases, and much other evidence also indicates nucleic acids to be the major target for uv radiation. It is immediately absorbed, however, by various materials relevant to food. The absorption of the shorter wavelengths by oxygen has already been noted. Absorption by water depends on its clarity; in clear water, intensity is reduced by about two-thirds for every 5 cm of depth, but in river water two-thirds of the energy is absorbed by only 1 cm of depth. Its activity is also affected by certain ions, e.g., Fe^{3+}. Proteins and bases

*Irradiation is the process of applying radiant energy to a target, such as a food.

also absorb uv strongly, so that in milk, for example, a layer only about 0.1 mm thick can absorb 90% of the incident energy. Absorption is roughly proportional to the concentration of such substances, while penetration is even less into solid foods.

Since the penetration of uv radiation is small into liquids and negligible into solid foods, its main value lies in killing microbes suspended in air or exposed on surfaces. The organisms, however, are still resistant if they are covered by films of protective substances, as in aerosols or on the surface of wet or greasy foods.

C. Methods of Measurement

The intensity of radiation at various wavelengths may be measured by physical methods, but there is no simple method for estimating effective dose, which must be judged arbitrarily in terms of time and distance from a given lamp. Attempts to measure microbiological effect directly are confused by the large influence of intrinsic and extrinsic factors. Nevertheless, with some care, it can be done (see Section II,A).

II. EFFECTS OF ULTRAVIOLET LIGHT ON MICROORGANISMS

A. General Lethality

When a uniform population of microbial cells, spores, or conidia, is irradiated at constant intensity, the number of survivors declines exponentially with time (i.e., with the total dose of radiation applied). Because of this, it is convenient to express the resistance of an organism as the dose needed to reduce the number of survivors to one-tenth (i.e., by 90%)—a quantity virtually independent of the absolute numbers involved and of the dose rate. This quantity, as in the corresponding situation with heating, can be called the *D* value or "decimal reduction dose." Typical *D* values are *Serratia marcescens,* 1500 μW sec; *Escherichia coli,* 2000 μW sec; spores of *Bacillus mesentericus,* 10,000 μW sec.

In establishing such *D* values for uv much greater care is needed than with ionizing radiation, because of ready absorption of the radiation by the medium, etc., as indicated above. The *D* value also depends on the physiological and other conditions of the cells.

B. Susceptibility of Different Species

There is lack of satisfactory information on the susceptibility of different microbial species to uv. Different strains of a single species may differ in resistance, and comparison of observations by different workers is confounded by differences in irradiation conditions, cell age, etc. Consequently, there are manifest discrepancies in the literature.

It has been stated (see Sykes, 1958, p. 173) that, broadly speaking, gram-negative, nonsporing rods are the most easily killed by uv irradiation, that staphylococci and streptococci require about five times as much, bacterial spores about 10 times; mold spores (dry) about 50 times, and viruses still more. Table 2.1 indicates that most gram-positive bacteria and spores are less resistant than this generalization suggests.

Nevertheless, in broad terms, the order of resistance to uv among bacteria resembles that to ionizing radiation. It is quite different in other major groups: yeasts are about as resistant as bacteria to uv, molds much more so; the converse holds for ionizing radiation (see Chapter 3).

C. Public Health Implications

In any practical use, uv radiation of the nature and intensity involved here, is immediately dangerous to operatives, to their skin and especially their eyes. Appropriate screening and/or limitations on period of exposure are essential.

While much concern has been expressed over the mutagenic effects of ionizing radiation, little has been expressed about hazard from such mutations when using uv, perhaps because long experience without special precaution, but without ill-effects, has indicated the hazard to be negligible.

If the uv lamp produces emissions below 200 nm, there is the possibility of production of ozone, which, besides further accelerating oxidation of fats, is toxic to man at concentrations below the odor threshold (cf. Chapter 10).

III. USE OF ULTRAVIOLET LIGHT IN THE TREATMENT OF FOOD

The greatest value of uv treatment lies in the decontamination of air, for which purpose it is widely used (e.g., in hospitals). It may also be applied to surface sterilzation of foods or food-handling equipment.

A. Air

In food microbiology laboratories, uv lamps have a use in cleansing the air in inoculating areas, which may be of special value where industrial conditions lead to heavily contaminated atmospheres.

Similarly, bakeries have found that uv light can help to control the occurrence on bread of mold spores, or to control the spread of microbes to the surface, but not the interior, of custard pies.

TABLE 2.1

Approximate Doses of Ultraviolet Radiation Needed to Destroy Various Microorganisms

Species observed[a]	Dose needed for one decimal reduction (*D* value) (μW sec × 10^3)
Gram-negative bacteria	
Facultative anaerobes	
Escherichia coli	3–4
E. coli	2.1–6.4
Branhamella (*Neisseria*) *catarrhalis*[b]	3–4
Proteus vulgaris	<3
Salmonella typhi (*Eberthella typhosa*)	3–4
Salmonella typhi[b]	2.1
Serratia marcescens	3–4
Serratia marcescens[b]	0.8–2.4
Shigella flexneri (*paradysenteriae*)	3–4
Aerobes	
Agrobacterium tumefaciens	3–4
Pseudomonas aeruginosa (*pyocyanea*)[b]	5.5
Pseudomonas fluorescens	3–4
Obligate aerobes	
Fusobacterium nucleatum (*Fusiformis fusiformis*)	<3
Phototrophic	
Rhodospirillum rubrum	5–6
Gram-positive bacteria	
Bacillus anthracis	5–6
Bacillus subtilis (vegetative)	6–8
Bacillus subtilis (spores)	8–10
Corynebacterium diphtheriae	5–6
Micrococcus piltonensis[c]	6–8
Micrococcus sphaeroides[c]	10–20
Micrococcus luteus (Sarcina lutea)	10–20
Staphylococcus aureus (albus)	4–5
Staphylococcus aureus	3–4
Staphylococcus aureus[b]	2.2
Streptococcus sp. (α-hemolytic)	3–4
Streptococcus sp. (β-hemolytic)	4–5
Streptococcus pyogenes[b]	2.2
Molds	
Aspergillus glaucus	50–100
Aspergillus flavus	50–100
Aspergillus niger	<200
Mucor racemosus	20–50
Geotrichum (*Oospora*) *lactis*	10–20
Penicillium digitatum	50–100
Penicillium expansum	20–50
Penicillium roquefortii	20–50
Rhizopus nigricans	>200

(*continued*)

TABLE 2.1 (*continued*)

Species observed[a]	Dose needed for one decimal reduction (*D* value) (μW sec × 10³)
Yeasts	
Saccharomyces cerevisiae (baker's yeast)	3–4
Saccharomyces cerevisiae (molasses yeast)	6–8
Saccharomyces cerevisiae (brewer's yeast)	<3
Saccharomyces ellipsoideus (wine yeast)	5–6
Saccharomyces ellipsoideus (wild)	8–10

[a] Data from Proctor and Goldblith (1951) (except as noted in footnote *b*). Dose calculated on the basis that the arbitrary "click" unit quoted represented 200 μW.

[b] Hollaender (1942). Dose calculated on the basis of 1 μW sec = 10 ergs.

[c] Deleted from Bergey's Manual, 7th and 8th Editions (Buchanan and Gibbons, 1974).

In large tanks filled with concentrated sugar syrups, otherwise microbiologically stable, condensation in the head space leads to diminished sugar concentration at the surface of the syrup, which permits superficial growth of airborne molds. This can be prevented by mounting uv lamps in the head space.

B. Liquids

Because of the limited penetration of uv radiation, liquids must be exposed to it in layers of thinness appropriate to their absorbing power.

Although uv light is in principle excellent for treatment of water, the requirement of exposure in layers only about 1 cm deep makes the equipment complex and costly, and the water must first be clarified. To kill significant numbers of microorganisms, relatively long exposure times are needed, so that throughput is low. These limitations make this method less practical than chlorination.

Irradiation of milk requires exposure as a very thin film passing the source, making control difficult and throughput very low. The exposure promotes oxidation of the fat and development of off-flavors. Hence, though bacterial counts can be reduced by 90–99% in appropriate conditions, the procedure has no practical value.

C. Surfaces

Ultraviolet irradiation can effectively destroy microbes on a surface, but only if the surface has been cleaned effectively beforehand and if a sufficient time of exposure is allowed. As a rule, only those sufaces directly irradiated will be sanitized, because even ordinary glass is relatively opaque to the germicidal wavelengths.

Ultraviolet irradiation can be used to sterilize packages where other methods, notably heat, are inapplicable. Usually the packages must be open to allow the radiation to penetrate, as, for example, treating the material used to form the containers in aseptic filling of ultrahigh temperature (UHT) milk. Only if the material is transparent to uv, eg., some thin plastic bags, can the interior be treated without opening the package. Sugar used for canning may contain thermophilic spores which are difficult to destroy except by uv.

D. Solid Foods

Numbers of about 10 spores of *Bacillus stearothermophilus* per gram of sugar may be destroyed by exposures of about 30 min, in layers of sugar 4 mm deep, penetration perhps being assisted by multiple reflections from the surfaces of the crystals. Exposure times can be shortened if the layer of sugar is turned by shaking but are greatly lengthened if the sugar is lumpy. Sugar is little affected by uv light.

Another use in the food industry is in chill rooms used for storage of meat to delay the growth of surface microbes. Reduction of aerial contamination seems unlikely to be important (cf. Chapter 15). Although the energy actually applied to the meat surface is relatively low, and the microbes are numerous and largely embedded in a protective substratum, there are persistent claims of useful prolongation of storage life. Generally speaking, microorganisms on only those solid surfaces directly exposed to uv will be affected. However, production of ozone, or reflectance of the uv rays can have a lesser effect on microorganisms lying on surfaces not directly exposed.

The problem here, however, is not sterilization so much as retardation of multiplication; moreover, the radiation is applied continuously. With an organism multiplying 10-fold in a period of about 10^4–10^5 sec, and a decimal reduction dose of about 10×10^3 μW sec (cf. Table 2.1), the multiplication could apparently be offset by irradiation at effective intensities of 1–0.1 μW. In fact, lag phases are longer, growth rates lower, and maximum cell densities smaller, even with such low irradiation intensities at the meat surface (Kaess and Weidemann, 1973). For example, uv irradiation of a spoilage *Pseudomonas* on meat reduced its growth rate to about 85% of the unirradiated control with an intensity at the surface of 2 μW/cm^2 and to about 75% with 24 μW/cm^2. An intensity of 0.2 μW/cm^2 sufficed to retard production of aerial hyphae by meat molds and to restrict the size of bacterial colonies. These data refer to 0°C and a near-saturated atmosphere [equilibrium relative humidity (ERH) 0.993]; with ERH 0.985 growth was reduced more. The appearance of improvement was greater than these data imply, because the irradiation tended to eliminate all growth above the meat surface. An intensity of 2 μW/cm^2 prevented production of aerial hyphae by molds, and 24 μW/cm^2 confined bacterial growth to capillary spaces below the meat surface. Colonies were visible only with difficulty and many of the bacteria developed as long filamentous cells. The intensity on the *shaded* side of a carcass, 2 m from a single 25 watt lamp, may be about 0.2 μW/cm^2, enough to be significant, and this level may be increased if the walls of the room have reflecting surfaces. Destruction of spoilage odors by uv may mislead a buyer to believe that the food is fresh and unspoiled. This effect could not be reproduced by the introduction, instead, of ozone (Kaess and Weidemann, 1973).

E. Damage to Foods

Ultraviolet irradiation is unsuitable for use with certain foods which contain fat, especially unsaturated fats, since it greatly accelerates development of rancid flavors owing to its strong catalytic effect upon lipid oxidation (cf. butter exposed to sunlight). Hence, its use with dairy products must be restricted, and it is much more applicable to mutton or beef than to pork in chill rooms. It damages green vegetables, producing discolored spots on the leaves.

IV. COMBINATION EFFECTS AND INTERACTIONS

There is some evidence that uv has less effect on dry microorganisms than on wet ones, but experiments with organisms suspended in air at different relative humidities may not have taken into account the effects of clumping, and there is much disagreement in detail. Up to tenfold greater resistance has been reported in "dry" as compared with moist atmospheres, and there are reports of a relatively abrupt change in resistance at relative humidities about 60%. There is no information about the corresponding relations to foods. Temperature, within the normal range 0°–40°C, has almost no effect on sensitivity to uv. Organisms exposed to uv become more sensitive to heat and vice versa as with ionizing radiations. The lethal effects of uv, unlike those of ionizing radiation, are are not much increased by the presence of oxygen.

Like ionizing radiation, uv inhibits cell division before growth in size; hence, sublethal doses lead to development of filamentous cells, which later either resume normal cell division or die.

Conditions after irradiation have a major influence on the recovery of irradiated cells. Though the subject is extensive, little has emerged that is important in applications to food: most noteworthy are the restorative effect of visible light ("photoreactivation") and suggestions that recovery is best on poor media and at temperatures suboptimal for growth.

3

Ionizing Irradiation

I. INTRODUCTION

Ionizing radiation is characterized by a very high energy content, great penetrating power, and lethality due to action at the cellular level. Until recently, the main applications of ionizing radiation have been in the field of medicine, in industrial processes, and as tools to examine the genetics and structure and function of microorganisms. Practical application of ionizing radiation to destroy microorganisms on commercial products (principally pharmaceuticals) has developed only within the past two decades.

The lethality to microorganisms of α, β, γ, and X rays was established early in this century. With the exception of X rays, methods for production of ionizing radiation in a controlled and economic way, having the intensity needed for large scale industrial use, were not available until after World War II when the intensive studies of atomic fission processes for military purposes had resulted in a much clearer understanding of the sources and origins of ionizing radiation and a great accumulation of technical knowledge in this field. Public attention was focused on the peaceful uses of atomic energy, and work on the preservation of foods by this means began in a number of countries during the early 1950s and has continued with variable intensity ever since. Initially, only the few countries possessing atomic capability were involved, with the bulk of the work in the United States, but in recent years interest in irradiation of foods has extended to many other countries.

The early work clearly established the potential usefulness of irradiation as a means of food preservation and indicated from a practical standpoint that the two most feasible methods would be machine-generated focused electron beams (β radiation) and γ rays derived from the decay of radioactive isotopes (such as ^{60}Co) available as a by-product of reactor technology. During the early 1960s a few governments permitted the use of ionizing irradiation for processing a limited range of foods. Subsequently this permission was withdrawn in the United States because experimental evidence suggested that irradiation could induce potentially mutagenic, teratogenic, or carcinogenic compounds in some foods. Investigators in a number of countries are examining the safety of a range of irradiated foods by animal feeding trials and by cellular tests of mutagenicity and teratogenicity of extracted material. Results to date support the safety of the process, and it seems likely that irradiation will again be approved as a food processing method.

The literature of effects of ionizing irradiation on microorganisms is extensive, and we do not intend that this chapter should cover more than those facts important to the microbiologist concerned with food products. Ionizing irradiation has certain advantages over other methods for destroying bacteria in foods.

1. It is highly lethal, but the dose may be adjusted to yield pasteurizing or sterilizing effects.
2. At lower levels (<0.5 Mrad) it produces virtually no organoleptically detectable change in the product.
3. Even at high dose levels (>1 Mrad) total chemical change in the food is small.
4. There are no residues of non-food material.
5. Negligible heat is produced so that raw products retain their raw food characteristics and foods may be processed in the frozen state.
6. Penetration of radiation is instantaneous, uniform, and deep, permitting precise process control (in contrast to heat).

There are also some disadvantages.

1. Enzymes are usually not inactivated at bactericidal dose levels and may remain active in food during storage.
2. Chemical changes, though small in total amount, may cause unacceptable organoleptic changes in sensitive foods or foods given high dose levels. These changes are commonly associated with free radical action, which may also be important as a secondary bactericidal factor.
3. Some studies have suggested that mutagenic, teratogenic, carcinogenic, or even toxic factors may be induced in foods by irradiation, although recent authoritative assessments now suggest that these were largely groundless (FAO/IAEA/WHO Expert Committee, 1977).
4. Dose levels used to destroy microorganisms are several orders of magnitude greater than those killing man and therefore very strict safety precautions must be taken to protect operators and food handlers. This involves heavy shielding of the source if radioisotopes are used, and continuous monitoring of the area and of personnel.

A. Definitions

Ionizing radiation: An emanation of sufficiently high photon energy to displace electrons from target molecules.

Irradiation: The process of applying radiant energy to a target, such as food.

β Radiation (cathode rays): Beams of very highly (negatively) charged electrons derived from decay of certain radioactive isotopes or, more commonly in practice, from machines such as linear accelerators. Electron beams may be focused, and their energy contact may be varied. They may produce high intensity effects over areas of a few square centimeters, but their penetration is limited to a few centimeters (e.g., 5 MeV β rays may penetrate 2.5 cm). They are most usefully applied for surface treatments or thin food layers.

X Radiation: Produced by focused cathode rays in an evacuated tube. They have high energy and good penetrating power and may be focused by shielding. They are uneconomic for industrial use except in some special circumstances (e.g., metal parts inspection) and are not used in food processing.

γ Radiation: Very high intensity radiation with great penetrating power produced during the decay of radioactive isotopes. The most widely used isotope is ^{60}Co which generates γ rays suitable for treatment of food up to 40 cm thick.

The sterilizing effect of radiation is related to the ionizing capacity and therefore to the energy of the radiation or more specifically energy absorbed by the food. Radiation is therefore quantitated in terms of "radiation dose," which may be expressed as dose delivered (roentgen) or dose absorbed (rep, roentgen equivalent physical or rad). In practice absorbed dose is the most useful measurement for foods. The rep, widely used in early studies of food irradiation, is defined as the quantity of ionizing radiation absorbed by soft tissues in gaining 83 ergs of energy per gram of tissue. This is roughly equivalent to the absorption of one roentgen. In more recent work the term rad is generally used. One rad is defined as the dose equivalent to the absorption of 100 ergs per gram of substrate. The terms kilorad (krad) and megarad (Mrad) are used to indicate doses of 1000 and 1,000,000 rads, respectively. The new SI system uses the Gray (Gy = 100 rads). Differences among these units are not important for present purposes. With a source of given intensity, the dose applied is directly proportional to the time of exposure.

Radiation source strength is measured in curies (Ci) where one Ci equals 3.7×10^{10} disintegrations per second.

B. Properties of Ionizing Radiation

When high energy radiation strikes atoms it excites them, causing ejection of electrons and the formation of ion pairs consisting of the negatively charged electron(s) and the positively charged residues. Ionization may proceed further as ejected electrons strike other atoms, giving rise to a chain effect that terminates when particle energies fall to a low level.

Water is readily ionized and may be the primary source of ionization in foods with secondary effects on other molecules, possibly more a result of water ionization than of direct hits. In the presence of oxygen, highly reactive compounds may be produced, such as H, H_3O^+, and H_2O_2.

Radiation at the energy flux levels used for foods (<2 MeV) does not induce radioactivity. However, radioactivity can be induced by very high energy radiation in excess of 10 MeV, which is unsuitable for use in food treatment.

It is characteristic of ionizing radiations that they penetrate matter instantly, to a degree depending on the nature of the radiation and the density of the material. For a given energy, γ rays penetrate much farther than electrons. The useful penetration of 3 MeV β rays into water is approximately 1 cm and the same for liquid or solid foods of roughly the same density; whereas the 1.1 MeV γ rays from ^{60}Co penetrate about 20 cm of food but are stopped by a similar thickness of lead. By irradiating from both sides, simultaneously or successively, it is possible to attain sufficiently uniform absorption throughout the whole depth of a piece of food of roughly twice these thicknesses. Further, given appropriate "geometry" and regularity in shapes, the distribution of dose throughout the food can be nearly uniform. When the molecular damage is in a vital part of a living cell, that cell dies, and there is much evidence indicating that damage to the nucleus or nuclear material is vital. In a small proportion of cases, the damage results in mutation.

Larger organisms are killed by much smaller doses of ionizing radiations, as illustrated in Table 3.1. Consequently, when killing bacteria using dose levels of the order of 100 krad, the radiation must be screened from human operators.

Fuller accounts of ionizing radiations are to be found in textbooks (e.g., Bacq and Alexander, 1961; Desrosier and Rosenstock, 1960) or in specialized reports (e.g., Hannan, 1955; International Atomic Energy Agency, 1970a).

C. Methods of Measurement

Dose is normally measured by dosimeters, while the source is monitored by modified ionization chambers. The most widely used dosimeters are cobalt glass, ferrous sulfate, or cerous sulfate. Cobalt glass changes color in proportion to the dose received. Ferrous and cerous salts are oxidized to ferric and ceric salts as a result of interaction with hydroxyl radicals produced in water by radiation.

The Fricke dosimeter, consisting of an acid solution of ferrous salts containing chloride to inhibit side reactions, is frequently used in food work. It has been determined that 15.45 ± 0.11 ferric ions are formed for every 100 eV of absorbed energy. Ferric ion concentration thus becomes a direct measure of energy absorbed. The ion yield is termed the *G* value.

TABLE 3.1

Approximate Lethal Radiation Doses for Various Organisms[a]

Type of organism	Dose (krad)
Higher animals and man	0.5–1
Insects	1–100
Vegetative bacteria and yeasts	50–1,000
Bacterial spores	1,000–5,000
Viruses	1,000–20,000

[a] After IAEA (1970a).

Dose distribution in a container of food exposed to a source of ionizing radiation may not be uniform. Variations between 100 and 125% are commonly encountered and should be anticipated in calculating food processing treatments.

Calculation of lethal doses for microorganisms depends on measurement of survivors and mathematical treatments similar to those used in thermal process calculations. This is described in the next section.

II. EFFECT OF IONIZING RADIATIONS ON MICROORGANISMS

A. General Effects

The death of microorganisms is a consequence of the ionizing action of high energy radiation. Most studies indicate that lethal damage to microbial DNA resulting in loss of ability to reproduce is a primary cause of lethality, but damage to other sensitive and critical molecules (e.g., in membranes) may also have an effect. Dose levels lethal for microorganisms do not cause inactivation of most enzymes or major change to proteins and other large molecules. Free radicals and other reactive molecules are formed, particularly in water, but it is not clear to what extent primary ionization (''direct hits'') and secondary effects are responsible for the lethal damage to microorganisms. Sublethal damage may interfere with recovery of particular bacteria from irradiated foods. Microorganisms differ greatly in their sensitivity to irradiation.

In a general sense, death of microorganisms from exposure of populations to ionizing irradiation is logarithmic in nature. Thus, if a culture or suspension of a single strain of a microorganism is exposed to a consistent flux of lethal ionizing irradiation, a constant fraction of the population will be killed at equal time intervals, regardless of the total numbers. In practice, this means that a graph of the logarithm of survivors against dose is a straight line over most of its length

(Fig. 3.1). A small "shoulder" may appear at the beginning of the curve in some cases. This may be due to phenomena unconnected with the actual irradiation sensitivity of the microorganisms such as aggregation of cells, or to there being at least two sensitive sites per cell which require inactivating before the cell loses its viability. The situation is similar mathematically to death due to heat, and the *D* value (decimal reduction time) concept of the thermobacteriologist has been applied usefully to irradiation death. The *D* value is defined as the dose necessary to reduce the population 10-fold (one $\log_{10}$ unit). Thus a 2*D* treatment would imply a dose sufficient to reduce the viable numbers 100-fold.

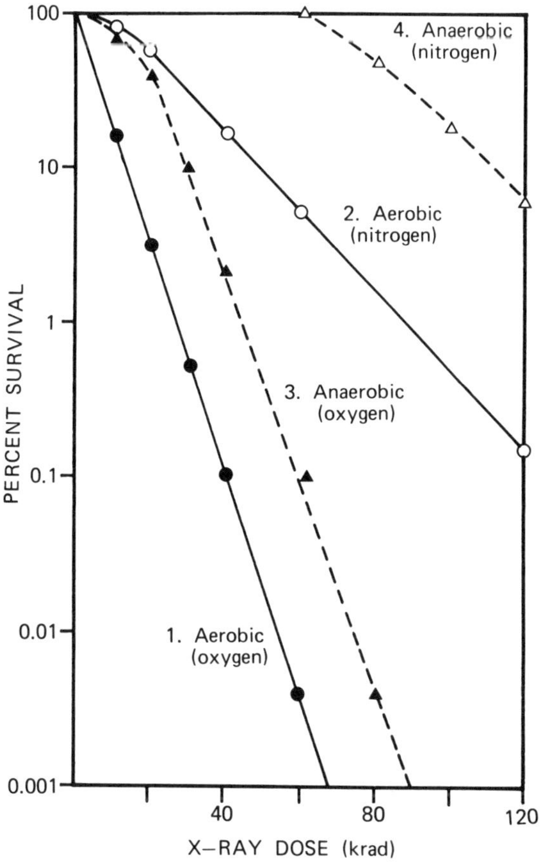

Fig. 3.1 Lethal effect of X irradiation on *Escherichia coli* B/r. (1) Cells grown aerobically, irradiated in oxygen. (2) Cells grown aerobically, irradiated in nitrogen. (3) Cells grown anaerobically, irradiated in oxygen. (4) Cells grown anaerobically, irradiated in nitrogen. Adapted from Hollaender *et al.* (1951).

B. Factors Affecting Sensitivity to Irradiation

Removal of oxygen increases the resistance of microorganisms to the lethal effect of irradiation (see Fig. 3.1), suggesting that oxidizing reactions occurring as a result of irradiation are important.

Removal of water also increases the resistance of microorganisms to ionizing irradiation. In fully dehydrated substrates, doses 2–3 times as high are required to obtain microbicidal effects equivalent to those in hydrated systems, due to the

TABLE 3.2

Approximate Radiation Doses Needed to Reduce Viable Numbers a Millionfold, for Various Bacteria Irradiated in Frozen Media[a]

Microorganisms	Dose (krad)
Gram-negative bacteria	
Facultative anaerobes	
Enterobacter aerogenes 2/3[b]	300[c]
Enterobacter aerogenes 1/3	300–600
Escherichia coli 5/6	300[c]
Escherichia coli 1/6	300–600
Proteus vulgaris	300[c]
Salmonella 4/6	300[c]
Salmonella 2/6	300–600
Serratia marcescens	300[c]
Aerobes	
Alcaligenes faecalis	600–900
Pseudomonas aeruginosa	300[c]
Gram-positive bacteria	
Lactobacillus plantarum (arabinosus)	300[c]
Lactobacillus casei	300[c]
Micrococcus radiodurans[d]	>3000
Staphylococcus aureus 6/6	300[c]
Streptococcus faecalis 1539	600–900
Streptococcus faecalis QM	1200–1500
Streptococcus faecium 2/4	900–1200
Streptococcus faecium 2/4	1200–1500
Streptococcus lactis	300–600

[a] Calculated from Anellis *et al.* (1973).

[b] Number of strains/total number tested.

[c] The lowest dose tested.

[d] Not recognized or classified *incertae sedis* in the 8th Edition of "Bergey's Manual" (Buchanan and Gibbons, 1974).

minimizing of the secondary ionization and free radical effects noted earlier. Freezing, which effectively immobilizes water molecules, yields similar protective effects presumably due to immobilization of reactive molecules within the ice crystal lattice until they decay (Table 3.2). Note that the effect is much less with spores, presumably because of the sequestering of water in the spore structure.

The suspending medium influences irradiation sensitivity. Thus different doses may be necessary to obtain the same microbicidal effect in different foods. Additionally, foods may create localized anaerobic conditions even when oxygen is present externally, and this can greatly affect irradiation resistance.

C. Effects on Bacteria

Radiation sensitivity of microorganisms differs with species and even with strain, although the range of resistance among strains of a single species is usually small enough to be ignored for practical purposes.

Table 3.3 tabulates radiation sensitivities for a number of common bacteria, yeasts, and molds as reported by various authors, whose studies were undertaken under a wide variety of conditions and using different irradiation procedures: it is therefore only a rough guide to the relative resistance of different groups of microorganisms.

Gram-negative bacteria, including common food spoilage organisms (e.g., *Pseudomonas*) and enteric species, including pathogens (e.g., *Salmonella, Shigella*) are generally more sensitive than are gram-positive bacteria. The fecal streptococci are rather resistant, bacterial spores are still more resistant, and *Micrococcus radiodurans* is exceptionally resistant.

The bacteria responsible for the spoilage of meat, milk, and poultry at low temperatures belong to the so-called *Pseudomonas–Achromobacter* association (members of the group *Achromobacter* now having been renamed either *Acinetobacter* or *Moraxella*). Whereas the *Pseudomonas* species of this association (e.g., *P. geniculata*) have low resistance to irradiation, there is a wide spread of resistance among the *Acinetobacter–Moraxella* species, so that strains belonging to this group are likely to survive and form a major element in the flora of perishable foods after exposure to moderate radiation doses. Most *Acinetobacter* isolates had low resistance, but among numerous other isolates from beef irradiated up to 270 krad, *D* values were about 100 krad for the most resistant, which were identified as *Moraxella* (Tiwari and Maxcy, 1972). This places the resistance of *Moraxella* on a par with that of the enterococci (Table 3.3).

Food poisoning organisms are no exception to the above general rule; i.e., most of the gram-negative pathogens, together with *Escherichia coli* and other members of the family Enterobacteriaceae, have low resistance to irradiation,

TABLE 3.3

Approximate Radiation Doses Needed to Reduce Viable Numbers a Millionfold for Various Microorganisms[a]

Microbial species	Reference[b]	Dose (krad)
Gram-negative bacteria		
Facultative anaerobes		
Enterobacter (Aerobacter) aerogenes	(3)	<50
Enterobacter (Aerobacter) aerogenes	(2)	100–150
Enterobacter (Aerobacter) cloacae	(2)	150–200
Escherichia coli	(3)	<50
Escherichia coli I	(2)	150–200
Escherichia coli	(10)	200–300
Escherichia coli II	(2)	200–300
Proteus vulgaris	(3)	<50
Proteus vulgaris	(10)	100–150
Salmonella cholerae-suis	(10)	300–500
Salmonella enteritidis	(7, 10)	300–500
Salmonella gallinarum	(2)	300–500
Salmonella paratyphi A	(10)	300–500
Salmonella paratyphi B	(10)	500–700
Salmonella pullorum	(10)	200–300
Salmonella typhi	(3)	<50
Salmonella typhimurium	(7)	300–500
Salmonella typhi (typhosa)	(10)	300–500
Salmonella wichita	(10)	300–500
Serratia marcescens	(3)	<50
Shigella dysenteriae	(10)	200–300
Shigella paradysenteriae	(10)	150–200
Shigella sonnei	(10)	150–200
Vibrio parahaemolyticus	(8)	<50
Vibrio parahaemolyticus	(8)	50–100
Obligate anaerobes		
Bacteriodes fragilis subsp. *ovatus*	(3)	<50
Bacteriodes fragilis subsp. *vulgatus*	(3)	<50
Aerobes		
Acinetobacter sp.	(11)	50–100
Alcaligenes ammoniagenes[c]	(3)	100–150
Alcaligenes viscosus[c]	(3)	50–100
Brucella abortus	(7)	200–300
Moraxella	(11)	500–750
Pseudomonas aeruginosa	(3)	<50
Pseudomonas aeruginosa	(3)	150–200
Pseudomonas fluorescens	(7)	50–100
Pseudomonas geniculata[c]	(4)	<50
Gram-positive bacteria		
Bacillus (vegetative)	(3)	200–300
Bacillus cereus	(6)	2000–3000

TABLE 3.3 (*continued*)

Microbial species	Reference[b]	Dose (krad)
Bacillus coagulans	(5)	500–750
Bacillus coagulans	(5)	1000–2000
Bacillus pumilus	(5)	100–2000
Bacillus pumilus	(5)	2000–3000
Bacillus stearothermophilus	(4)	1000–2000
Bacillus subtilis	(5)	1000–2000
Clostridium sp.	(4, 5)	1000–2000
Clostridium botulinum A	(2, 4)	2000–3000
Clostridium botulinum E	(2)	1000–2000
Clostridium perfringens	(9)	2000–3000
Lactobacillus (heterofermentative)	(4)	200–300
Lactobacillus (homofermentative)	(4)	500–750
Leuconostoc	(4)	200–300
Leuconostoc dextranicum	(3)	50–100
Leuconostoc mesenteroides	(3)	50–100
Micrococcus sp.	(4)	300–500
Micrococcus conglomeratus [c]	(3)	200–300
Micrococcus radiodurans [c]	(4)	>3000
Micrococcus roseus	(4)	>3000
Propionibacterium acnes	(3)	100–150
Micrococcus luteus (*Sarcina lutea*)	(3)	300–500
Staphylococcus aureus (*albus*)	(3)	<50
Staphylococcus aureus	(4)	<50
Staphylococcus aureus	(3)	50–100
Staphylococcus aureus	(2)	200–300
Staphylococcus aureus	(2, 7)	300–500
Staphylococcus aureus	(10)	1000–2000
Streptococcus faecalis	(4)	300–500
Streptococcus faecalis	(2, 10)	500–750
Streptococcus faecium	(4)	1000–2000
Actinomycetes		
Microbacterium thermosphactum [c]	(4)	200–300
Mycobacterium tuberculosis	(2)	100–150
Molds and Yeasts		
Aspergillus fischeri	(1)	200–300
Aspergillus flavus	(1)	200–300
Aspergillus fumigatus	(1)	150–200
Aspergillus niger	(1)	300–500
Aspergillus oryzae	(1)	150–200
Aspergillus terreus	(1)	200–300
Aspergillus versicolor	(1)	200–300
Candida sp.	(4)	300–500
Candida albicans	(1)	750–1000

(*continued*)

TABLE 3.3 (*continued*)

Microbial species	Reference[b]	Dose (krad)
Candida krusei	(1)	1000-2000
Candida parakrusei	(1)	1000-2000
Candida pseudotropicalis	(1)	300-500
Cryptococcus neoformans	(1)	750-1000
Debaromyces kloeckeri	(4)	500-750
Kluyreromyces fragilia	(1)	750-1000
Penicillium camemberti	(1)	50-100
Pencillium notatum	(1)	150-200
Saccharomyces cerevisiae	(1)	750-1000
Saccharomyces ellipsoideus	(1)	500-750
Schizosaccharomyces octosporus	(1)	300-500
Torulopsis candida	(4)	200-300
Torula rosei	(1)	300-500
Viruses	(4)	>3000

[a] The values are taken from the various references indicated, sometimes directly, sometimes by calculation of *D* values, on the basis of the likely populations in the irradiated cultures if no precise information was given. Where two entries appear for the same species and author, they represent the range among several strains examined.

[b] Key to references: (1) Bridges *et al.* (1956); (2) Erdman *et al.* (1961a,b); (3) Koh *et al.* (1956); (4) Niven (1958); (5) Pepper *et al.* (1956); (6) Lewis *et al.* (1973); (7) Maxcy and Tiwari (1973); (8) Matsuyama (1973); (9) Roberts (1968); (10) Quinn *et al.* (1967); (11) Tiwari and Maxcy (1972).

[c] Not recognized, or classified *incertae sedis* in the 8th Edition of "Bergey's Manual" (Buchanan and Gibbons, 1974).

although *Salmonella* serotypes differ widely in their resistance. Hence *E. coli* is unlikely to be a useful "indicator" for *Salmonella* in irradiated foods. There is also a wide spread of resistance reported for *Staphylococcus aureus* and *Streptococcus* spp., especially *S. faecium* which is among the most resistant of vegetative forms. Spores of some strains of *Bacillus cereus* are among the most resistant *Bacillus* spp.; and *Clostridium botulinum* type A includes strains long established to have the most radiation resistance of all spores, although spores of some food-poisoning strains of *C. perfringens* are almost as resistant.

To prevent damage to flavor, it is sometimes recommended to irradiate in the frozen state, which increases bacterial resistance by about twofold, more or less equally with all vegetative bacteria (cf. Tables 3.2 and 3.3). Hence, the general order of comparative resistance is not much changed by freezing before irradiation, though the scale or dosage is perhaps doubled. The same probably applies to drying, though this has been investigated less systematically.

Since resistance of bacterial spores is little affected by freezing, when the aim is to destroy spores it is advantageous to irradiate a food in the frozen state, for deleterious change in the food is roughly halved. Similarly, in the dry state, the resistance of the more resistant vegetative organisms, e.g., *Streptococcus faecium* is raised to equal that of spores, as just noted.

The use of irradiation has revealed bacterial species unusually resistant to it; although these must have been present before, they were so inconspicuous as to be unknown. *Micrococcus radiodurans* was the first to be isolated from meat, but similar species are apparently widely distributed in the environment. Matsuyama (1973) recorded about 10 highly resistant isolates from various natural radioactive sources, although there is no evidence of correlation between background radioactivity and the occurrence of resistant strains in nature. More recently, Kristensen (1974) obtained an isolate about as resistant as *M. radiodurans* from the flora of three reactor pools. Although he assumed that these bacteria (gram-positive non-sporing rods) were mutants characterized by their high radiation resistance, 15 clones isolated from the three locations were apparently identical, and it is more probable that the radiation field was simply selecting a resistant organism normally present in water in low numbers.

D. Effects on Molds and Yeasts

In general, the radiation resistance of molds is of the same order as that of the vegetative bacteria except for those more sensitive. For example, Sommer (1973) quotes the following doses for 0.01% survival: *Penicillium expansum,* 200 krad; *Mucor,* 250 krad; *Botrytis cinerea,* 300 krad; *Rhizopus,* 350 krad; *Alternaria citri,* 400 krad; *Cladosporium herbarum,* 450 krad; and *Alternaria tenuis,* 550 krad, which figures indicate a range from 300 to 840 krad for the doses corresponding to the millionfold inactivation which Table 3.3 represents.

Such figures are only approximate because the dose levels may be modified by relevant circumstances. For example, as with bacteria, sensitivity to irradiation is greatest at early stages of growth, while doses required in air are only about half as great as those required under anaerobic conditions.

The yeasts are distinctly more resistant than the molds are and as resistant as the more resistant bacteria. Especially resistant species, such as *Candida krusei,* are as resistant as the more sensitive bacterial spores.

There seem to be no records yet of outstandingly resistant molds or yeasts corresponding to the highly resistant bacteria.

E. Effects on Viruses

Viruses are even more resistant than bacteria, so that treatments which destroy virtually all bacteria will not reliably eliminate viruses.

F. Public Health Implications

The extreme lethality of radiation to human beings makes strict safety precautions essential in the irradiation process. Even after three decades of intensive research, sale of irradiated foods is rarely permitted, and there remains in the minds of health authorities concern that some of the chemical changes might produce compounds that are toxic, carcinogenic, or mutagenic.

Some microbiological problems peculiar to irradiation processing of foods are nevertheless clear (Thatcher, 1963; Ingram and Farkas, 1977).

There has been concern that pathogens may sometimes survive irradiation whereas harmless organisms do not, e.g., by some tests salmonellae are more resistant than the supposed indicator organisms.

There are several reports of the isolation from strongly irradiated foods of bacteria with a very high resistance, but these species, typified by *Micrococcus radiodurans,* are nonpathogenic, slow-growing, and do not take part in spoilage. Many grow poorly or not at all at 10°C and below, in the presence of as little as 2% NaC1, and are killed by mild heating. Though they presumably occur in unirradiated foods, their presence has never been detected. Though these resistant bacteria may survive in irradiated foods, their numbers are likely to remain small, and they will be harmless and easy to control, if control is necessary. Radiation pasteurization would be likely to leave many surviving virus particles, and even radiation "sterilization" might leave some. Nevertheless, irradiation reduces the number of infective particles, up to 10-fold for a dose of about 500 krad, which is better than refrigeration, which tends to preserve them. Moreover, the significance of viruses in foodborne disease is still uncertain, and there are as yet no specific requirements that viruses shall be absent from food, in which they do not multiply and are killed on cooking.

However, irradiation could not be used, as is heat processing, to permit export of meat from countries where dangerous virus diseases (e.g., rinderpest) are endemic, but the details have yet to be established.

There has been special concern over the possibility that the irradiation of food might produce mutations in its microorganisms (Thatcher, 1963; Ingram and Farkas, 1977), which might conceivably be undesirable, in the following respects:

1. Existing pathogens might become more virulent.
2. Organisms initially harmless might become pathogenic.
3. Vital characterisitcs might be changed, making it difficult to establish identity, especially important with pathogens.
4. Organisms might become more resistant to irradiation.
5. Combinations of these possibilities could be unusually hazardous: e.g., if a mesophilic pathogen became at the same time more resistant to irradiation and more virulent or more capable of growing readily at refrigeration temperatures.

6. Mutagenic effects would be expected to increase with the irradiation dosage; on the other hand, for such mutation to be expressed, the mutated organism must survive and multiply. One might therefore expect this type of change to be most serious with intermediate-dose pasteurization treatments, which leave appreciable populations of cells which have survived considerable irradiation doses.

In the few experiments with viruses, no mutations were detected (FAO/IAEA/WHO Expert Committee, 1977), but such possibilities have been explored with particular reference to salmonellae in relation to radicidation treatments of food. The general results may be summarized as follows:

1. *Acquired virulence*. Pathogens do not become more virulent; if virulence is changed, it is generally diminished. However, with *Clostridium botulinum*, a small increase has been demonstrated under realistic conditions; with *Aspergillus flavus* and *A. parisiticus*, only under highly artifical conditions. The effective doses reduced the numbers of viable toxigenic cells to a degree far greater than they enhanced toxin production of the surviving cells, indicating that even with such mutation the overall effect would be to reduce greatly the likelihood and level of toxin production.

2. *Induced pathogenesis*. The only known report of pathogenic properties arising apparently *de novo* refers to nontoxigenic strains of *A. flavus*, a few of which, months after irradiation, developed small amounts of aflatoxin. It is conceivable that this was stimulation of toxin formation from a level below the detection threshold to one above and did not represent acquisition of the property *de novo*. There is no known record of the acquisition by a nonpathogenic species of pathogenic properties as a result of irradiation, even from the medical field where ionizing irradiation has long been widely used.

3. *Modification of key characteristics*. In considering effects on critically important characteristics, it is necessary to distinguish between cells that have been irradiated once or twice, and cultures that have been subjected to repeated cycles of irradiation and growth many times. With isolated irradiation treatments, a proportion of surviving cells may be "damaged" and may fail to grow in unfavorable conditions or on selective media. However, this need present no unusual problem in recognition or enumeration because heated or dried cells behave similarly, and such cells can be restored by a short period of resuscitation on a favorable medium, after which their properties are normal.

With repeated cyclic irradiation, significant changes occur: morphology may become unrecognizable; a proportion, usually small, of the survivors may lose a particular biochemical reaction; and serological reactions may be weakened. In the most extensively investigated instance (that of salmonellae), general identity was not made doubtful and (though some reactions were weaker) serological typing remained possible even with recycled cells. Changes of phage pattern, with loss and acquisition of particular responses, have been reported with recycled *Staphylococcus aureus*. Perhaps the most troublesome changes so far recorded are biochemical changes in a minor proportion of cells of *Escherichia coli* type I, changing the diagnosis to type II or Intermediate II—a small but significant change in that species.

4. *Acquired resistance to irradiation*. Repeated irradiation cycles sometimes diminish, but more often increase, irradiation resistance. Under favorable circumstances not wholly understood, the increase may be large. In the extreme example, 84 cyclic treatments increased the resistance of *Salmonella* to the level of *Micrococcus radiodurans*, a change some argue could make *Salmonella* radicidation with a dose about 500 krad wholly ineffective.

The significance of the possibilities of modification of key characteristics or of acquired resistance clearly depends on the likelihood of repeated cyclic irradiation under favorable conditions; most important of these conditions is that, following each irradiation, there should be opportunity for sufficient multiplication to restore the high degree of inactivation before the next irradiation. Recycling experiments have naturally been made under optimal conditions from this point of view, but such conditions seem unlikely in practice. Moreover, if such regrowth did occur, it could be prevented by application of hygienic measures and/or temperature control.

5. *Hazardous combinations of acquired characters*. At present, there are no indications of unfavorable combinations, rather the reverse. Unusually resistant mutant strains tend to have weaker growth potential and diminished virulence, and the minimum growth temperature seems much more likely to be raised than lowered by irradiation. Such questions, however, deserve further investigation.

The general conclusion, as regards possible microbiological hazard due to the mutagenic action of ionizing irradiation, seems to be that such risk is negligible. An exception might conceivably arise with repeatedly recycled cells, under circumstances which in practice seem improbable and easy to avoid. This conclusion is strengthened by two further general considerations. First, heating can be mutagenic; but none of the postulated hazards is known to have arisen. Second, given the wide medical use of X and γ rays in nonsterile situations, one would suppose similar hazards to arise there, yet no microbial novelties have been ascribed to that cause. Experience thus suggests that the conceivable difficulties are not, in practice, likely to become important; and these hypothetical risks seem trivial, in comparison with the large and repeatedly demonstrated advantages which irradiation could bring, in the elimination of undesirable microorganisms.

In the few experiments with viruses, no mutations were detected (FAO/IAEA/WHO Expert Committee, 1977).

III. USE OF IONIZING RADIATIONS TO CONTROL MICROORGANISMS IN FOODS

Two kinds of process can be broadly defined. The first aims to produce a food which is safe and can be kept indefinitely free from microbially induced spoilage: all the microorganisms are killed, the food has been sterilized. Alternatively, the few surviving microbes remain inactive as if dead under the given conditions, a state called "commercial sterility."

A second, more modest, aim is to improve microbiological quality short of "sterilization": i.e., to improve safety by virtual elimination of pathogens; or, to prolong storage life by reducing numbers of spoilage organisms. Heat treatments designed to attain such ends are called "pasteurization."

These two types of objective have long been distinguished in irradiation processing under the general descriptions "radiation pasteurization" for lower dose treatments leaving numerous survivors, and "radiation sterilization" for high-dose treatments aimed at stable products. Different names are now preferred, however. Low-dose treatments are distinguished according to objective: if this is to extend storage life, the process is called *radurization;* if it is to remove some particular pathogen, *radicidation.* The distinction is arbitrary, for a particular irradiation treatment produces the same effect whatever it is called. Treatment aimed at radurization will also kill pathogens and improve safety, while radicidation has the side effect of killing many spoilage organisms and extending storage life. The same occurs, of course, with the corresponding heat treatments.

With the higher dose treatments, the aim is often to produce the semblance of sterility rather than real sterility, hence the term *radappertization* is preferred to "radiation sterilization," appertization being the well-known French name for the corresponding type of heat process after its originator, Louis Appert.

Both types of procedures presuppose some form of packaging. Where the food has been sterilized, its continued integrity demands freedom from reinfection and long-term protection against chemical alteration. This usually requires exclusion of air by means of a "hermetic" package equivalent to a can. The requirements are less strict for pasteurized foods. Wrapping at least sufficient to prevent gross recontamination is necessary to retain the benefits of radurization, spoilage organisms being nearly ubiquitous. Foods treated against pathogens like *Salmonella* seem less likely to pick up equivalent recontamination, and wrapping may sometimes be unnecessary on this account.

The relative resistances illustrated in Tables 3.2 and 3.3 bear several important resemblances to those relating to heat resistance.

1. The common spoilage bacteria of proteinaceous foods under cool conditions (e.g., *Pseudomonas*) are relatively sensitive, hence large numbers of them

can be destroyed, and the life of refrigerated perishable foods correspondingly prolonged, by relatively small irradiation doses (of the order of 100 krad) as by mild heat treatments.

2. Much the same applies to the molds, e.g., *Mucor* or *Botrytis,* which commonly spoil dry or acid foods.

3. Important bacteriological pathogens, e.g., *Salmonella* or *Staphylococcus,* can be virtually eliminated by slightly higher doses (of the order of 250–500 krad); especially as they occur on normally acceptable foods only in relatively small numbers. As in heating, the bacterial groups likely to survive such intermediate treatments are spore formers and the fecal streptococci.

4. The same applies to pathogenic fungi like *Aspergillus flavus* and *A. parasiticus,* which are no more resistant than many common saprophytic molds.

5. The species with the most resistant spores is *Clostridium botulinum,* hence "sterilizing" treatments are concerned first with this species, for foods where it is a hazard. The 12*D* concept can be applied as a safety criterion to irradiated as to heat processed food. The species of higher irradiation resistance than *C. botulinum* (e.g., *Micrococcus radiodurans*) are not known to have any significance in foods.

6. In foods where *C. botulinum* is controlled otherwise, the safe irradiation dose can be lower. This is the situation with foods of pH 4.6 or below, where any surviving *C. botulinum* spores are inhibited from developing by the acidity.

7. Similarly, in cured foods most spores surviving irradiation are unable to multiply in the presence of salt and nitrite, hence doses can be lower than if salts are absent, by an amount presumed to depend on the concentration of salts and the temperature of storage. There follows a discussion of examples of these different types of process.

A. Radurization

Processes aiming to prolong storage life can be divided into those designed for near neutral or "low acid" foods that are spoiled by bacteria and those acid or dry foods spoiled by fungi.

1. Foods Spoiled by Bacteria

Various flesh foods, which are spoiled in similar ways under chill conditions by similar *Pseudomonas–Acinetobacter–Moraxella* associations, respond in a similar manner to radurization treatments.

At relatively low dose levels, about 100 krad, most of the typical spoilage organisms are destroyed. Hence, during chilled storage, a longer time must elapse (usually up to twice as long) before numbers rise again to the levels necessary to cause spoilage. When spoilage occurs, it takes a normal putrefactive form.

With relatively high doses, approaching 500 krad, the normal spoilage organisms are eliminated, and the more resistant and metabolically less active species remain, e.g., resistant *Moraxella* spp., lactic acid bacteria (especially fecal streptococci), and yeasts. During chilled storage, this surviving flora grows less rapidly than the normal spoilage flora, and the storage life is correspondingly further extended to three or four times normal. Eventually a quite different kind of spoilage results.

As with the unirradiated foods, the nature of the spoilage flora is strongly influenced by the nature of the packaging. In anaerobic (oxygen impermeable) packages, lactic acid bacteria and yeasts predominate.

The above relations, which have been extensively investigated in a variety of foods (International Atomic Energy Agency, 1963, 1966, 1967, 1968, 1970c, 1973a, 1973b) depend on the relative sensitivity to irradiation of the bacterial association responsible for spoilage at chill temperatures. If storage is at higher temperatures, spoilage involves more resistant bacteria (e.g., Enterobacteriaceae, which are facultatively anaerobic), and the benefits of irradiation and effects of anaerobic packaging are not so great.

In all such treatments, some spore-forming pathogens would be expected to survive, but since they are at least equally present in normal foods, they would create unusual problems only in special circumstances. In fish irradiated with doses >100 krad and stored above 5°C, *Clostridium botulinum* type E is sometimes able to grow and produce toxin before detectable spoilage occurs; hence storage below 3°C is prescribed.

2. *Foods Spoiled by Fungi*

A rather similar situation to that of bacteria exists among acid or dry foods spoiled by fungi. The common spoilage fungi, e.g., *Alternaria, Botrytis, Mucor, Rhizopus,* are about as sensitive as vegetative bacteria (excluding the most sensitive of the latter). Consequently, radurization doses in the same range 100–500 krad have a similar effect in extending shelf-life of foods like cereals, bread, fruit, and vegetables. Because the bacteria usually associated (mainly *Erwinia*) with strawberries were virtually eliminated by such treatment, it has been suggested by Tamminga *et al.* (1975) that the absence of such bacteria could be used as an indicator of adequate treatment against fungi.

When fruit products are liquid, e.g., juices, they tend to be spoiled by yeasts rather than molds. Attempts have been made to control these yeasts by radurization doses of irradiation, but without much success. This is because yeasts are relatively resistant and occur in considerable numbers in juices produced under industrial conditions, while the flavor of juices is relatively easily damaged by irradiation. The necessary dose of radiation can be reduced about one-third by adding a light heat treatment (of the order of 10 min at 50°C).

Consequently there have been various proposals to combine irradiation with mild heating in the treatment of foods spoiled by fungi or yeasts. Such products correspond closely, from the microbiological viewpoint, with those that have been heat pasteurized.

B. Radicidation

As regards elimination of pathogens, it is again convenient to separate foods where the problems are bacterial from those where the problems are fungal.

1. Destruction of Bacteria

Treatment with doses of about 250 krad or more greatly reduces numbers of vegetative bacterial pathogens and can be used virtually to remove them from flesh foods (IAEA, 1968): e.g., *Vibrio* spp. from fish (inactivation approaching 10^{10}), or *Salmonella* from poultry or meat (inactivation about 10^5). Dose levels about twice as great, of the order of 500 krad, are needed if the food is frozen or dried. Unlike corresponding heat treatments, such doses do not destroy enzymes, the absence of which consequently cannot be used as tests for effective treatment. Hence, treatment must be controlled by bacteriological examination, if not by process control (see Section III,D). Because of the difficulty of detecting small numbers of surviving *Salmonella,* it has been suggested that better control would be exercised by examining for Enterobacteriaceae; no corresponding indicator has yet been suggested for *Vibrio parahaemolyticus* in fish.

A smilar treatment could conceivably be used to remove *Staphylococcus aureus* from foods but the treatment cannot readily be applied to dairy products, where it would seem especially useful, because milk products suffer unusual chemical and organoleptic damage on irradiation.

2. Destruction of Fungi

Selective treatments against pathogenic fungi, on a basis similar to those used against pathogenic bacteria, seem possible in principle. The effect of irradiation is to diminish the proportions, in the total fungus flora, of the genera *Aspergillus* and *Penicillium,* especially important in production of mycotoxins in stored foods.

There is special interest attached to *Aspergillus flavus* and *A. parasiticus,* which form carcinogenic aflatoxins. *Aspergillus flavus* is among the more sensitive molds; a dose of 400 krad destroys 10^6 conidia. There are reports that irradiation may increase aflatoxin production, but decreases are more frequent and the increases reported have been trivial compared with the reduction in number of viable conidia. It therefore appears unlikely at present that this effect will create special problems.

Despite these encouraging indications, no clear proposals for a radicidation type of process against fungi have yet emerged.

C. Radappertization

1. *Clostridium botulinum type E*

The main problem with high dose treatments is control of resistant bacterial spores. Strains of *C. botulinum* possess the most irradiation-resistant spores. Hence the problem is to control this dangerous species, either directly by the use of "sterilizing" doses of irradiation, or with lesser doses assisted by auxiliary factors like acidity or curing salts.

By analogy with the 12*D* concept in thermal processing, an inactivation of 10^{12} is the aim; and the requisite dose of radiation, for the most resistant strains of *C. botulinum,* is estimated at about 4500 krad.

Control by bacteriological examination is inconceivable at such levels in inactivation, and control will have to be exercised chiefly by strict regulation of the process, as in thermal processing. In thermal processing, however, there is an additional safeguard, because if the process becomes marginal, containers are likely to be "blown" by gas-forming harmless clostridia (e.g., *Clostridium sporogenes* and the so called "putrefactive anaerobes") which are more heat-resistant than *C. botulinum.* At present, no corresponding organisms of higher irradiation resistance are known; but experience might reveal them as it did putrefactive anaerobe PA 3679, a highly heat-resistant organism often used as a standard culture for heating studies.

However, as there are no known thermophiles more irradiation-resistant than *C. botulinum,* foods safe from *C. botulinum* should not be liable to thermophilic spoilage under tropical conditions—i.e., there should be no class of "tropical conserves" requiring extra treatment, as there is with heat-processed products.

2. *Acid Foods*

In foods of pH 4.6 or below, where *C. botulinum* cannot grow, it has been supposed that dose levels could be substantially reduced. The procedure has still not been thoroughly investigated, however, and several questions remain unanswered.

1. The critical pH level might not be exactly 4.6. This value is based on the most heat-resistant form of *C. botulinum,* type A, with the additional safeguard that the heat resistance is increasingly reduced, as the pH becomes more acid. It is not known whether exactly the same relations apply for irradiation.
2. The foregoing problem apart, the process-determining species would be one combining acid-tolerance with relatively high irradiation resistance. A likely species would seem to be *Bacillus coagulans (B. thermoacidurans),* which is relatively resistant to irradiation (*D* value ca. 200 krad), as well as to acid and heat. In this case, spoilage should be the familiar flat-souring, and process control could be by pH measurements as with the thermal process.

3. Alternatively, yeasts might be important, being acid-tolerant, and relatively more resistant to irradiation than to heat.

4. In any case, relatively large irradiation doses, of the order of 2000 krad, seem certain to be needed, and these would be sufficient to cause organoleptic damage in many fruit products.

Byssochlamys (Paecilomyces) would apparently not present problems in irradiated fruit products: the *D* value even for ascospores is only about 50 krad, and less if irradiation is at 50–72°C.

3. Cured Foods

Experience shows that cured meats are safe and stable with irradiation treatments much less than those required for uncured meats, despite the fact that viable microorganisms often remain. In heat processing, this occurs because the bacterial cells that survive a heat process are "damaged" so that they are unable to grow in the presence of relevant concentrations of NaCl (ca. 2–4.5%) and $NaNO_2$ (ca. 50–150 mg/kg), especially if temperatures are below about 25°C. In the laboratory, irradiated spores behave like heated ones in this respect, suggesting that similar combinations might be effective for irradiated foods.

In fact, canned ham can be made safe from *C. botulinum* and stable with irradiation doses of about 2500 krad, compared with about 4500 krad for uncured meat. If irradiation is at temperatures below 0°C, to minimize radiation-induced off-flavor, the dose required to control *Clostridium botulinum* in ham will be higher (ca. 3.7–4.7 Mrad), but the amount of nitrite can be reduced (Wierbicki and Heiligman, 1973; Wierbicki *et al.,* 1976).

Part of the heat process in less strongly pasteurized hams could be replaced by irradiation, but the relative amounts of heat and irradiation and the contribution of salt and nitrite have yet to be determined.

D. Control

In certain heat pasteurization processes, effectiveness is routinely checked by simple tests for the absence of residual enzyme activity (phosphatase in milk, amylase in egg pulp). This type of control cannot be used for corresponding irradiation processes because enzymes are more resistant to irradiation than are microorganisms.

Instead it appears that effectiveness will have to be checked either by (a) bacteriological tests on the processed food; and/or (b) control of the process itself. This is the situation now existing with heat-processed foods for which no suitable enzyme test has been devised; and where bacteriological control is statistically too unreliable, as in heat "sterilized" foods, one has to rely mainly on process control. The irradiation process is, for physical reasons, easier to control and check by dosimetry, than a corresponding heat process; further, regulatory control of irradiation-processing establishments is likely to be much more strict than for heat processing.

IV. COMBINATION EFFECTS AND INTERACTIONS

The problems of the surviving microflora clearly differ according to the nature of the food and its associated species. Foods too dry to permit growth of microorganisms, or frozen foods, present no problems; they simply carry fewer microorganisms of all kinds after irradiation than before.

Similarly, partially dry or acid foods which support molds but not bacteria will not give rise to any problems related to bacteria, whether of food poisoning or spoilage. If they do not normally support hazardous species of fungi, there is no reason to expect that these will appear after irradiation, for these fungi are not especially resistant to irradiation, and as far as is known, the food is not rendered more likely to support them by an irradiation treatment. Thus, no public health problems can be foreseen with the majority of fruits, for example.

The semi-dry grains or pulses, which are normally liable to support toxigenic fungi if mishandled (Chapter 23), may present the same problem after irradiation, but the indications are that irradiation would diminish it. All fungi are greatly reduced in number. The nontoxigenic types appear to be relatively favored by irradiation. Toxin production by the toxigenic species is generally diminished, despite a few reports to the contrary. The possibility of growth and toxin production is avoided if the dry product is kept under storage conditions which prevent mold growth, as is also necessary for the unirradiated dry product if risk is to be avoided.

The wet, perishable, high-protein foods that normally support bacterial growth are those most likely to raise microbiological problems when irradiated. Because the vegetative food poisoning bacteria are relatively sensitive, irradiation effects large reductions in their numbers, so that they are not found in irradiated normal foods which, in this respect, are much improved by irradiation. Even among the more radiation-resistant vegetative species, none has yet been found associated with a pathogenic risk in irradiated foods.

The foreseeable bacteriological problems arise from the resistant spores of a few food poisoning species, notably *Clostridium botulinum, C. perfringens, Bacillus cereus* (and perhaps *B. licheniformis*). *Clostridium botulinum* type E is especially troublesome, despite its relatively sensitive spores, because it grows well at most food storage temperatures, but it is fortunately restricted to fish and seafoods. The corresponding problems with vacuum-packaged and hot-smoked fish have been signalled by outbreaks of food poisoning, which led to restrictions on those foodstuffs; and it seems clear that restrictions would similarly be desirable with radurized seafoods. The simplest solution already suggested would be a requirement to hold the food below 3°C, which prevents growth of *C. botulinum* type E.

Problems of spore survival, exactly analogous in principle, arise with pasteurized foods other than fish, whether heated or irradiated; but they are less serious where spores of the mesophilic species are concerned because these require relatively high temperature to multiply freely. Thus, cooked meat is handled without temperature restrictions in many countries despite its well-known tendency, if it is kept warm, to cause outbreaks of food poisoning from *C. perfringens* with the implication of risk from *C. botulinum* also. The risk from irradiated meat would be similar, although conceivably less, because spores of *C. perfringens* tend to require heat activation before they will germinate. Similar considerations probably apply to the mesophilic *B. cereus*. The purist might argue that to avoid any risk, all irradiated meats should, for example, be stored at temperatures at which *C. perfringens* or *C. botulinum* types A and B cannot grow, below 10°C (cf. Chapter 1). With the exactly similar hazard from cooked meats, sometimes higher temperatures clearly suffice to reduce the risk to proportions that are generally tolerable; hence it may be suggested that no special temperature restrictions would be necessary for irradiated meats, at least in temperate climates. The spores of *C. perfringens* and *C. botulinum* are so common in soil and dust that they may be expected to occur even in normal foods.

These effects of radurization on the associated microbiological risks to public health may be summarized as follows. (a) It produces obvious and large improvements as regards the various vegetative species significant in food. (2) It raises familiar problems through the survival of spores, in the case of fish and seafoods sufficient to justify limitations in processing and storage temperature, in the case of meats apparently not so, at least in temperate climates. There is a balance of advantage to be evaluated here; in many applications (e.g., radicidation of meats) the balance seems to be in favor of using ionizing irradiation.

Clearly, the foregoing generalizations might be invalidated if a food initially contained unusually large numbers of some pathogen, thereby making its survival in greater numbers more likely. Irradiation processes are based on assumptions of normal hygienic quality in the foods treated. In this respect they are no different from other food processes. It is well understood by manufacturers, for example, that packaging will not restore spoiled food, or that safe canning procedures are impossible with very heavily contaminated raw materials.

The likelihood that spoilage will take unusual forms has been understood for some time. If, for example, irradiation causes lactic acid bacteria to replace a normally gram-negative flora, spoilage will take the form of souring rather than putrefaction; and, since souring is less objectionable than putrefaction, a larger chemical change with more bacteria may have to develop before spoilage is recognized. Exactly similar situations arise with cured or packaged foods, which often carry very large numbers of bacteria without being regarded as spoiled. With irradiated foods, there is no indication that the surviving flora is of significant public health interest, nor that it constitutes a spoilage problem of unusual significance. While organoleptic quality may be a criterion of acceptability from a commercial standpoint, it is not necessarily so from the public health point of view; and there is clearly no reason to expect that new criteria of spoilage would be unacceptable to consumers, especially if it were understood that they are associated with a diminished risk to public health.

In these effects of irradiation on the bacterial flora of foods, there have been many parallels with the effects of heating, despite occasional differences in detail. The species likely to be significant in irradiated foods are much the same as in familiar heat-processed foods, especially as regards those significant for public health. Correspondingly, it is anticipated that generally similar criteria and methods of examination should be applicable to irradiated and to the corresponding heat-processed foods (IAEA, 1970b). This conclusion is, however, undesirably theoretical, because experience under commercially realistic conditions is still lacking.

4

Reduced Water Activity

I. INTRODUCTION

The growth and metabolism of microorganisms demands the presence of water in an available form. The most useful measurement of the avilability of water is water activity (a_w). The a_w in a food may be reduced by increasing the concentration of solutes in the aqueous phase of the food, either by removing water or by adding solutes. Some water molecules are oriented about the solute molecules, and others become absorbed onto insoluble food constituents. In both instances the water becomes less available to enter into reactions.

Dehydration is a method of food preservation that depends upon the reduction of a_w by water removal. In curing or salting and in syruping or sugaring, it is addition of solute which lowers a_w and preserves the foods. A small reduction in a_w often has sufficient effect to preserve a food when combined with other factors such as nitrite in many cured meats, and smoke ingredients in smoked, salted, and dried foods. Curing is discussed in Chapter 8, while dehydration procedures are described later in this chapter.

A. Definitions of Water Activity and Related Terms

1. Definition of Water Activity

The a_w of a food or solution is the ratio of the water vapor pressure of the food (p) to that of pure water (p_0) at the same temperature:

$$a_w = p/p_0$$

When a solution becomes more concentrated, vapor pressure decreases and the a_w falls from a maximum value of 1 for pure water (in the absence of capillary or adsorptive forces). The a_w is related to freezing point, boiling point, equilibrium relative humidity (ERH), and osmotic pressure.

2. *Relation of Water Activity to Equilibrium Relative Humidity and Osmotic Pressure*

Many earlier studies of microbial response to concentration and available moisture were reported in terms of ERH or osmotic pressure.

ERH, like a_w, is the ratio of vapor pressures of solution and pure water, but is expressed as a percentage. Hence

$$\text{ERH (\%)} = a_w \times 100.$$

ERH refers strictly to the atmosphere in equilibrium with a solution or food and is a less appropriate expression than a_w as a measure of available water.

Osmotic pressure is related to a_w by the expression

$$\text{Osmotic pressure} = \frac{-RT \log_e a_w}{\bar{V}}$$

where R is the gas constant, T the absolute temperature, $\log_e a_w$ the natural logarithm of a_w and $\bar{V}$ the partial molar volume of water. The use of osmotic pressure presupposes the presence of a membrane of appropriate permeability properties.

3. *Relation of Water Activity to Solute Concentration*

Raoult's law can be expressed in the form

$$\frac{p}{p_0} = \frac{n_2}{n_1 + n_2}$$

i.e., the vapor pressure of the solution relative to that of the pure solvent is equal to the mole fraction of the solvent, where p and p_0 are the vapor pressures of the solution and the solvent, respectively, and n_1 and n_2 are the number of moles of solute and solvent, respectively. For a 1 molal solution of an ideal solute,

$$a_w = \frac{p}{p_0} = \frac{n_2}{n_1 + n_2} = \frac{55.51}{55.51 + 1} \quad \text{or} \quad 0.9823$$

However, solutes are not ideal. On the one hand, interactions between molecules may reduce the effective number of particles in solution, while, on the other, dissociation may greatly increase the number. Such factors lead to departures from the ideal which are large for nonelectrolytes at concentrations above 1 molal and for electrolytes at all concentrations. Experimentally determined molal osmotic coefficients are used to compensate for nonideal behavior and a_w values are calculated from the expression

$$\log_e a_w = \frac{-\nu m \phi}{55.51}$$

where m is the molal concentration of the solute, ν is the number of ions generated by each molecule of solute ($\nu = 1$ for nonelectrolytes), and ϕ is the molal osmotic coefficient.

Scott (1957) gives molalities (gram molecules of solute per kilogram of water) for various solutes at a range of a_w levels and compares them with ideal values. These are particularly useful in preparing media of known a_w levels for studies of microbiological water relations. Data on the a_w of solutions of direct relevance to food preservation [namely, sodium chloride, sucrose, glucose, invert sugar, and incompletely hydrolyzed starches (glucose syrups)] are given in Tables 4.1 and 4.2. Concentrations in °Salometer, °Brix, and °Baumé, terms widely used in the food industries, are given where appropriate.

TABLE 4.1

Concentrations of Sodium Chloride and Sucrose at Various Values of Water Activity (a_w) at 25°C

	NaCl			Sucrose		
a_w	m	%,w/w	°Salometer	m[b]	%, w/w (°Brix)	°Baumé
1.000	0	0	0	0	0	0
0.995	0.150	0.88	3.32	0.272	8.52	4.75
0.990	0.300	1.74	6.57	0.534	15.45	8.59
0.980	0.607	3.43	12.94	1.03	26.07	14.43
0.960	1.20	6.57	24.79	1.92	39.66	21.79
0.940	1.77	9.38	35.40	2.72	48.22	26.34
0.920	2.31	11.90	44.91	3.48	54.36	29.57
0.900	2.83	14.18	53.51	4.11	58.45	31.69
0.880	3.32	16.28	61.43	4.93	62.77	33.90
0.860	3.80	18.18	68.60	5.58	65.63	35.36
0.840	4.26	19.94	75.25			
0.820	4.71	21.59	81.47			
0.800	5.15	23.13	87.28			
0.753	6.16	26.5	100.00			

[a] Based on data of Robinson and Stokes (1959).

[b] Molality, the number of gram moles per kilogram of water.

TABLE 4.2

Concentrations of Glucose and Glucose-Containing Syrups at Various Levels of Water Activity (a_w) at 25°C

	Glucose[a]		Inverted sugar[b]
a_w	m[c]	%,w/w	%,w/w
0.995	0.259	4.45	2.05
0.990	0.542	8.90	4.11
0.980	1.04	15.74	8.22
0.960	2.21	28.51	16.43
0.940	3.38	37.83	24.65
0.920	4.31	43.72	32.87
0.900	5.24	48.54	41.09
0.880	6.27	53.05	49.30
0.860	7.81	58.45	57.52
0.850	9.00	61.84	61.63

[a] Based on data of Taylor and Rowlinson (1955).
[b] Based on data of Grover (1947).
[c] Molality, the number of gram moles per kilogram of water.

4. *Relation of Water Activity to Water Content*

Data such as those just referred to enable one to predict the a_w of a simple solution of known composition, but the relationship between the composition of a food and its a_w is much more complicated. To obtain information on the latter it is customary to measure the a_w values of the food at a range of water contents; these are plotted to obtain a water sorption isotherm (Fig. 4.1).

The factors in foods that reduce water vapor pressure and hence a_w include adsorption of water molecules onto surfaces, capillary forces, and the solution of components already referred to. In the isotherm it is generally accepted that the steep lower section represents the adsorption of water to form a monolayer of molecules on the adsorption sites of the solid material. With the addition of more water, the isotherm rises steeply again as solutes dissolve and capillary spaces are filled. These phenomena overlap and some may, indeed, not be represented in the isotherm. There may be no obvious monolayer in foods containing little structural material, and there may be little capillary influence.

For many foods the isotherm obtained by adsorption of water to a dry preparation differs from that obtained by drying a wet food (Fig. 4.2). This effect is termed hysteresis and the divergent isotherms the hysteresis loop.

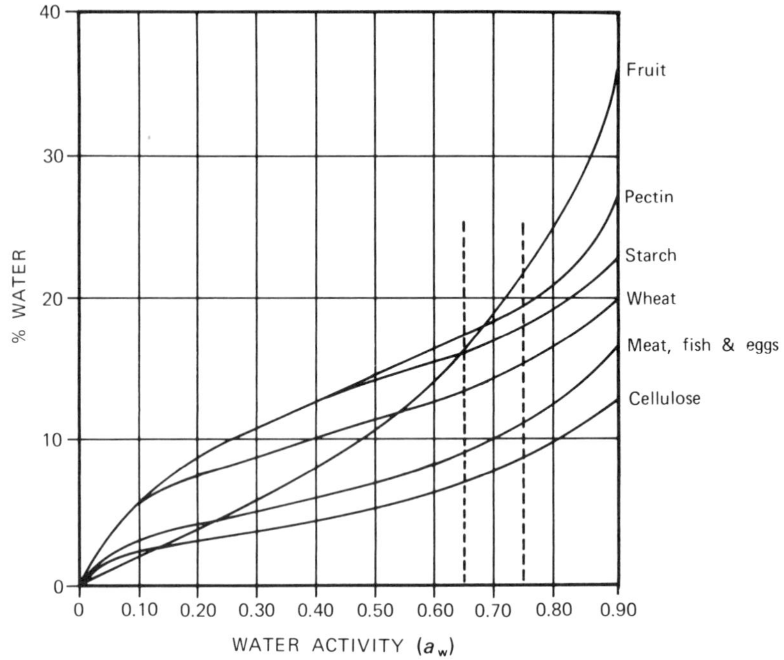

Fig. 4.1 The water vapor sorption isotherm of some foods at approximately 20°C. From Mossel (1975).

In practice, most low a_w foods are produced by desorption of water, i.e. drying, and it is the water relations of the desorption isotherm that are important. Here hysteresis does not pose a problem. However the situation may be more complicated with formulated, intermediate moisture foods (0.60–0.90 a_w) in which ingredients of a range of a_w levels are mixed together.

In foods showing marked hysteresis at high levels of a_w, microorganisms may grow more rapidly in systems adjusted by a desorption process than in those prepared by an adsorption process at the same a_w (Acott and Labuza, 1975b).

5. *Relation of Water Activity to Temperature*

As the water vapor pressure of solutions is affected by temperature, so is the water activity, but the effect is small with most solutes unless the solutions are saturated. In such cases the amounts of some substances in solution, and hence the a_w, may change markedly with temperature.

At temperatures below the freezing point of a solution or a food, the vapor pressures of liquid water and of ice decrease with decrease in temperature. The a_w of a frozen food or solution is a function of the temperature and will be 0.953

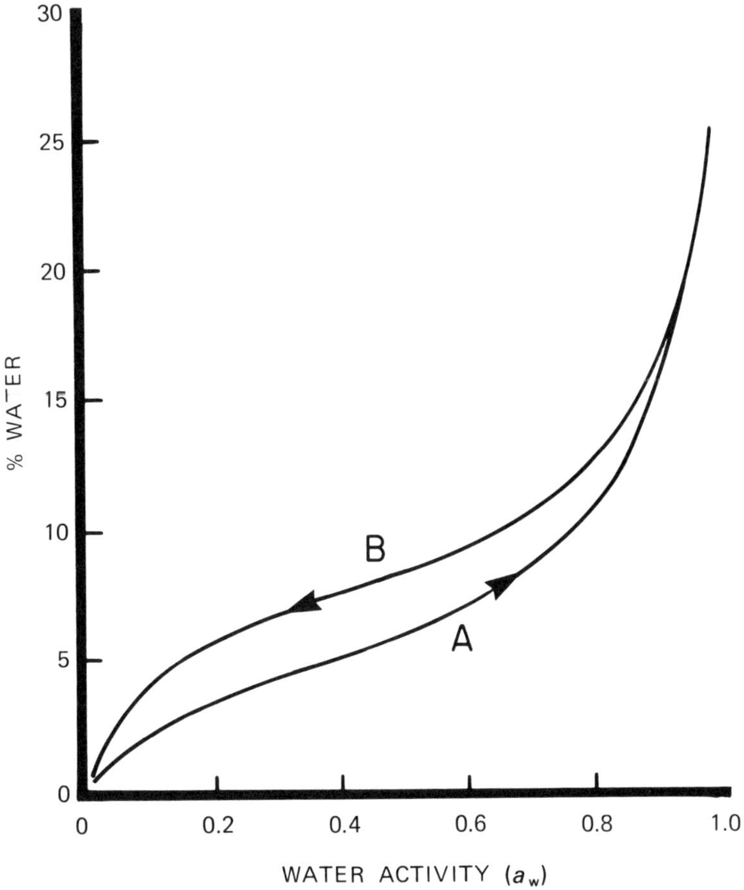

Fig. 4.2 Water sorption isotherms showing hysteresis. Curve A, adsorption; curve B, desorption.

at −5°C, 0.907 at −10°C, 0.864 at −15°C, and 0.823 at −20°C. Scott (1962) discusses the limits of microbial growth in frozen media in terms of temperature and a_w.

B. Measurement of Water Activity

The measurement of a_w is commonly performed in the food laboratory either by commercially available hair or electric hygrometers or by graphical interpolation. In the latter technique, samples of the food are held at constant temperatures in containers (preferably evacuated) containing solutions (often saturated) of known a_w levels. The changes in weight occurring over a period of several hours are plotted against the a_w of the controlling solution. The a_w at which the curve indicates no change in weight is the a_w of the food.

There are several other approaches to a_w measurement, including direct measurement of vapor pressure and the use of isopiestic techniques. Accuracy of measurement greater than ± 0.005 a_w is unnecessary for most food-related applications with ± 0.02 a_w adequate in many instances. A comparison of the advantages, disadvantages, and accuracies of seven methods of a_w determination has been made by Labuza *et al.* (1976).

In the construction of water sorption isotherms, the accurage determination of water *content* as well as of water *activity* is important, and the same method should be used when obtaining data for use in conjunction with an isotherm as was used in establishing the isotherm. Methods for measurement of water activity and content of foods are discussed by Troller and Christian (1978).

II. EFFECTS OF REDUCED WATER ACTIVITY ON MICROORGANISMS

A. General Effects

1. Growth and Related Processes

Many microorganisms, including the pathogenic bacteria, grow most rapidly at levels of a_w in the range 0.995–0.980. (The a_w of most laboratory media is 0.999–0.990.) Below this a_w, the growth rate and the stationary population or ultimate cell mass decrease, and the length of lag phase increases. At a sufficiently low a_w, which is difficult to define precisely, the lag phase becomes infinite, i.e., growth does not occur (Scott, 1957).

The growth of most bacteria and fungi is confined to the a_w range above 0.90. However, the organisms of greatest importance in food preservation include many which can grow at much lower levels of a_w. These are variously termed halophiles, xerophiles, and osmophiles.

Halophiles are unable to grow in salt-free media and frequently require substantial amounts of sodium chloride for growth. They are usually bacteria and are commonly more salt tolerant than nonhalophilic organisms. Xerophiles are organisms variously defined as those growing most rapidly under relatively dry conditions or capable of growth at a_w below 0.85. All known xerophilic microorganisms are yeasts or molds. Osmophiles are organisms capable of growth in environments of high osmotic pressure. The term is most commonly applied to sugar-tolerant yeasts, and is synonymous with xerophile, which is perferred.

In Fig. 4.3, relative rates of growth of five microorganisms are compared as

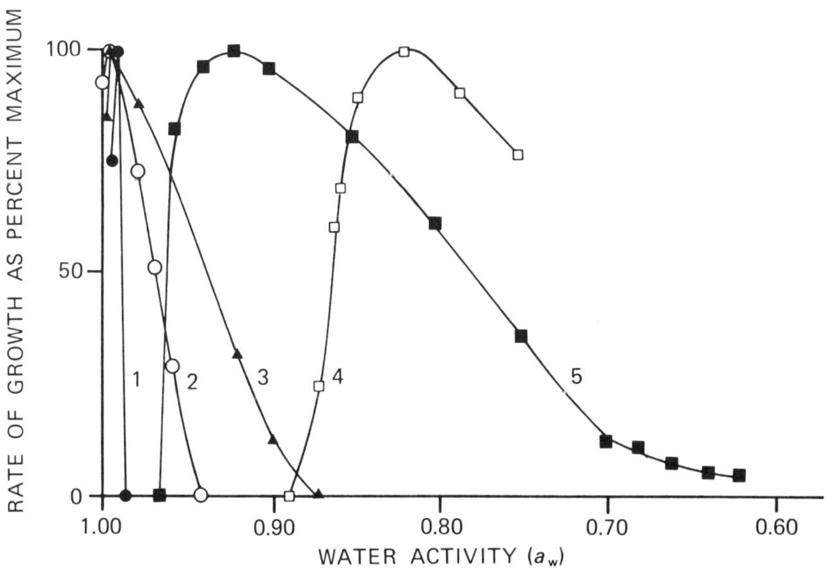

Fig. 4.3 The water relations of growth of five microorganisms. (1) *Mycoplasma gallisepticum* (Leach, 1962); (2) *Salmonella oranienburg* (Christian and Scott, 1953); (3) *Staphylococcus aureus* (Scott, 1953); (4) *Halobacterium halobium* (Christian, unpublished); (5) *Xeromyces bisporus* (Scott, 1957).

functions of a_w. Curves 1, 2, and 3 cover the range of responses most commonly seen in nonhalophilic bacteria. Curves 4 and 5 depict, respectively, the responses of an extremely helophilic bacterium and a xerophilic mold.

A knowledge of the lowest a_w at which an organism can grow is of major importance in the microbial ecology of foods. Table 4.3 gives the approximate minimum a_w levels permitting growth of a selection of microorganisms relevant to foods and food processing.

For many species, the effect of a_w upon growth is largely independent of the nature of the solute which controls the a_w. In such cases it can be concluded unequivocally that it is a_w or a related property of aqueous environment that is responsible for the effect (Scott, 1957). In other cases, marked solute effects are evident (Marshall *et al.*, 1971). The reasons for solute specific effects are not clear, except in the halophilic bacteria, which have a high requirement for sodium ions. Microorganisms respond to reduced a_w by accumulating potassium, amino acids, or polyols (Christian and Waltho, 1962; Christian and Hall, 1972; Brown, 1974; Measures, 1975). These are referred to by Brown (1974) as "compatible solutes," since at high intracellular concentrations they function as protectors rather than inhibitors of enzyme activity.

The values for limiting a_w listed in Table 4.3 were obtained for the most part under conditions in which factors other than a_w (e.g., temperature, pH, E_h, nutrition) were near optimum for growth. When these and other environmental agents depart from the optimum for a particular organism, its resistance to reduced a_w decreases and the minimum a_w permitting growth increases (see Section IV).

2. *Survival*

Like growth, microbial survival is affected by the a_w of the environment. Rates of microbial death are also reduced at ambient temperatures (Scott, 1958) and under refrigeration (Ingram, 1951) when a_w is lowered or solute concentration increased. The protection afforded by low a_w can be substantially reduced under acid conditions (Christian and Stewart, 1973).

In food microbiology, microbial survival is most important in the context of heat processing. Survival at elevated temperatures is generally poorest at high a_w. For bacterial spores, the protection resulting from reduction of a_w is maximal in the range 0.2–0.4 a_w when heated in the absence of solutes (Murrell and Scott, 1966). Survival of vegetative bacteria in the presence of solutes or food materials shows no such consistent pattern, being markedly influenced by the nature of substances present (Corry, 1973).

B. Effects on Specific Groups of Organisms

1. *Gram-Negative Bacteria*

Important gram-negative bacteria include *Pseudomonas* spp. and the Enterobacteriaceae, which can usually grow only above 0.96 and 0.93 a_w, respectively. Exceptions are some salt-tolerant strains of *Enterobacter* which grow well at 0.90 a_w. Both moderately and extremely halophilic gram-negative bacteria exist (see Section II, B, 4). The important food-poisoning *Vibrio parahaemolyticus* is also halophilic, with a strict requirement for a small amount of NaCl. Its limit for growth is about 0.94 a_w (Rödel *et al.*, 1973).

2. *Gram-Positive Non-Spore-Forming Bacteria*

This group of organisms includes many that are more tolerant than gram-negative bacteria are to low levels of a_w. Many Lactobacillaceae, like the Enterobacteriaceae, have limits for growth near 0.94 a_w, but the Micrococcaceae has representatives capable of growth below 0.90 a_w. *Staphylococcus aureus*, the most drought-tolerant of the pathogenic bacteria, has a lower limit variously reported on different substrates as between 0.86 and 0.83 a_w. Production of staphylococcal exterotoxin in food slurries has not been observed below 0.93 a_w (Troller and Stinson, 1975).

TABLE 4.3

Approximate Minimum Levels of Water Activity (a_w) Permitting Growth of Microorganisms at Temperatures Near Optimal[a]

	a_w
Molds	
Alternaria citri	0.84
Aspergillus candidus	0.75
A. conicus	0.70
A. flavus	0.78
A. fumigatus	0.82
A. niger	0.77
A. ochraceous	0.77
A. restrictus	0.75
A. sydowii	0.78
A. tamarii	0.78
A. terreus	0.78
A. versicolor	0.78
A. wentii	0.84
Botrytis cinerea	0.93
Chrysosporium fastidium	0.69
C. xerophilum	0.71
Emericella (Aspergillus) nidulans	0.78
Eremascus albus	0.70
E. fertilis	0.77
Erotum (Aspergillus) amstelodami	0.70
E. carnoyi	0.74
E. chevalieri	0.71
E. echinulatum	0.62
E. herbariorum	0.74
E. repens	0.71
E. rubrum	0.70
Monascus (Xeromyces) bisporus	0.61
Mucor plumbeus	0.93
Paecilomyces variotii	0.84
Pencillium brevicompactum	0.81
P. chrysogenum	0.79
P. citrinum	0.80
P. cyclopium	0.81
P. expansum	0.83
P. fellutanum	0.80
P. frequentans	0.81
P. islandicum	0.83
P. martensii	0.79
P. palitans	0.83
P. patulum	0.81

(*continued*)

TABLE 4.3 (*continued*)

	a_w
P. puberulum	0.81
P. spinulosum	0.80
P. viridicatum	0.81
Rhizopus nigricans	0.93
Rhizoctonia solani	0.96
Stachybotrys atra	0.94
Wallemia sebi (Sporendonema epizoum)	0.75
Yeasts	
Debaryomyces hansenii	0.83
Saccharomyces bailii	0.80
S. cerevisiae	0.90
S. rouxii	0.62
Bacteria[b]	
Bacillus cereus	0.95
B. megaterium	0.95
B. stearothermophilus	0.93
B. subtilis	0.90
Clostridium botulinum type A	0.95
C. botulinum type B	0.94
C. botulinum type E	0.97
C. perfringens	0.95
Enterobacter aerogenes	0.94
Escherichia coli	0.95
Halobacterium halobium	0.75
Lactobacillus viridescens	0.95
L. plantarum	0.94
Microbacterium sp.	0.94
Paracoccus (Micrococcus) halodenitrificans	0.86
Micrococcus luteus (lysodeikticus)	0.93
Pediococcus cerevisiae	0.94
Pseudomonas fluorescens	0.97
Salmonella sp.	0.95
Staphylococcus aureus	0.86
Vibrio costicolus	0.86
V. parahaemolyticus	0.94

[a] From Troller and Christian (1978).
[b] a_w adjusted with salts.

3. *Spore-Forming Bacteria*

Most *Bacillus* spp. grow at a_w levels of 0.94 to 0.89 (Marshall *et al.*, 1971). The most common limit is 0.90–0.91 a_w. Germination and growth of food-poisoning strains of *B. cereus* are prevented at 0.97–0.93 a_w, depending upon the nature of the bulk solute (Jakobsen *et al.*, 1972).

Types A, B, and E of *Clostridium botulinum* differ in their water relations. The lower limits in salt-adjusted media for growth from spore inocula are 0.95 a_w for type A, 0.94 a_w for type B, and 0.97 a_w for type E (Ohye and Christian, 1967). Toxin production occurred at levels of a_w closely approaching these growth minima. Growth of all three types proceeded at lower levels when the a_w was controlled by glycerol instead of by salt (Baird-Parker and Freame, 1967). For *C. perfringens,* lower limits near 0.95 a_w in NaC1 have been reported by a number of workers, again with lower minima in glycerol-adjusted media.

4. Halophilic, Xerophilic, and Osmophilic Microorganisms

Strictly, only organisms that grow well at some reduced levels of a_w but fail to grow at high levels should be included in these groups. The halophilic bacteria meet this criterion, all having a specific and substantial requirement for NaC1. Contrasting examples are the moderate halophile, *Vibrio costicolus,* growing over the NaCl range 0.3–3.6 *M* (0.990–0.855 a_w), and the extreme halophile, *Halobacter halobium* growing from 3 *M* NaC1 to saturation (0.885–0.75 a_w) (see Fig. 4.3).

Fungi grow more slowly than most bacteria at high a_w, but frequently grow at much lower levels of a_w. Many are particularly well adapted to growth in such conditions. The term "xerophile" is commonly restricted to fungi capable of growth at an a_w below 0.85 (Pitt, 1975). Figure 4.3 includes as an example of such an organism data for the most xerophilic mold studied (*Xeromyces bisporus*). Note the requirement for, as well as the tolerance of, physiologically dry conditions.

Osmophilic yeasts are typically those growing in the presence of high concentrations of sugars, again without necessarily requiring a greatly reduced a_w. Many of these are also capable of growth in concentrated saline environments (Onishi, 1963) and so would be more accurately described as xerophilic.

Although many of the mycotoxigenic molds of concern in foods are xerophilic, little is yet known of the influence of a_w on toxin production. There are no reports of the production of aflatoxins below 0.83 (Northolt *et al.,* 1976).

C. Spoilage

1. 0.98 a_w and Above

Most microorganisms of concern in foods have optimal a_w levels for growth above 0.98. Although many can achieve more than half of their maximum growth rate at 0.98 a_w, some are substantially inhibited at this level, so that reducing a_w to 0.98 can have a marked effect on the composition of the flora of some foods.

For proteinaceous foods, particularly meat and fish, held at refrigeration temperatures in air, *Pseudomonas* spp. are important spoilage agents, producing volatile amines, hydrogen sulfide, and esters of organic acids. Under anaerobic conditions, lactic acid bacteria predominate, causing souring.

The addition of small amounts of salt (ca. 2%) to meat under aerobic conditions is sufficient to retard growth of pseudomonads, which have lower limits near 0.97 a_w, and encourage a lactic fermentation similar to that occurring anaerobically in salt-free meat at higher a_w levels. Both pseudomonads and certain lactic acid bacteria can form substantial amounts of slime.

Fresh fruit and vegetables are spoiled by a wide range of bacteria and fungi. Bacterial attack on the less acid fruit and vegetables by *Erwinia carotovora* and related species which ferment pectins and cause soft rots is common. Low concentrations of salt often select a lactic fermentation, as in the production of sauerkraut from cabbage. A wide variety of molds cause spoilage exemplified chiefly by softening of tissues, discoloration, and visible mycelium.

In unconcentrated fruit juices, fermentation by wild yeasts is common, although at low temperatures acid fermentations due to *Leuconostoc* spp. may predominate.

All nonacid canned foods are subject to spoilage by spore-forming bacteria, should heat processing be inadequate. Spoilage takes many forms but is predominantly acid, gas, putrefactive, or sulfide.

In general, although most spoilage bacteria and fungi have optima for growth in this a_w range, the former grow more rapidly and will predominate unless restrained by unfavorable pH, temperature, or substrate.

2. *Below 0.98 to 0.93* a_w

Spoilage by gram-negative bacteria gives way to gram-positive types—Lactobacillaceae, Bacillaceae, Micrococcaceae, but salt-tolerant coliforms are also important, e.g., in cucumber pickles. Lightly salted foods of about 0.97 a_w, such as sausages (frankfurters and bologna), are susceptible to spoilage by lactic acid bacteria. At lower levels of a_w, micrococci grow profusely on surfaces but have little adverse effect on the quality of sausage. Molding is a problem on bread and high moisture prunes.

Most canned, cured meats have a_w near the top of this range. The reduced a_w is important; in combination with salt and nitrite it prevents the outgrowth of bacterial spores. Among acid canned foods there is an a_w effect on stability of canned pears in syrup to spoilage by butyric anaerobes (Jakobsen and Jensen, 1975). Those with a_w below 0.98 may be stabilized by a milder heat process than is required when the syrup is more dilute.

3. *Below 0.93 to 0.85* a_w

Spoilage organisms of concern are gram-positive bacteria (predominantly cocci), yeasts, and molds. A variety of molds may grow on the surfaces of hard cheeses and bakery goods, in some cases causing discoloration and off-flavors. Dry sausages may support surface growth of micrococci or molds, neither of which appear to be detrimental and may, in the case of some molds, be considered an indication of correct aging. Sweetened condensed milk with an a_w close to 0.85 is spoiled by gas formation from sucrose-fermenting yeasts and mold colonies ("buttons") may grow on the surface of condensed milk in cans with air in the headspace.

4. *Below 0.85 to 0.60* a_w

Xerophilic fungi and osmophilic yeasts predominate in the spoilage of foods in this a_w range. Common foods in this category are jams and jellies preserved by high concentrations of sucrose and invert sugar. Molds will develop on the surface if aerobic conditions exist and fermentative yeasts grow well in the depths of these products, with attendant carbon dioxide evolution. Heavily salted fish, if exposed to air, suffers proteolytic attack from extreme (predominantly red) halophilic bacteria and may support growth of the "dun" mold *Wallemia sebi,* which appears to cause no undesirable flavor changes.

5. *Below 0.60* a_w

Unless rehydration occurs, microorganisms do not grow.

D. Pathogens

1. *0.98* a_w *and Above*

Within this range, all food-borne bacterial pathogens grow rapidly, in many cases with little organoleptic change to the substrate. Processes presenting hazard include those that selectively destroy competitors while actively favoring growth of common pathogenic contaminants, e.g., the cooking and holding at warm temperatures of meat containing spores of *Clostridium perfringens.*

Salmonella spp. grow well in many high a_w foods; they, like most other Enterobacteriaceae, are good competitors in a mixed flora at high a_w, unless competing organisms reduce the pH rapidly. *Clostridium botulinum* is a hazard of high a_w foods inadequately heat processed. Putrefactive and rancid odors accompany growth in most proteinaceous foods, but foods more acid or lower in protein can support extensive growth and toxin production with little evidence of putrefaction.

Vibrio parahaemolyticus, generally confined to seafoods, is sensitive to both heat and cold but grows very rapidly under favorable conditions of pH, temperature, and a_w. It does not cause marked organoleptic changes to seafoods, even at infectious concentrations.

Staphylococcus aureus competes poorly with the natural microflora of most foods at high levels of a_w. Most mycotoxigenic molds grow well, but unless foods are acid, bacterial spoilage will usually precede molding and toxigenesis.

2. *Below 0.98 to 0.93* a_w

The growth of most food-borne bacterial pathogens (*Bacillus cereus, Clostridium botulinum, C. perfringens, Salmonella* spp., *Vibrio parahaemolyticus*) is prevented at levels of a_w within this range (see Table 4.3). Salmonellae can divide at only one-half of their maximal rate at 0.97 a_w and at about one-quarter at 0.96 a_w. If other environmental factors (pH, temperature) are suboptimal, growth of all of the above mentioned bacteria will be very substantially inhibited at 0.96 a_w. In contrast, staphylococci grow at one-half of their maximum rate at 0.94 a_w and are likely to be of major concern within this range, in which are found the lightly salted meat products commonly incriminated in staphylococcal food poisoning.

The bacterial spoilage agents of nonacid foods are much inhibited in the range 0.98–0.93 a_w, and fungi may become a relatively greater problem in many instances. The risk of mycotoxin production can thus be greater here than in foods of higher a_w.

3. *Below 0.93 to 0.85* a_w

As the only bacterial pathogen growing within this range, *Staphylococcus aureus* is the major public health hazard. Many of its competitors grow slowly, if at all, while *S. aureus* maintains a growth rate at 0.94 a_w about one-half of that at optimal a_w (Scott, 1953). While enterotoxin production has not been reported in foods below 0.93 a_w (Troller and Stinson, 1975), enterotoxin A formation has been demonstrated at 0.867 a_w (30°C) and 0.887 a_w (25°C) within 2 weeks in a salt mixture broth (Lotter and Leistner, 1978). The formulation of intermediate moisture foods is designed to give an a_w of about 0.85 in order to inhibit this organism, with antimycotics added to control the more drought-resistant fungi. Some molds can produce mycotoxins at all a_w levels within this range (Troller and Christian, 1978).

4. *Below 0.85 to 0.60* a_w

Many xerophilic molds (i.e., those growing at a_w below 0.85) are mycotoxigenic (Pitt, 1975). However there are no reports of mycotoxin production at 0.80 a_w or below.

5. *Pathogens Surviving but Not Growing in Food of Low* a_w

Pathogenic microorganisms remain viable for long periods in many foods of low a_w and in dried foods in particular. The only practical method of destruction in such foods is heat, as exemplified in the hot-room pasteurization of spray- or pan-dried egg white. The enhanced heat resistance at low a_w has been referred to in Section II, A, 2. In contrast, highly salted foods are lethal to many pathogens. Salmonellae, for example, die rapidly in salted sausage casings. They also die in brines at room temperature, but survive well in refrigerated brines.

The pathogens are often of greatest concern after rehydration of such foods. It is essential that they be controlled by appropriate heating (e.g., in dried soups) and subsequent refrigeration at a temperature below 7°C.

III. USE OF REDUCED a_w TO CONTROL MICROORGANISMS IN FOODS

Microbial stability is conferred on many foods by removal of water (dehydration) or by adding solutes to achieve an adequately low level of a_w. In dehydration, energy is applied to the food as heat, which raises the vapor pressure of the water present to the extent that it evaporates from the food surface. Evaporative cooling lowers the surface temperature and additional heat input is required to maintain the vapor pressure at the desired level. As the water at the surface is removed it is replaced from within the food by such processes as diffusion, convection, capillary flow, and shrinkage.

The evaporation or vaporization of moisture from the food results from the difference in vapor pressure between the atmosphere and the food surface. As drying proceeds, the rate of water loss decreases, because the rate of water movement to the surface becomes limiting, the surface layers become less permeable, and the increasing concentration of solutes reduces the vapor pressure at the surface. To reach the desired degree of dryness, it may thus be necessary to reduce the ambient vapor pressure or raise the temperature of the food.

In general, the gentler the drying process, the greater is the likelihood that contaminating microbes will survive. Freeze drying, commonly used to preserve microbial cultures, may have relatively little effect on microbial viability, while roller drying may destroy most of the vegetative organisms present.

Although adequately dried foods are stable, the various processes pose different microbiological problems, and for this reason will be described individually (see Section III, A, following).

The solutes commonly added to foods to reduce a_w are salt and sugar. The syruping of fruits to produce jams, jellies, and similar preserves usually involves some concomitant removal of water (concentration) by heating. Salting is applied predominantly to meat and fish and to some vegetables. Depending upon the nature of the food it may be dry salted or soaked in brine. Rapid penetration of salt can be effected by "pumping," i.e., injecting the curing brine into the deep tissues of large portions of meat. In cured meats, the growth-inhibiting effect of salt is augmented by the preservative action of nitrite (see Chapter 8).

A. Specific Processes

1. *Solar Drying*

The foods most commonly dried by solar radiation are fruit, fish, rice and other grains, coffee, and copra. The drying of the first two will be taken as examples.

Vine fruits are dipped into or sprayed with a solution containing alkali and fatty acid esters to facilitate water loss through the skin. Stone fruits are halved and pitted. Sulfur dioxide is sometimes added to preserve color and flavor. The fruits are then spread on racks or trays in the open, protected where possible from dust and birds. Solar drying is most successful in regions where a rain-free period of low humidity coincides with the harvest. Wetting of the food or slow drying may allow rapid growth of fungi.

In contrast, the sun-drying of fish is practiced most commonly in the tropics where humidity is usually high. Before exposure to the sun, split fish are frequently salted to inhibit microbial growth. The likelihood of rain demands provision for rapidly covering the fish. Insect infestation and spoilage by halophilic bacteria and molds are common problems.

2. *Tunnel and Belt Drying*

In tunnel drying, subdivided wet food, commonly fruit and vegetables, is spread on trays which are stacked on trolleys and moved progressively through a tunnel. Hot, dry air is blown along the tunnel supplying heat to the food and carrying away the evaporated water vapor.

For belt drying, the food, again commonly fruit or vegetables, is placed in a thin even layer on a porous belt through which hot air moves vertically as the belt passes through the drier. In both systems, but particularly in tray drying, overloading causes slow drying that promotes rapid microbial growth in the warm, wet product.

3. *Fluid Bed Drying*

Rapid drying is obtained by this method in which wet, particulate foods, commonly vegetables, are passed continually onto a porous plate. Hot, dry air passing through this plate moves the material slowly along the bed, suspending it gently in the air stream and thus facilitating water removal.

4. *Pan Drying*

This is a batch process, used for certain egg products, which involves the blowing of hot air over the surface of the food until it is dry. If drying is too slow, microbial growth is likely below the surface of the product.

5. *Foam Mat Drying*

A stable foam is prepared by mixing emulsifier with blended raw material, commonly fruit juice, spread onto a belt and dried with hot air in a tunnel. The greatly increased surface area produced by foaming speeds up the removal of water.

6. *Drum or Roller Drying*

Drum drying is a relatively inexpensive process which is used to dry a wide range of slurried foods. A purée or suspension of the food is spread continuously onto a heated metal drum which rotates at 1–5 revolutions per minute. The material dries in the course of less than one revolution and is scraped from the drum surface in a continuous sheet by a metal blade. Although the surface of the drum, steam heated from within, is about 150°C, evaporative cooling keeps the product temperature to a maximum of about 90°C. Some foods acquire ''cooked'' flavors, but there are no microbiological problems during dehydration.

7. *Spray Drying*

This widely used process is applied particularly to milk, coffee, and egg albumen. Liquefied food is sprayed in the form of very small droplets into a stream of air at about 200°C. The droplets dry rapidly with little off-flavor or discoloration and form hollow particles which rehydrate readily. Foamed material may also be spray dried. Although microorganisms cannot multiply during the rapid drying process, there is a risk of bacterial growth during preheating, the holding stage, and when the hot dry product passes over cold surfaces and condensation occurs.

8. *Freeze Drying*

Heat is applied to frozen food under vacuum to increase the vapor pressure. Moisture sublimes from the surface and condenses as ice on a cold surface. The food does not thaw during the process so that shrinkage does not occur. No microbial growth is possible during freeze drying.

9. *Concentration*

This term applies to the reduction of water content in liquid foods without conversion to a dry state. It is used for the preparation of meat extract, tomato paste, fruit juice concentrate, condensed milk, etc. In some cases, e.g., meat extract, the result is a microbiologically stable product with an a_w below 0.70. Meat extract is concentrated by heat under vacuum in jacketed vessels, and there is little that can be done to prevent some multiplication of thermophilic bacteria during the early stages of the process. Control of the final numbers is effected by ensuring a low initial population of thermophilic spores.

For most liquid foods, however, the a_w of the concentrate remains high—

above 0.95—and an additional process (canning, freezing) is required to confer stability. In such cases, concentration is simply a means of conserving space. Many other liquids are concentrated as a cost-saving preliminary to dehydration by several of the methods discussed earlier in this section.

B. Principal Groups of Foods Based on a_w

1. 0.98 a_w *and Above*

Fresh meats and fish
Fresh fruits and vegetables
Milk and other beverages
Canned vegetables in brine
Canned fruit in light syrup

These are very moist foods, including those containing less than 3.5% sodium chloride or 26% sucrose in the aqueous phase. Food-poisoning bacteria and common spoilage microorganisms, with the exception of extreme xerophiles and halophiles, grow almost unimpeded at levels of a_w within this range.

2. Below 0.98 to 0.93 a_w

Evaporated milk
Tomato paste
Lightly salted fish, pork, beef products
Canned cured meats
Fermented sausages (not dried)
Cooked sausages
Processed cheese
Gouda cheese
Canned fruits in heavy syrup
Bread
High moisture prunes

Maximum concentrations of salt or sugar in the aqueous phase of these foods will be near 10% and 50%, respectively. All known food-poisoning bacteria can grow at least in the upper part of this range.

Below 0.93 to 0.85 a_w

Dry or fermented sausage (Hungarian, Italian types)
Dried beef
Raw ham
Aged cheddar cheese
Sweetened condensed milk

This group includes foods with up to 17% salt or saturated sucrose in the aqueous phase. Only one bacterium known to cause food-poisoning (*Staphylococcus aureus*), but many molds that produce mycotoxins, can grow in this a_w range.

4. *Below 0.85 to 0.60* a_w

Intermediate moisture foods
Dried fruit
Flour
Cereals
Jams and jellies
Molasses
Heavily salted fish
Meat extract
Some aged cheeses
Nuts

No pathogenic bacteria grow within this range. Spoilage by xerophilic, osmophilic, or halophilic organisms can occur.

5. *Below 0.60* a_w

Confectionery
Chocolate
Honey
Noodles
Biscuits
Crackers
Potato chips
Dried egg, milk, vegetables

Microorganisms do not multiply below 0.60 a_w, but can remain viable for long periods.

C. Problems of a_w in Food Mixtures

It should not be assumed that the water content and hence the a_w of an initially homogeneous food remains homogeneously distributed during transport and storage. A food in which the *average* water content is sufficiently low to ensure microbiological stability may contain areas of much higher (and much lower) a_w. The gradients that lead to this unsafe situation result most commonly from temperature differences arising from climatic changes during transport, diurnal variations, and exposure to sunshine. In extreme cases, condensation on overhead surfaces results in liquid water dropping back onto the foodstuff. Although such problems are most commonly encountered in silos, conveyors, and in ships' holds, foods in much smaller containers can also be subjected to a_w gradients (see Chapter 23).

Moisture gain or loss will also introduce a_w gradients in improperly packaged or unpackaged foods. Condensation is common on the surfaces of foods removed from chilling rooms, while evaporation may be rapid when warm food is placed under refrigeration in circulating cold air. Carcass meat provides examples of both condensation and evaporation, where the a_w of the surface on which microbial growth occurs may not be the a_w of the bulk of the muscle.

When ingredients are mixed in formulating a food, the migration of water between the components will depend not on their relative moisture contents but on their levels of a_w. Thus a component of lower moisture content will lose water to another of higher moisture content if the latter has the lower a_w. The amount of moisture that will be transferred and the equilibrium value of a_w reached can be determined from knowledge of the relative amounts of the different ingredients and from their water sorption isotherms.

IV. INTERACTIONS OF a_w AND OTHER FACTORS

In the preceding sections, the limitations on microbial activities by a_w were for the most part established in menstrua with optimal pH, temperature, E_h, and nutrition. When any one of these is less than optimal, the range of a_w favorable to growth is narrowed.

Freezing to sufficiently low temperature or holding at sufficiently high temperature will prevent growth of microorganisms in all foods, independent of a_w. Otherwise, combinations of reduced temperature and pH with appropriate levels of a_w are commonly used, although in many cases the object is an extended shelf-life but not a shelf-stable product. The storage categories of meat products listed in Table 4.4, although only approximate in terms of a_w and pH, are clear

TABLE 4.4

Storage Categories of Meat Products Based on the a_w and pH Values of the Product, with Recommended Storage Temperatures[a]

Category	pH and a_w	Temperature
Easily perishable	pH >5.2 and a_w >0.95	≤+5°C
Perishable	pH 5.2–5.0 (inclusive) or a_w 0.95–0.90 (inclusive)	≤+10°C
Shelf-stable	pH ≤5.2 and a_w ≤0.95 or only pH <5.0 or only a_w <0.90	No refrigeration required

[a] Modified from Leistner and Rödel (1975).

examples of this approach. Note, however, that these include traditional products containing intrinsically inhibitory substances that may also contribute to stability. These categories cannot necessarily be extrapolated to other foods.

A mild heat treatment that inactivates vegetative cells will provide a stable product if outgrowth of surviving spores can be prevented either by curing salts as in meats or by controlled pH and a_w as demonstrated for canned pears (Jakobsen and Jensen, 1975). Microorganisms in general are more resistant to destruction by heat in dry foods than in moist foods (see Section II, A, 2).

Although yeasts grow in liquids with a_w between 0.75 and 0.62, they are not reported as growing on solids with the same a_w (Scott, 1957). It follows that the spoilage agents for the latter foods will be molds and hence anaerobiosis will enhance preservation. At slightly higher a_w, heavily salted fish are spoiled by extremely halophilic bacteria that are obligate aerobes and anaerobic conditions again extend shelf-life.

The action of preservatives is enhanced when microbial growth is retarded by reduced a_w and combinations of controlled a_w and preservative also have a place in food preservation.

5

pH and Acidity

I. INTRODUCTION

In the unprocessed state, most foods such as meat, fish, and vegetables are slightly acidic, most fruits are moderately acidic, and a few foods such as egg white are alkaline. For thousands of years, the acidity of foods has been increased, either naturally by fermentation or artificially by the addition of weak acids, to enhance microbiological stability and preserve the foods. Acidity may be the primary factor in preservation of fermented foods such as yogurt, sauerkraut, and pickles, or may be an ancillary factor combined with other factors such as chemical preservatives, heat, and water activity (a_w).

A. Definitions

The relationship between pH and the hydrogen ion concentration $[H^+]$ that it represents is

$$pH = -\log_{10} [H^+] \qquad \text{or} \qquad pH = \log_{10} (1/[H^+])$$

Thus pH is the negative logarithm of the hydrogen ion or proton concentration. In foods, the acids of interest are almost always weak acids (HA), which dissociate into H^+ and A^-. At equilibrium, the ratio $[H^+][A^-]/[HA]$ is a constant (K_a); i.e., $K_a = [H^+][A^-]/[HA]$. This can be rewritten

$$[H^+] = K_a [HA]/[A^-]$$

and, taking the negative logarithm of both sides of the equation,

$$-\log H^+ = -\log K_a - \log[HA] + \log [A^-]$$

or

$$pH = pK_a + \log [A^-]/[HA]$$

If $[A^-]$ and [HA] are in equal concentration, the log of the ratio is zero, and pH = pK_a. In other words, pK_a = pH when the concentrations of the dissociated acid and undissociated acid are equal. pK_a values for various acids are tabulated in Table 7.1 and in handbooks of chemistry and physics. Knowing the concentration of the acid, the pH, and the pK_a or K_a, one can calculate the amount of undissociated acid present (see Table 7.3.). Strong acids have very low pK_a, and therefore are almost entirely dissociated in solution; that is; they contribute $[H^+]$ in higher proportion than a weak acid does. For example, the pH (at 25°C) of a 0.1 *M* solution of HCl is 1.08, whereas the pH of a 0.1 *M* solution of acetic acid is 2.87.

Titratable acidity is the quantity of standard alkali (usually 0.1 *M* NaOH) required to neutralize an acid solution. It measures the amount of free hydrogen ion, as well as the amount of hydrogen ion released from undissociated acid during the titration.

B. Properties of pH and Acidity

The pH of a food is one of several principal factors that determine survival and growth of microorganisms during processing, storage, and distribution. The various factors described elsewhere in this volume usually are affected by pH. It is sometimes difficult to separate the effect of pH per se, from that of other factors which in turn are affected by pH.

Hence, microorganisms are affected by the level of free H^+ ions (i.e., pH, per se) and by the concentration of undissociated weak acid which is in turn affected by pH. The anions of some weak acids (e.g., acetiç or lactic) are metabolized within the bacterial cell, so that H^+ is released, acidifying the cell interior to inhibitory levels. Other anions are not so metabolized, and therefore do not acidify the cell interior.

C. Measurement of pH and Acidity

The pH is most commonly determined with an electronic pH meter, with an accuracy of ±0.01 pH over the range 0–14. The pH meter comes equipped with a glass membrane electrode and a reference electrode, or with the increasingly popular single probe combination electrode. The reference electrode is usually the calomel type, utilizing a salt bridge filled with a saturated solution of potassium chloride.

A number of different methods utilizing electronic pH meters for the measurement of pH in intermediate moisture foods have been compared (Acott and Labuza, 1975a).

1. AOAC method 14.022 (Horwitz, 1975) using glass and calomel electrodes.

2. Direct measurement using either a semimicro combination electrode or a nonaqueous electrode, specially designed to correct for osmotic gradients (e.g., Beckman 39142).

3. Slurry dilution method, using equal weights of water and product, tested after equilibration for 24 hr at 4°C.

4. Gran plot extrapolation in which serial dilutions are made and the measured pH is plotted on special paper (e.g., Gran plot paper, Orion catalog No. 900093). The pH is estimated by extrapolation to zero dilution. The pH is limited to one decimal place of certainty because of the inaccuracy of extrapolation.

The methods gave similar results, and the AOAC direct reading method is as reliable as any other method, provided the food is sufficiently plastic to allow contact with the sensing tip of the electrode. If the food is somewhat dry, the Gran plot technique is best; the buffering capactiy of the food may affect the measurement.

For standard methods for measuring the pH of various foods and the use of reference buffer solutions, see the Twelfth Edition of the "Official Methods of Analysis of the Association of the Official Analytical Chemists" (Horwitz, 1975). Additional information is presented by Munro (1970) as well as by manufacturers of laboratory equipment. A recent and particularly helpful procedure for determining the pH or acidity of acidified foods is provided in the U.S. Good Manufacturing Regulations for Acidified Foods (U.S. Department of Health, Education and Welfare, 1979).

The physical characteristics of the product determine the most appropriate method for preparing the sample for measurement of pH. For liquids and moist homogeneous products it is sufficient to immerse the tip of the electrode(s) into the sample and ensure that there is adequate contact. Foodstuffs containing solids may be homogenized before a sample is taken, so that a bulk pH can be measured accurately. Some foodstuffs are nonhomogeneous, having gradients of pH from the surface to core (e.g., vegetables during the pickling process). It is useful to know the differences among the pH values measured at several points that reflect the nature and size of the sample. To measure the pH of a moist surface film (e.g., fish fillet or beef steak) it is sufficient to apply the two electrodes firmly enough to obtain a reading on the meter. The meter reads the pH at the surface of the glass electrode. The calomel electrode merely establishes electrical contact (Elliott, 1947).

Under certain circumstances, e.g., with heavily buffered or very acid foods, titratable acidity may provide a more useful method of assessing acidity. A substantial proportion of a weak acid in solution is undissociated and thus does not contribute directly to the pH, since pH is a measure of only proton concentration and not of total concentration of acid.

Because the antibacterial activity of weak acids depends primarily on the concentration of the undissociated molecule (Chapter 7) it is first necessary to estimate the pH of the food, then to estimate the total concentration of the acid by titration, usually with 0.1 *M* NaOH. During the titration, the dilution of the acid and the removal of H^+ ions will cause dissociation of the undissociated acid, freeing the remaining H^+ ions so they can be titrated. For fruit and fruit products use AOAC method 22.060; for milk, use AOAC method 16.023 (Horwitz, 1975). Titratable acidity measures not only the level of preservative acids, but also in some products, the amount of microbial spoilage that has taken place (Chapters 18, 19, 23, and 25).

II. EFFECTS OF pH ON MICROORGANISMS

A. Mechanisms

Many microbes can grow over a wide range of pH values. One might assume from this that microbial cells possess efficient methods to stabilize their internal pH. However, there is little direct evidence to support this assumption. Instead, investigation has shown the internal pH to be considerably affected by the pH of the menstrum.

This question has been evaluated by measuring the weak acid 5,5,-dimethyl-2,4-oxazolidinedione (DMO) across the cell membrane. In the pH range 5–7 the internal pH of *Escherichia coli* was usually more alkaline than that of the external menstruum; above pH 7, it was more acid (Kashket and Wong, 1969). The effect of a strong inorganic acid (phosphoric) and a weak organic acid (acetic) on the internal acidification of yeast cells is illustrated in Table 5.1. The weak organic acid acidified the cell interior more effectively than the strong inorganic acid did; at pH 4, formic, acetic, and butyric acids inhibited glycolysis in intact cells (Neal *et al.*, 1965).

A variety of weak acids at or below their pK_a values are potent inhibitors of amino acid transport in *Penicillium chrysogenum*. The effective compounds include sorbate, benzoate, and propionate; and it has been suggested that the undissociated form of these weak acids may diffuse freely through the cell membrane and ionize in the cell to yield protons that acidify the alkaline cell interior (Hunter and Segel, 1973). Similar effects have been found using a variety of microbes and the effectiveness of undissociated acids as preservatives has been noted for many years (Ingram *et al.*, 1956).

A number of experiments have suggested the underlying mechanism for the activity of weak acids. Freese *et al.*, (1973) determined the ratio of pH values inside the microbial cell to those outside, using weak lipophilic acids as preser-

TABLE 5.1

Influence of the Type of Acid (Buffer) and pH of the Medium on the Apparent Internal pH of Yeast Cells[a]

Buffer	pH of medium	Apparent internal pH 1 min	Apparent internal pH 10 min
Acetate	6.2	6.1	6.2
Phosphate	6.2	6.2	6.3
Acetate	4.2	4.9	5.1
Phosphate	4.1	5.7	5.8
Acetate	3.0	4.3	4.3
Phosphate	3.0	6.5	5.5–6.5

[a] From Neal *et al.* (1965).

vatives, and concluded that it is the rate of proton leakage into the cell relative to the proton-rejecting capacity of the cell that determines the extent to which an environment is inhibitory.

There are three types of acid preservatives.

1. *Strong acids* (e.g., hydrochloric and phosphoric acids) rely on very low pH, and hence high external proton concentrations, to acidify the cell interior. Such conditions are usually unacceptable in foods, but acceptable in carbonated beverages where phosphoric acid is the acidulant.
2. *Weak lipophilic acids* cause leakage of hydrogen ions across the cell membrane, acidifying the cell interior, and inhibiting nutrient transport. Some acids will dissociate to give anions (e.g., lactate, citrate) which the cell can transport and whose presence does not therefore inhibit energy yielding metabolism. Other acids, e.g., acetic and formic, are very effective preservatives since they are not only proton conductors but also may yield inhibitory concentrations of their anions within the cell.
3. *Acid-potentiated ions* such as sulfite and nitrite (the salts of weak mineral acids) become highly inhibitory at low pH.

The cells of various microbial species have different tolerances toward internal acidification or anion accumulation, and their membranes have different permeability characteristics for lipophilic acids. In a mixed flora, acidity may select a particularly tolerant component of the initial population. Yeasts and lactobacilli are often selected by low pH (see Table 5.2).

TABLE 5.2

Some Types of Spoilage Typically Associated with Foods of Certain pH Values

pH and type of food	Examples	pH range	Additional factors	Spoilage organisms	Reference
Alkaline pH >7.0	Egg albumen	Up to 9.6	Conalbumin, lysozyme	Predominantly gram negative, e.g., *Pseudomonas, Aeromonas, Proteus*	Board (1969)
Neutral pH 7.0–6.5	Milk	7.0–6.8	Antimicrobial agents	Homo- and heterofermentative lactobacilli, *Bacillus* spp., *Micrococcus*, coliforms	
	Ham	6.5	Lactic acid	*Pseudomonas, Sarcina*, micrococci	Ingram (1962a)
			Nitrite, sodium chloride	*Alcaligenes, Acinetobacter–Moraxella*	
	Bacon	6.0–5.9	Sodium chloride 4% w/v	Micrococci, *Vibrio*	Gardner (1971)
	Chicken	6.7–6.2	1°C storage	*Pseudomonas putrifaciens* [b]	Barnes and Impey (1968)
	Packaged chopped meat	6.8–6.5	Sulfite	*Microbacterium thermosphactum* [b]	Dowdell and Board (1971)
	Carcass meat	7.2–5.2		*Acinetobacter–Moraxella, Pseudomonas, Lactobacillus*	
	Vacuum-packaged meat	6.6	Dark cutting meat	*Micrococcus, Bacillus*, Lactobacillaceae, psychrotrophic Enterobacteriaceae	Patterson and Gibbs (1977)
Low acid pH 6.5–5.3 [a]	Raw beef	5.8–5.4	Storage 2°–4°C	*Staphylococcus, Micrococcus*	Labadie *et al.* (1975)
	Raw veal		Acidified, HCl or lactic acid		
		5.4		Yeasts, *Pseudomonas*	Labadie *et al.* (1975)
		5.4		*Pseudomonas*	Labadie *et al.* (1975)
	Bacon	5.6	Sodium chloride 4%	*Micrococcus*, yeasts (*Candida*)	Gardner (1971)
	Canned vegetables	6.4–5.4	Heat	Flat-sour organisms, *Bacillus stearothermophilus*	Jay (1970)

(*continued*)

TABLE 5.2 *(continued)*

pH and type of food	Examples	pH range	Additional factors	Spoilage organisms	Reference
	Marinades with mayonnaise	6.8		Gram-negative oxidative bacteria, *Bacillus* spp., *Serratia*, yeasts, molds	Bartl (1973)
	Vacuum-packaged fresh meat	5.4	4°C storage	Heterofermentative lactobacilli, gram-negative fermentative rods	Patterson and Gibbs (1977)
	Aerobic-packaged fresh meat	5.4	4°C storage	Homo- and hetero-fermentative lactobacilli, *Proteus*, *Pseudomonas* *Aeromonas*, psychrotrophic Enterobacteriaceae	Patterson and Gibbs (1977)
Medium acid pH 5.3–4.5 [a]	Canned products including meat and vegetable mixtures, soups, sauces		Heat	Thermophilic anaerobes, *Clostridium saccharolyticum*, [b] putrefactive anaerobes	Jay (1970)
	Fermented vegetables	5.1–3.9	Weak organic acids	*Leuconostoc mesenteroides*, yeasts	Orillo *et al*. (1969)
	Pickled cucumbers	7.0–4.5	Acetic acid	Yeasts, coliform bacteria, heterofermentative-lactic acid bacteria	Etchells *et al*. (1975)
	Cottage cheese	4.5		Gram-negative oxidative bacteria, molds	
Acid pH 4.5–3.7 [a]	Pickled cucumbers	4.0	Acetic acid	*Bacillus nigrificans*, [b] pectinolytic molds *Enterobacter*, (*Aerobacter*), lactobacilli, pediococci	Jay (1970)
	Mayonnaise	4.1–3.0		*Zygosaccharomyces*, *Saccharomyces*, *Lactobacillus brevis*	Jay (1970)
	Tomatoes, fruits	4.0		*Clostridium pasteurianum*, *Bacillus coagulans*, Lactobacilli, butyric anaerobes	Jay (1970)
	Juices	4.5		*Acetobacter–Gluconobacter*, yeasts, Lactobacillaceae, Catalase-negative cocci	Mossel (1971)

	Fruits	4.5		Lactobacillaceae, molds	Mossel (1971)
	Fruit juice	5.5–3.5		*Lactobacillus buchneri*	Eskin *et al*. (1971)
	Dried fruits	4.5	a_w 0.90, sulfite	Molds, yeasts	Mossel (1971)
	Yogurt	4.2–3.8	Reduced a_w fruit-supplemented natural yogurt	Yeasts, molds	Davis (1975)
	Fermented vegetables	4.3–3.9		*Pediococcus cereviseae*	Orillo *et al*. (1969)
		4.5–3.5		Lactobacillaceae	Orillo *et al*. (1969)
	Orange juice	4.0		*Leuconostoc mesenteroides*	Rushing *et al*. (1956)
		4.0		Lactobacillaceae	Rushing *et al*. (1956)
	Sour vegetable preserves	4.2	Acetic acid	*Bacillus* spp., lactobacillaceae *Saccharomyces bailii, Pichia, Moniliella*	Dakin and Radwell (1971)
	Marinated herring	4.0	Cooked	*Lactobacillus plantarum*	Blood (1975)
High acid pH <3.7[a]	Pickles, fruit juice	3.9–3.5	Canned	Molds, yeast, some lactic acid bacteria	
	Grapefruit segments	3.5	Heat, sugar, syrup	*Lactobacillus brevis*	Juven (1976)
	Sauerkraut	3.7–3.1		*Lactobacillus brevis, L. cucumeris, L. plantarum, Torula* spp.	Jay (1970)
	Citrus fruits	3.5–3.0		*Penicillium italicum, P. digitatum*	Eskin *et al*. (1971)
	Salad dressing	3.9–3.2		Yeasts, lactobacilli, *Bacillus* spp.	Smittle (1977)

[a] pH values used in canning.
[b] Either not listed or classified *incertae sedis* in the 8th Edition of Bergey's Manual (Buchanan and Gibbons, 1974).

B. Growth

The pH limits for growth differ greatly among microorganisms within the range of about 1–11. Many microorganisms grow optimally near pH 7, and grow well between pH 5 and 8. However, there are some exceptions: the acetic acid bacteria grow optimally between pH 5.4 and 6.3; and lactic acid bacteria grow optimally between pH 5.5 (or below) and 6.0. Table 5.3 lists the minimum and maximum pH values for growth of a wide variety of microorganisms. In general, yeasts and fungi grow at much lower pH than do bacteria, whereas the maximum pH values for growth are similar among bacteria, yeasts, and molds.

Data in Table 5.3 were obtained in experiments in which all conditions other than pH were optimum. The table shows growth at wider ranges of pH than those that would occur in foods, where less than optimum conditions would reduce the extent of the pH range of growth, e.g., competing microbial species, unfavorable oxygen tension, adverse storage temperature, reduced water activity, or heat-damage during processing. Figure 5.1 illustrates such interaction between pH and salt concentration as they affect the lag, growth rate, and maximum cell population of *Staphylococcus aureus*. The minimum pH permitting growth of this bacterium in laboratory medium is 4.5 in 8–10% NaC1 and 6.0 in 16% NaCl, while at no pH is toxin synthesized at NaCl concentrations above 10% (Riemann *et al.*, 1972).

The pH tolerance of microorganisms may also be greatly modified by the nature of the acid used for acidification. Thus *Lactobacillus brevis* grew in APT broth to pH 3.0 acidified with citric, hydrochloric, phosphoric, or tartaric acid, to pH 3.7 acidified with lactic acid, and to pH 4.0 acidified with acetic acid (Juven, 1976). The acid tolerance of *Staphylococcus aureus* in pasteurized milk is also altered by the type of acid used to acidify the milk. A 90% reduction in growth rate was achieved at pH 5.2 with acetic, at pH 4.9 with lactic, at pH 4.7 with phosphoric or citric acid, and at pH 4.6 with hydrochloric acid (Minor and Marth, 1970). In skim milk over the pH range 6.7–4.0 citric acid was most inhibitory to *Salmonella typhimurium*, lactic acid was somewhat less so, and hydrochloric acid, least inhibitory (Subramanian and Marth, 1968). At various pH values in a variety of low-acid foods, acetic acid as an acidulant was usually more inhibitory than citric acid against the growth of *Clostridium botulinum* (Townsend *et al.*, 1954).

It is generally held that the pH relations of microorganisms are fairly stable. Chung and Goepfert (1970), who established that the minimum pH for growth of several serotypes of *Salmonella* was 4.05 ± 0.05, were unable to reduce this limit by sequential subculture at low pH. However, Huhtanen (1975) found it possible to increase the pH tolerance of 32 *Salmonella* serotypes by 0.8–1.0 pH units using a pH gradient agar plate technique. The increased tolerance was rapidly lost after transferring to trypticase soy agar, and he considered that a

TABLE 5.3

The Limits of pH Allowing Initiation of Growth by Various Microorganisms, in Laboratory Media Adjusted with Strong Acid or Alkali[a]

	Minimum pH	Maximum pH
Gram-negative bacteria		
Acetobacter acidophilum[b]	2.8	4.3
Alcaligenes faecalis	6.4	9.7
Escherichia coli	4.4	9.0
Klebsiella pneumoniae (aerogenes)	4.4	9.0
Proteus vulgaris	4.4	9.2
Pseudomonas aeruginosa	5.6	8.0
Salmonella paratyphi	4.05	7.8
Salmonella schottmuelleri	4.05	8.0
Salmonella typhi	4.0–4.5	8.0–9.6
Thiobacillus thiooxidans	1.0	9.8
Vibrio parahaemolyticus	4.8	11.0
Gram-positive bacteria		
Bacillus cereus	4.9	9.3
B. subtilis	4.5	8.5
B. stearothermophilus	5.2	9.2
Clostridium botulinum	4.7	8.5
Clostridium sporogenes	5.0	9.0
Enterococcus spp.	4.8	10.6
Bifidobacterium bifidum (Lactobacillus bifidus)	3.8	7.2
Lactobacillus spp.	3.8–4.4	7.2
Serratia marcescens	4.0	9.0
Micrococcus sp.	5.6	8.1
Staphylococcus aureus	4.0	9.8
Streptococcus faecium	4.4–4.7	9.2
Streptococcus lactis	4.3–4.8	9.2
Streptococcus pyogenes	6.35	9.2
Yeasts		
Candida pseudotropicalis	2.3	8.8
Hansenula canadensis	2.15	8.6
Saccharomyces cerevisiae	2.35	8.6
Saccharomyces fragilis	2.4	9.05
Saccharomyces microellipsoides	2.2	8.8
Saccharomyces pastori	2.1	8.8
Saccharomyces exiguus	1.5	—
Schizosaccharomyces octosporus	5.45	7.05
Candida krusei	1.5	—
Hanseniaspora melligeri	1.5	—
Rhodotorula mucilaginosa	1.5	—
Molds		
Aspergillus oryzae	1.6	9.3
Penicillium italicum	1.9	9.3
Penicillium variabile	1.6	11.1
Fusarium oxysporum	1.8	11.1
Marasmius foetidus	2	6.8
Phycomyces blakesleeanus	3.0	7.5

[a] Assembled by Dr. Martyn Brown.

[b] Not listed, or classified *incertae sedis* in the 8th Edition of Bergey's Manual (Buchanan and Gibbons, 1974).

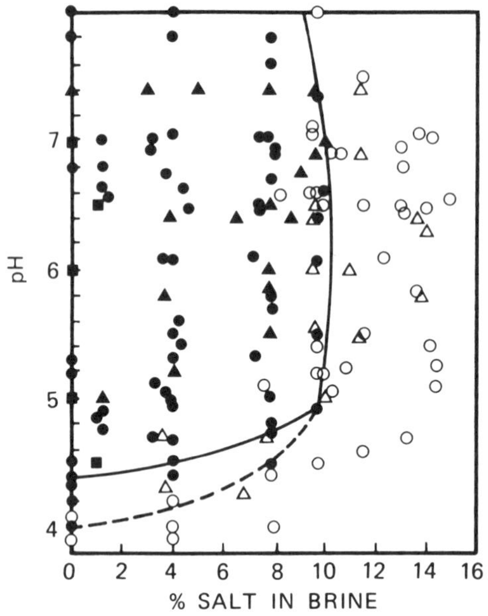

Fig. 5.1 Combination of pH values and brine concentrations which supported (closed symbols) or did not support (open symbols) production of enterotoxin A (squares), B (triangles), and C (circles) aerobically in laboratory media and foods at inoculum levels up to 10^7 cells/ml(gm)(solid line) or 10^8 cells/ml(gm)(broken line). (From Riemann *et al.*, 1972.)

liquid medium at pH values lower than those giving visible growth was bactericidal, not bacteriostatic, for both acid-conditioned and wild-type strains.

From a study of interactions between staphylococci and saprophytic microbes over a range of environmental conditons, including pH, Peterson *el al.*, (1964a), concluded that the primary effect of pH on staphylococcal growth was direct rather than through its influence on saprophyte growth and competition.

Thus the significant effects that pH can exert on microbial growth in foods may be influenced by many other environmental factors whose effects will, in turn, be modified by pH.

C. Survival

When acidity or the concentration of preservatives is sufficiently high to halt microbial growth, the length of time that microbes are able to remain viable becomes critical in assessing their potential contribution to spoilage or food poisoning, should conditions again become favorable for growth. Under adverse conditions microbes may be deprived of the maintenance energy needed to regulate their internal ionic environment and to turn over their cellular components; such conditions lead to loss of viability. When microbes can transport sufficient nutrients for maintenance, but not for growth, they can remain viable for long periods.

As previously noted for microbial growth, a number of food preservation factors may interact with pH to affect the survival of microorganisms. Survival of *Staphylococcus aureas* (10^8 cells/ml) in trypticase soy broth acidified with acetic, citric, hydrochloric, lactic, and phosphoric acids was examined by Minor and Marth (1972). When the organism was exposed to the medium adjusted with hydrochloric acid to pH 5.2–3.6, between 90 and 99.99% of the cells were inactivated within 48 hr. Acetic, lactic, and phosphoric acids were more active against *S. aureus* than were hydrochloric or citric acids. Mixtures of lactic and hydrochloric acids inactivated more cells than did either acid alone. The killing action was attributed to the undissociated acid molecule, since the anions of the acids had no effect on survival.

The survival of *Staphylococcus aureus* and *Salmonella newport* in low water activity foods (0.43 a_w), stored for 9 weeks at 25°C emphasizes the importance of the interaction among food preservation factors (Fig. 5.2). The foods were pro-

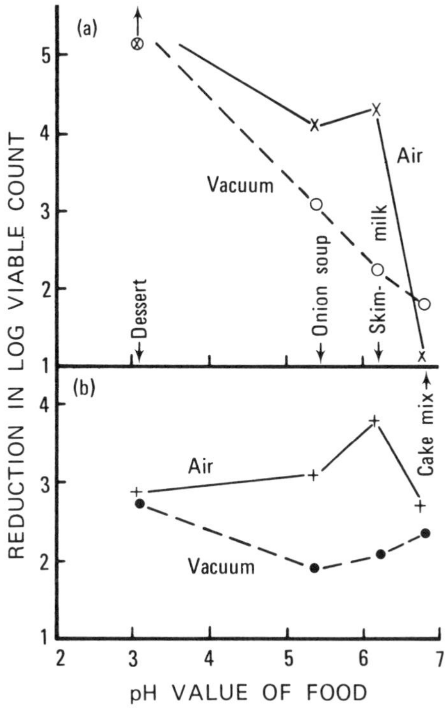

Fig. 5.2 Relation of pH value of foods to the reduction in viability of bacteria during storage for 9 weeks at 25°C and 0.43 a_w in air and in vacuum. (a) *Staphylococcus aureus:* (b) *Salmonella newport*. (From Christian and Stewart, 1973.)

gressively more destructive with decreased pH. The a_w has a major effect on the survival of pathogens in dried foods and implicit factors such as pH may modify this effect (Christian and Stewart, 1973). The survival of both organisms tested was greatest in cake mix (pH 6.8) and least in dessert (pH 3.1); the staphylococci were particularly susceptible to acidity, whereas the salmonellae were not consistently susceptible.

III. EFFECTS OF pH ON FOOD SPOILAGE MICROORGANISMS

A. Non-Spore-Forming Bacteria

Non-spore-forming bacteria figure prominently among the organisms associated with the spoilage of foods in all pH ranges, as shown in Table 5.3. Meats with relatively high pH (6.2) putrify from gram-negative members of the *Pseudomonas–Acinetobacter–Moraxella* group upon storage at or below 10°C. However, they fail to grow at or below pH 5.3 (Labadie *et al.*, 1975; Bem *et al.*, 1976).

Despite considerable differences in the composition of the cell wall or membranes of gram-positive and gram-negative organisms, their limits of pH tolerance are only slightly different (Table 5.3). However, the bacteria responsible for the spoilage of acidic (pH <4.5) foods are all gram-positive. Most commonly involved are species of *Lactobacillus,* which are important both as spoilage agents for some foods and as useful aids in the production of fermented dairy products and vegetables. The gram-positive organisms include some that are especially resistant to the presence of high concentrations of undissociated acids in the medium; e.g., lactobacilli are resistant to lactic or acetic acid. In addition, some lactobacilli may produce weak lipophilic acids in sufficient quantities to inhibit enteric bacteria. The growth of *L. acidophilus* in laboratory media inhibits or destroys (at pH 3.8) *Salmonella enteritidis, S. typhimurium,* and *Shigella sonnei* present in mixed culture (Cieplinska and Zychowicz, 1974).

Egg albumen, among the most alkaline of foods (pH >8.6) is not colonized by gram-positive bacteria, because it contains the enzyme lysozyme (*N*-acetyl muramidase) which is most active at alkaline pH. Lysozyme acts by rupturing the walls of gram-positive microbes causing lysis of the cells (see Volume II, Chapter 19).

Interest in the effect of pH on the heat resistance of microorganisms centers on bacterial spores. However, the resistance of vegetative cells to heat is also affected and diminishes rapidly under acid or alkaline conditions. Weak acids are usually more effective than strong acids at lowering the heat resistance of vegetative cells, and acid may be added to offset the protective effect of sugar on microbes exposed to heat (Hansen and Riemann, 1963).

Many environmental factors in addition to lysozyme and heat interact with pH in their effects on vegetative bacteria. Table 5.2 lists examples of food spoilage in which the effects of salt, temperature, preservatives, and a_w are related to pH.

B. Spore-Forming Bacteria

Spores produced during growth of vegetative cells in unprocessed foods can survive the minimal heat processes designed to kill the vegetative cells. However, in many canned foods, thermal processing is designed to destroy both spores and vegetative cells. Where a milder thermal process does not kill all the spores present, the function of the specific environment created within the food by the processor is primarily to inhibit the germination of the surviving spores. Its second function is to arrest the growth of any vegetative cells derived from these germinated spores or to inhibit the growth of post-processing contaminants. Acidification often is used in conjunction with curing salts, sugar, or salt to inhibit germination or outgrowth from germinated spores, and canned foods may be classified on the basis of their acidity (Hersom and Hulland, 1964; see also Table 5.2).

Most of the data relevant to the influence of pH on spores have been obtained with vegetative pathogenic bacteria and are discussed in Section II, C. Such organisms are fortunately acid sensitive, but some spore-formers are spoilage agents of canned acid foods. These include the butyric anaerobes, such as *Clostridium pasteurianum*. In foods such as mandarin oranges this organism is inhibited by pH values below 3.5. When the pH is above this level, the incidence of spoilage increases; between pH 3.67 and 3.75, none of the samples spoiled; between 3.90 and 3.97, 33% spoiled; between 4.12 and 4.24, 92% spoiled; and above pH 4.42 all the samples spoiled (Ikegami *et al.*, 1970). The same organism grew in canned pears at pH 3.8 if the a_w was ≥ 0.985, but not at pH 3.8–4.0 when the a_w was in the range 0.985–0.975 (Jakobsen and Jensen, 1975).

Glucono-δ-lactone (GDL) has been suggested as an acidulant for fermented sausage (Mol and Timmers, 1970). In a pasteurized meat mix, GDL at 0.2%, w/v, decreased the pH from 6.2 to 5.95. When the meat contained salt alone (4.3%, w/v), the lower pH did not improve stability, but with nitrite (64 ppm) and brine (salt 4.1% w/v) the growth of spoilage clostridia was inhibited. The size of the inoculum, storage conditions before and after pasteurization, and the interaction of the brine, nitrite, and pH all affect the storage stability of products.

There is conflicting evidence on the effect of pH on the thermal inactivation of bacterial spores. However, spores are often easily inactivated at low pH, probably because the environmental pH alters the ionic environment within the spore coat. Alderton and Snell (1963) have suggested that spores act as weak cation exchangers so that low pH, i.e., high $[H^+]$, permits replacement of other bound ions in the spore wall by H^+, thus lowering the stability of the spore.

The pH of the recovery medium exerts considerable effect on the number of spores apparently viable after heat treatment (Roberts, 1970).

C. Yeasts and Molds

Yeasts and molds are usually acid tolerant, grow well below pH 4, and include some of the most acid-tolerant microbes found in foods (see Tables 5.2 and 5.3).

In acid foods, yeasts when present in a heterogeneous population will usually outgrow the bacteria. The ability to grow at low pH may depend on energy-requiring systems that prevent acidification of the cell interior. The type of acid present in the medium will determine pH tolerance. *Saccharomyces bailii* grows in the presence of high concentrations of sorbic and benzoic acids (600 mg/l) or in 2% acetic acid at pH values below the pK_a of weak organic acids (Warth, 1977). Starved cells concentrate these acids intracellularly, but on addition of glucose the intracellular concentration is considerably reduced. In *S. bailii,* resistance to acid preservatives results primarily from an inducible, energy-requiring system that transports preservatives from the cell. Since the pump requires energy, the organisms will be resistant only when a sufficient energy source is present, i.e., in foods with a high concentration of glucose. In such foods fermentation will be stimulated and growth yield diminished (Warth, 1977).

The pH is an important factor governing the survival of yeasts that have been subjected to a sublethal stress. For a range of yeasts, low pH (<3.5) in the recovery medium inhibits growth.

The pH and acid resistance of some yeasts isolated from spoiled carbonated beverages, sauce, and other liquid acid food products have been examined by Pitt (1974); these results are included in Table 5.3. There is little information on the effect of pH on spoilage by molds; although they are tolerant of low pH values, they grow only aerobically.

IV. EFFECTS OF pH ON PATHOGENS

Gram-negative pathogens such as the salmonellae may be controlled by pH below 4.0, but even lower pH values or a combination of low pH and other factors, such as low temperature, are needed to control coliforms. The type of acid used for acidification is critical. The imposition of conditions which kill the pathogens is important if foods are likely to suffer temperature abuse during storage. In many meat products the growth of salmonellae is restricted by the growth of acid-producing organisms (lactobacilli). Because of the high buffering capactiy of such foods, large amounts of acid are produced without a corresponding drop in pH. In poorly buffered foods, such as pickles, much less activity by similar acid producers will produce inhibitory concentrations of weak acids. The minimum pH values permitting growth of salmonellae in various acids are summarized in Table 5.3.

The effect of pH on the bactericidal activity of formic, acetic, propionic, and butyric acids against *Salmonella typhimurium* was determined (Goepfert and Hicks, 1969). The effectiveness of the acid diminished as the pH of the medium was increased and as the fatty acid chain length increased. The rate of death of cells in the presence of volatile fatty acids at a given pH and temperature was concentration dependent; an increase in temperature increased the killing rate at fixed pH and concentration (see also Chapter 7).

Investigation of the survival of salmonellae (*S. oranienburg*, *S. senftenberg*, and *S. pullorum*) in liquid egg white with pH values ranging from 6.0 to 10.0 showed that the number of organisms surviving diminished only during storage at pH 9–10 (Banwart and Ayres, 1957). *Vibrio parahaemolyticus* is more sensitive than *Escherichia coli* to acidity, and growth is completely inhibited at pH 4.5–5.0. The vibrio is killed in 0.5% (v/v) acetic acid in a few minutes (Sakazaki, 1973).

Most meat products do not have sufficiently low pH values and high salt concentrations to inhibit *Clostridium botulinum* completely (reviewed in Riemann *et al.*, 1972). However, *C. botulinum* did not grow in pickles containing acetic acid (0.9–1.8%, v/v); and outgrowth of *C. botulinum* spores was inhibited at pH 4.8, but not at pH 5.0. A concentration of 0.9% acetic acid in the brine with cucumbers was sufficient to prevent outgrowth from spore inocula as large as 10^6 per cucumber (Ito *et al.*, 1976). The outgrowth of spores, which had been heated at 80°C for 20–30 min, was inhibited by sodium nitrite, and this effect depended on pH (Roberts and Ingram, 1973). Evidence was presented that undissociated nitrous acid was the inhibitory agent and that it was effective at pH 6, the pH of cured meat. Other workers have also shown that in meat products pH is a most important factor in preventing or delaying the outgrowth of *C. botulinum* spores.

Some data on pH and salt relations for growth of *Staphylococcus aureus* are in Table 5.4. Many workers have concluded that toxin production by *S. aureus* can be inhibited by various combinations of salt and pH and that under certain conditions toxin production does not accompany growth (Riemann *et al.*, 1972; Genigeorgis *et al.*, 1971a). Toxin production by clostridia may also be inhibited be adverse pH; the lowest pH for toxin production by *C. botulinum* depends on the food product. In laboratory media the organism is usually less acid tolerant than in foods, and sodium chloride and curing salts reduce its tolerance; type E is less tolerant to acid than types A and B. Combinations of salt and pH required to halt growth of and toxin production by *C. botulinum* type E have not been established with certainty (Fig. 5.3), but growth of *C. botulinum* without toxin production has not been demonstrated (Riemann *et al.*, 1972).

TABLE 5.4

Effects of pH Value and Salt Concentration on Growth Phases of *Staphylococcus aureus* MF31 in Trypticase Soy Broth at 37°C[a]

	NaCl (w/v): 0.5%			4.0%			8.0%		
pH	lag (hr)	μ^b	y^c	lag (hr)	μ^b	y^c	lag (hr)	μ^b	y^c
5.0	6	1.6	8.25	6	2.0	7.83	17	0.8	6.84
6.0	3	2.2	8.95	5	2.4	8.11	6	1.5	7.64
7.0	1	2.9	9.04	4	3.9	8.68	6	1.7	7.64
8.0	2	2.6	8.41	4	3.1	7.64	9	0.9	7.48

[a] From Iandolo *et al.* (1964).
[b] μ, maximum specific growth rate (divisions per hour).
[c] y, log maximum population.

Some have reported that *C. botulinum* grows in foods at pH values below 4.8, but there is no firm evidence that toxin was actually formed at the measured pH. If, as appears likely, toxin can be formed early in any fermentation or acidification process, it may be false to assume that toxin was formed at the terminal pH. Botulism toxin once formed is stable at chill temperature and at acid pH (Sakaguchi, 1969).

When *Bacillus cereus* was inoculated into canned foods including fruit, vege-

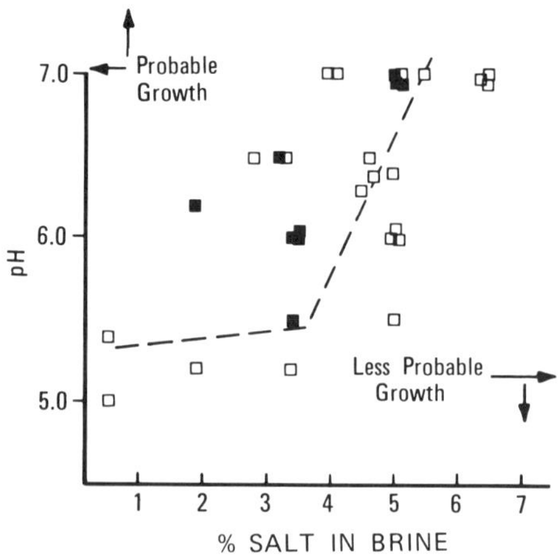

Fig. 5.3 Growth or toxin formation (closed symbols) and no growth or toxin formation (open symbols) by *Clostridium botulinum* type E. (From Riemann *et al.*, 1972).

TABLE 5.5

Normal pH Ranges of Selected Commercially Canned Foods[a]

Kind of food	Approximate pH range	Kind of food	Approximate pH range
Apples, whole	3.4–3.5	Duck	6.0–6.1
Apples, juice	3.3–3.5	Figs	4.9–5.0
Asparagus, green	5.0–5.8	Frankfurters	6.2
Beans		Fruit cocktail	3.6–4.0
Baked	4.8–5.5	Gooseberries	2.8–3.1
Green	4.9–5.5	Grapefruit	
Lima	5.4–6.3	Juice	2.9–3.4
Soy	6.0–6.6	Pulp	3.4
Beans, with pork	5.1–5.8	Sections	3.0–3.5
Beef, corned, hash	5.5–6.0	Grapes	3.5–4.5
Beets, whole	4.9–5.8	Ham, spiced	6.0–6.3
Blackberries	3.0–4.2	Hominy, lye	6.9–7.9
Blueberries	3.2–3.6	Huckleberries	2.8–2.9
Boysenberries	3.0–3.3	Jam, fruit	3.5–4.0
Bread		Jellies, fruit	3.0–3.5
White	5.0–6.0	Lemons	2.2–2.4
Date and nut	5.1–5.6	Juice	2.2–2.6
Broccoli	5.2–6.0	Lime juice	2.2–2.4
Carrots, chopped	5.3–5.6	Loganberries	2.7–3.5
Carrot juice	5.2–5.8	Mackerel	5.9–6.2
Cheese		Milk	
Parmesan	5.2–5.3	Fresh	6.4–6.8
Roquefort	4.7–4.8	Evaporated	5.9–6.3
Cherry juice	3.4–3.6	Molasses	5.0–5.4
Chicken	6.3–6.4	Mushrooms	6.0–6.5
Chicken with noodles	6.2–6.7	Olives, ripe	5.9–7.3
Chop suey	5.4–5.6	Orange juice	3.0–4.0
Cider	2.9–3.3	Oysters	6.3–6.7
Clams	5.9–7.1	Peaches	3.4–4.2
Cod fish	6.0–6.1	Pears (Bartlett)	3.8–4.6
Corn		Peas	5.6–6.5
On-the-cob	6.1–6.8	Pickles	
Cream style	5.9–6.5	Dill	2.6–3.8
Whole grain		Sour	3.0–3.5
Brine-packed	5.8–6.5	Sweet	2.5–3.0
Vacuum-packed	6.0–6.4	Pimento	4.3–4.9
Crab apples, spiced	3.3–3.7	Pineapple	
Cranberry		Crushed	3.2–4.0
Juice	2.5–2.7	Sliced	3.5–4.1
Sauce	2.3	Juice	3.4–3.7
Currant juice	3.0	Plums	2.8–3.0
Dates	6.2–6.4		

(continued)

TABLE 5.5 (*continued*)

Kind of food	Approximate pH range	Kind of food	Approximate pH range
Potatoes		Noodle	5.6–5.8
White	5.4–5.9	Oyster	6.5–6.9
Mashed	5.1	Pea	5.7–6.2
Potato salad	3.9–4.6	Tomato	4.2–5.2
Prune juice	3.7–4.3	Turtle	5.2–5.3
Pumpkin	5.2–5.5	Vegetable	4.7–5.6
Raspberries	2.9–3.7	Spinach	4.8–5.8
Rhubarb	2.9–3.3	Squash	5.0–5.3
Salmon	6.1–6.5	Strawberries	3.0–3.9
Sardines	5.7–6.6	Sweet potatoes	5.3–5.6
Sauerkraut	3.1–3.7	Tomatoes	4.1–4.4
Juice	3.3–3.4	Juice	3.9–4.4
Shrimp	6.8–7.0	Tuna	5.9–6.1
Soups		Turnip greens	5.4–5.6
Bean	5.7–5.8	Vegetable	
Beef broth	6.0–6.2	Juice	3.9–4.3
Chicken noodle	5.5–6.5	Mixed	5.4–5.6
Clam chowder	5.6–5.9	Vinegar	2.4–3.4
Duck	5.0–5.7	Youngberries	3.0–3.7
Mushroom	6.3–6.7		

[a] U.S. Department of Health, Education and Welfare (1969).

tables, soups, and meat with pH values ranging from 3.64 (apple and apricot puree) to 6.3 (beans), no growth was found in foods with a pH below 5.0, while at pH values above 5.15 the extent of growth was proportional to pH (Pivovarov and Kukhto, 1970). In laboratory media *B. cereus* may grow over the pH range 4.9–9.3 (Goepfert *et al.*, 1972). However, meat and rice products are less inhibitory than laboratory media, supporting growth at pH 4.35 (inoculum 10^8 cells/ml) (Raevuori and Genigeorgis, 1975). The effects of pH and NaCl on the growth of both *B. cereus* and *S. aureus* vary with the strain and growth medium used. At decreasing pH values a larger inoculum is needed to initiate growth, while both high NaCl concentrations and extreme pH values are needed to inhibit it (Raevuori and Genigeorgis, 1975; Genigeorgis *et al.*, 1971a).

V. USE OF pH TO CONTROL MICROORGANISMS IN FOODS

Low pH can assist in the preservation of foods in two ways: by directly inhibiting microbial growth and by reducing microbial heat resistance in foods that will subsequently be heat processed.

Foods preserved by lactic acid fermentation are derived predominantly from milk (cheese; Chapter 18), meat (fermented sausage; Chapter 15), and vegetables (sauerkraut, pickles, etc.; Chapter 21). Fermentation follows inoculation with lactic acid-producing bacteria or natural selection of the lactic acid bacteria from the normal flora of the food. Growth of the desired flora is assured by an appropriate level of salt, anaerobiasis, and control of temperature. Microbial spoilage may be inhibited also by direct acidification, i.e., acetic acid to fish or vegetables, or glucono-δ-lactone (which breaks down to lactic acid) to meat.

The overriding consideration in the processing of canned foods is the control of *Clostridium botulinum;* a pH of 4.6 or below completely inhibits the growth of this bacterium. Hence canned foods are divided into acid foods (pH 4.6 or below) and low-acid foods (pH above 4.6). The normal ranges of pH in a selection of commercially canned foods are given in Table 5.5. Note that in almost every case the range is entirely above or below 4.6.

The heat processing of canned foods is discussed in Chapter 1. In brief, most fruits, being naturally acid, may be given mild heat processes in which the temperature does not exceed 100°C; i.e., they need not be processed under pressure. Vegetables are predominantly low acid and require a severe heat process to destroy all spores of *Clostridium botulinum*. For a few vegetables (e.g., beetroot), an acid (usually acetic) may be added to permit the use of the milder heat treatment. It is essential that the pH of such acidified foods be permitted to equilibrate thoroughly *before* the heat process is applied. This requires sufficient acid, sufficient stirring, and sufficient time for the pH to fall to 4.6 or below at the center of solid materials.

6

Oxidation–Reduction Potential

I. INTRODUCTION

While the pH of foods is readily measured and the implications of those measurements understood, it is much more difficult to measure oxidation–reduction (redox) potential reproducibly (particularly between laboratories), and the relevance of the values obtained to the microbiology of the product is not clear.

Redox potential is believed to be an important selective factor in all environments, including food, and probably influences the types of microbes found and their metabolism. Differences in end-products of metabolism, discernible by the consumer as differences in color or taste, may be in some cases the consequence of redox differences. In particulate foods (e.g., meat products) or nonhomogeneous products (e.g., emulsions), redox potential may vary greatly from site to site due to localized high concentrations of various redox couples or nutrients such as glucose, furmarate, or malate. Gradients of redox potential may exist from the atmosphere to the depths of food when gaseous diffusion toward the center is restricted. Redox potential can denote the oxygen relations of living microbes and may be used to specify an environment in which an organism is capable of generating energy and synthesizing new cells without recourse to molecular oxygen. Aerobic microbes require positive redox values for growth, while anaerobes often require negative redox values. In different microbial cultures the redox value can range from an anaerobic figure below about −420 millivolts (mV) to an aerobic figure of about +300 mV.

Oxidation and reduction processes are defined in terms of electron migrations between chemical compounds. Oxidation is the loss of electrons, whereas reduction is the gain of electrons. When one substance is oxidized (i.e., releases electrons) another is always simultaneously reduced (i.e., takes up the electrons released). This electronic concept has suggested methods of studying quantitatively those reversible oxidation–reduction processes vital to living cells.

Measurement of electrode potential enables the degree of reduction or oxidation of a particular foodstuff to be estimated, and such measurements suggest the possibility of the grading of oxidizing and reducing systems, in foods, according to their intensity.

A. Definitions

Any reversible redox process can be expressed as

$$[\text{oxidant}] + [H^+] + n_e = [\text{reductant}] \tag{6.1}$$

in which n_e is the number of electrons transferred in the process.

The redox potential (E_h) of such a process is given by the equation devised originally by Nernst (see Pirt, 1975):

$$E_h = E_o + \frac{RT}{nF} \ln \frac{[\text{oxidant}]}{[\text{reductant}]} \tag{6.2}$$

in which E_o is the standard redox potential at pH 0, but with other solute components present at 1 *M* concentration (usually measured as the midpoint redox potential at pH 0, its value is assumed to equal the theoretical value for redox couples in dilute aqueous solutions); *R*, the molar gas constant; *T*, the temperature in °K; *F*, the faraday quantity of electricity; *n*, the number of electrons transferred in the process; and ln, the natural logarithm.

B. Measurement of Redox Potential

In many foods it is difficult to obtain a true measure of redox potential. For the redox potential of a sample to mimic that of the food from which it was taken it is essential that the gaseous environment be kept unchanged whether the food is vacuum-packaged, packed under an anoxic atmosphere, or packed in contact with the air. Because redox potential is lowered by metabolic activity, transportation should either be under chilled conditions (0°–4°C) or at the storage temperature of the food. Transportation temperature and time should always be specified to aid interpretation of the results. Oxygen is the main agent interfering with the measurement of redox potential. In general, withdrawn samples should not be used for the determination of redox potential, because oxygen may enter a sample during sampling. To prevent contamination of samples with oxygen, the top layer of the food sample must be removed under an inert gas before measurement.

Since the potential values measured depend on pH, each measurement of redox potential should be accompanied by a statement of pH. The pH may change the actual redox potential, but also, at the same E_h value the pH may create conditions favoring different types of metabolism. Roberts (1970) has demonstrated the influence of pH on the redox potential limiting the growth of *Clostridium perfringens* (Fig. 6.1) by interpreting the data of Hanke and Bailey (1945).

The concept of rH was introduced to eliminate this pH dependence by calculation. While rH can be calculated accurately when all hydrogen ions are fully dissociated at all pH values, inaccuracies will result when the degree of dissociation is altered by pH. Additionally, monovalent acids alter the redox potential by 30 mV per unit pH, bivalent acids by 57.7 mV, and acids with higher valency by up to 120 mV. To convert E_h (in volts) to rH, the following formula may be used.

$$E_h = 0.03\,(rH - 2 \times \text{pH}) \quad \text{or} \quad rH = \frac{E_h}{0.03} + 2 \times \text{pH} \qquad (6.3)$$

Redox potential may be measured using an inert metal (usually platinum) electrode in a circuit with a reference electrode. Alternatively, dyes that change color at specific redox potentials may be used, but these may interact with microbes and foods resulting in false values.

1. Use of Electrodes

The platinum electrode will respond to any system that gives an electrochemically reversible reaction at the electrode surface, provided that the rate of reaction is rapid compared with the electron drain or supply from the measuring electrode.

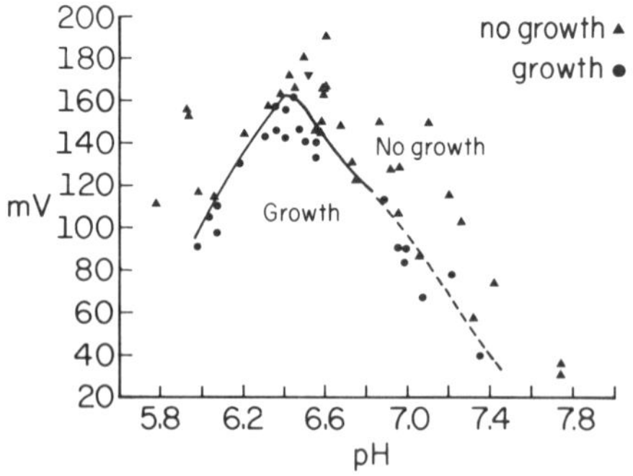

Fig. 6.1 The effect of pH and E_h on limitation of growth of *Clostridium perfringens*. [Data of Hanke and Bailey (1945), as interpreted by Roberts (1970).]

With modern high impedence (10^{13} ohms) measuring systems, reliable potentials can be obtained from redox couples giving very low exchange currents. For a fairly rapid reacting system such as Fe^{2+}/Fe^{3+}, measurements may be obtained when the concentration of the couple is as low as $10^{-5}\,M$ (Harrison, 1972). The calibration of redox electrodes is theoretically related to the standard hydrogen electrode, whereas in practice a calomel or other reference electrode is used. If the measured potential is E_m (mV) and the potential of the reference electrode is E_r (mV), then

$$E_h = E_m + E_r \tag{6.4}$$

Typical reference electrode values are given by Jacob in his detailed review of redox measurement (Jacob, 1970).

Equations (1) and (2) above show that E_h is pH dependent; they may be rewritten (Pirt, 1975):

$$E_h = E_o + 2.30 \frac{RT}{nF} \log \frac{[\text{oxidant}]}{[\text{reductant}]} - 2.30 \frac{bRT}{nF} \text{pH} \tag{6.5}$$

where b is the molar hydrogen ion concentration. If E_h is plotted against pH, the slope of the resultant line will be $-2.30\ bRT/nF$, which will represent the change in E_h for unit change of pH. At 30°C when the activities of the reductant and oxidant are constant, this is 58 mV (Pirt, 1975). The E_h reading (or apparent redox potential) depends both on the type of electrode (illustrated in Fig. 6.2) and on the electrode surface (e.g., polished, blackened; see Jacob, 1970).

For potentiometric measurement of the redox potential of liquid food samples, standard platinum and calomel electrodes are used. For measurements in solid foods it is necessary to use special thick-walled, slim, short-pointed platinum-

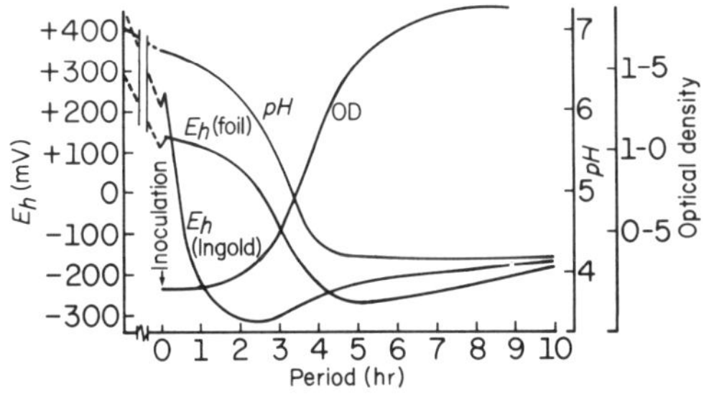

Fig. 6.2 Changes in optical density (OD), pH, and E_h (measured using two types of redox electrodes: simple platinum foil or Ingold combined with Pt, Ag/AgCl) during incubation of a culture of *Clostridium butyricum* spores in yeast-peptone-glucose medium at 37°C, sparged continuously with oxygen-free nitrogen containing CO_2 (5% v/v). (From Douglas and Rigby, 1974.)

tipped electrodes and KCl–agar bridges in pointed tubes with asbestos fibers melted into their tips (see Table 6.1.)

Unlike pH electrodes, redox electrodes require a certain time—depending on the type of sample—to reveal the redox value. Redox electrodes may be decontaminated using hydrogen peroxide. The electrode is kept for 5 min in 5% H_2O_2 solution, after which it is rinsed with sterile distilled water.

2. *Use of Dyes*

E_h may be estimated by means of indicator dyes, but the dyes themselves are likely to interact with foodstuffs and microbes (Jacob, 1970). The range of redox potentials over which any dye can provide a visible color change may be considered with reference to methylene blue (MB) (Pirt, 1975):

$$MB^+ \text{ (colored)} + H^+ + 2e = MBH \text{ (colorless)} \tag{6.6}$$

and

$$E_h = E_0 + 2.30 \frac{RT}{2F} \log \frac{[MB^+]}{[MBH]} - 2.30 \frac{RT}{2F} \text{pH} \tag{6.7}$$

or (in mV)

$$E_h = E_0 + 29 \log \frac{[MB^+]}{[MBH]} - 29 \text{ pH (at 30°C)} \tag{6.8}$$

The standard redox potentials of dyes (E_0; i.e., [reduced] = [oxidized] form) are often quoted (see Jacob, 1970). Some common redox indicators, their E_0 values and the influence of pH on E_0 are listed in Table 6.2. The useful range of an indicator is the E_h range over which the proportion of the oxidized form of the

TABLE 6.1

Potential Values of Various Reference Electrodes at 25°C

Reference electrode	Concentration of conducting salt	Value of constant potential (mV)
Standard hydrogen	—	0
Silver chloride	Normal	235
Calomel	Saturated	241
	Normal	280
	0.1 *N*	338
Sulfate	Normal	615

[a] From Jacob (1970).

TABLE 6.2

Oxidation–Reduction Potential Indicators[a,b]

pH	Phenol-*m*-sulfonate indo-2,6-dibromophenol	*m*-Chlorophenol indo-2,6-dichlorophenol	*o*-Chlorophenol indophenol	Phenol indophenol	*o*-Bromophenol indophenol	2,6-Dichlorophenol indophenol	Phenol indo-2,6-dibromophenol	*m*-Cresol indophenol	*o*-Cresol indophenol	2,6-Dichlorophenol-indo-*o*-cresol	Toluylene blue	Thymol indophenol	1-Naphthol-2-SO $_3$H-indophenol	1-Naphthol-2-SO $_3$H-indo-3,5-dichlorophenol	Thionine	Methylene blue	Indigo tetrasulfonate	Indigo trisulfonate	Indigo disulfonate
5.0	+0.390	+0.391	—	—	—	+0.366	—	—	—	+0.355	+0.221	—	—	+0.262	+0.138	+0.101	+0.065	+0.032	−0.010
5.2	0.378	0.378	—	—	—	0.352	—	—	—	0.322	0.208	—	—	0.249	0.124	0.088	0.053	0.020	−0.022
5.4	0.366	0.366	—	—	—	0.339	—	—	—	0.307	0.196	—	—	0.236	0.112	0.077	0.041	+0.008	−0.034
5.6	0.354	0.351	—	—	—	0.325	—	—	—	0.292	0.184	—	—	0.223	0.106	0.066	0.029	−0.004	−0.045
5.8	0.342	0.339	—	—	+0.308	0.310	—	—	—	0.277	0.173	+0.244	—	0.210	0.100	0.056	0.017	−0.016	−0.057
6.0	0.330	0.326	+0.301	—	0.296	0.295	—	—	—	0.261	0.162	0.233	+0.183	0.196	0.094	0.047	+0.006	−0.028	−0.069
6.2	0.319	0.312	0.288	+0.276	0.284	0.279	—	+0.259	+0.243	0.245	0.151	0.222	0.171	0.181	0.087	0.039	−0.006	−0.039	−0.081
6.4	0.307	0.297	0.275	0.265	0.272	0.263	—	0.246	0.230	0.228	0.141	0.211	0.159	0.166	0.080	0.031	−0.017	−0.051	−0.092
6.6	0.295	0.282	0.262	0.254	0.259	0.247	—	0.233	0.217	0.212	0.132	0.198	0.147	0.150	0.074	0.024	−0.027	−0.061	−0.104
6.8	0.284	0.268	0.248	0.241	0.245	0.232	—	0.221	0.204	0.196	0.123	0.185	0.135	0.134	0.068	0.017	−0.037	−0.072	−0.114
7.0	0.273	0.254	0.233	0.227	0.230	0.217	+0.218	0.208	0.191	0.181	0.115	0.174	0.123	0.119	0.062	0.011	−0.046	−0.081	−0.125
7.2	0.262	0.240	0.218	0.212	0.215	0.203	0.204	0.197	0.180	0.166	0.108	0.160	0.111	0.103	0.056	+0.004	−0.055	−0.091	−0.134
7.4	0.251	0.226	0.203	0.200	0.200	0.189	0.190	0.185	0.168	0.152	0.101	0.148	0.099	0.088	0.050	−0.002	−0.062	−0.099	−0.143
7.6	0.240	0.213	0.187	0.186	0.184	0.175	0.176	0.173	0.156	0.138	0.094	0.136	0.087	0.073	0.043	−0.008	−0.070	−0.107	−0.152
7.8	0.229	0.200	0.170	0.170	0.167	0.162	0.163	0.160	0.143	0.125	0.088	0.123	0.074	0.060	0.037	−0.014	−0.077	−0.114	−0.160
8.0	0.218	0.188	0.155	0.155	0.152	0.150	0.151	0.148	0.130	0.112	0.082	0.110	0.062	0.046	0.030	−0.020	−0.083	−0.121	−0.167
8.2	0.207	0.176	0.139	0.139	0.137	0.137	0.138	0.134	0.116	0.099	0.075	0.097	0.049	0.034	0.025	−0.026	−0.090	−0.127	−0.174
8.4	0.197	0.164	0.124	0.124	0.120	0.125	0.126	0.120	0.102	0.087	0.069	0.084	0.026	0.021	0.020	−0.032	−0.096	−0.134	−0.180
8.6	0.187	0.153	0.109	0.110	0.105	0.113	0.114	0.105	0.086	0.075	0.063	0.069	0.023	+0.010	0.014	−0.038	−0.102	−0.140	−0.187
8.8	0.177	0.143	0.095	0.096	0.092	0.101	0.102	0.091	0.072	0.063	0.057	0.056	+0.010	−0.002	+0.008	−0.044	−0.108	−0.146	−0.193
9.0	0.168	0.133	0.082	0.083	0.079	0.089	0.090	0.076	0.057	0.051	0.051	0.041	−0.003	−0.012	−0.001	−0.050	−0.114	−0.152	−0.199

[a] Hewitt (1950).
[b] E_o values at different pH (30°).

indicator changes from 90 to 10%; that is, log [MB]/[MBH] changes by 2. This means that the measured E_h ranges over about 60 mV and hence the E_h range of methylene blue is fixed by the E_0 value at a particular pH value, i.e., + 170 mV at pH 4, + 11 mV at pH 7, and −50 mV at pH 9.

The methylene blue reduction test is used in some countries to grade ice cream, milk, and meat, based on the assumption that there is a general correlation between the reduction time of methylene blue and numbers of microbes present. In practice, the correlation is poor (Anderson and Whitehead, 1974). Triphenyltetrazolium chloride (TTC) has also been suggested as a compound to estimate numbers of microbes on the surface of meat, but like methylene blue, is affected by the types as well as numbers of bacteria present.

3. *Problems of Sampling and Interpretation*

a. Poising. The redox potential of a food depends on the relative amounts of oxidized and reduced substances it contains. Foods containing large amounts of such substances are resistant to changes in redox potential and termed "heavily poised." Hence interpretation of the output from a redox electrode inserted into a food is not simple since that food may or may not be heavily poised. In a weakly poised food a small microbial population ($<10^5$/gm) may possibly cause a large change in redox potential, whereas in a heavily poised food a large microbial population ($>10^8$/gm) may scarcely affect the redox potential. To judge the significance of any change in the redox potential of a food the metabolic potential and size of the microbial population must be considered in relation to the redox buffering capacity of the food.

b. Side Reactions. Redox measurements are most useful when reversible couples with known components exist, and an equilibrium between them is attainable. Any food may contain redox couples—some not in equilibrium and others irreversible. The redox probe has been suggested as a suitable monitor of spoilage or desirable changes during storage in vacuum-packaged, anoxic gas-flushed or canned foods. It has been shown to indicate accurately such changes deep in *post-rigor* meat (see Barnes and Ingram, 1956), where empirical, and repeatable, relationships between metabolic events and redox potential are found. Under these circumstances the main substrates or microbial products have a high activity with the electrode and can mask all interfering or side reactions, so that the measured redox potential is a true representation of the state of the dominant redox couple.

c. Oxygen. All redox couples in a food will tend toward the fully oxidized state when oxygen is present. Although dissolved oxygen does not influence a platinum probe in pure water, in salt solutions (including foods such as ham and brined vegetables) electrolytic reactions may render the probe sensitive to dissolved oxygen (Harrison, 1972) and make it indicate artificially high (positive) redox values. Under such circumstances the apparent redox potential will be a function of the logarithm of the dissolved oxygen concentration, and large changes in redox potential will represent small changes in dissolved oxygen tension. Jacob (1970) reported a 60 mV fall per 90% decrease in dissolved oxygen concentration. At very low dissolved oxygen tensions, redox probes may show empirical relationships between oxygen and metabolic changes (Wimpenny, 1969). However, at these low oxygen levels the biochemistry of microbes and biochemically active foods undergoes metabolic changes that unpredictably alter these relationships. Redox potential and pH can be controlled in culture media, and probably in foods, by adjusting the concentration of gaseous O_2 and CO_2 in the head space above the medium (Dobson and Bullen, 1964). Although in theory a stable equilibrium can be reached between gas and water phase concentrations (relative to the solubility of the gases in the culture medium and the temperature), the uptake of O_2 and the release of CO_2 by microbial cultures will often exceed these concentrations.

II. EFFECTS OF REDOX POTENTIAL ON MICROORGANISMS

Though it is virtually impossible to define what we measure as the E_h of a bacterial culture (Morris, 1976), it is accepted that shifts in this value reflect changes in the balance between contending oxidizing and reducing agencies that are its prime determinants.

The fall in redox potential during storage of a food has been attributed to the liberation of gaseous H_2 and of reducing metabolites by food enzymes or actively growing microbes.

The most rapid fall in E_h as measured by platinum electrodes often coincides with the early logarithmic phase of growth of the microbes, when the metabolic activities are most intense (Hewitt, 1950; Oblinger and Kraft, 1973; Douglas *et al.*, 1973). Redox potential may be rapidly reduced during spore germination and outgrowth, when, although only small proportions of spores germinate initially, those that do germinate and grow produce reduced nicotinamide adenine dinucleotide (NADH) and other reductants which reduce the oxygen concentration to a level that is noninhibitory to the remaining spores. This reduction in oxygen level and the increased H_2 tension produced by cellular metabolism is shown as a large drop in redox potential. The environments provided by foodstuffs may be roughly categorized as those with restricted and those with unrestricted access to oxygen. In those with restricted access to oxygen, successful microbes produce $CO_2 + H_2O$ as their major endproducts; there is no significant production of end products such as organic acid, but secondary metabolites

TABLE 6.3

Bacterial Response to Redox Potential[a]

Bacterial category	Examples	Range from the highest redox potential for initiation of growth to the lowest redox potential found during growth	References
1. Aerobic bacteria Minimum tolerated oxygen pressure = 10^{-8} atm	*Bacillus subtilis*	+135 to −100 mV	Jacob (1970)
	Pseudomonas fluorescens	+500 to +100 mV	Oblinger and Kraft (1973)
2. Facultative anaerobes Minimum tolerated oxygen pressure = 10^{-12} atm	*Staphylococcus aureus*	+180 to −230 mV[b]	Jacob (1970)
	Staphylococcus aureus	+30 to −200 mV[c]	Jacob (1970)
	Proteus vulgaris	+150 to −600 mV[b]	Jacob (1970)
	Proteus vulgaris	+90 to −300 mV[c]	Jacob (1970)
3. Anaerobic bacteria	*Clostridium paraputrificum*	−30 to −550 mV	Jacob (1970)
	Clostridium perfringens (*welchii*)	+216 to −230 mV	Barnes and Ingram (1955)

[a] From Jacob (1970).
[b] Redox potential determined using a polished electrode.
[c] Redox potential determined using an unpolished electrode.

may include amines, etc. Where access to oxygen is restricted, fermentation occurs, and flavor changes may be imparted to the food before it is spoiled. Large numbers of microbes ($>10^6$/gm) must be present before such changes will be apparent to the consumer.

Some microbes, the strict aerobes or anaerobes, have only one terminal metabolic system for obtaining energy and are active over only a relatively narrow range of redox potential. Others, the facultative anaerobes, have alternative systems that may be switched on or off by the redox potential or the presence or absence of oxygen (see Table 6.3).

A. Strict Aerobes

Strict aerobes in the food environment (e.g., *Bacillus subtilis, B. megaterium,* micrococci, pseudomonads, acinetobacters, and moraxellae) use oxygen as the terminal electron acceptor in respiration. Although they may be proteolytic or lipolytic, their waste metabolites include water and carbon dioxide. They are an important fraction of the microflora only when O_2 is readily available, as is the case at the surface of meats and other foods stored in air. For example, slime may be caused at the surface of meat by growth of pseudomonads (Chapter 15) and ropiness of bread by *Bacillus subtilis* (Chapter 23).

B. Facultative Anaerobes

Facultative anaerobes, e.g., Lactobacillaceae, Enterobacteriaceae, Corynebacteriaceae, can use oxygen as a terminal electron acceptor, but in its absence they can use a variety of electron acceptors (e.g., NO_3^-, SO_4^{2-}). They grow on the surface and in the interior of foods; some possess proteolytic or lipolytic activity. Their waste products are often organic acids. Because of their ubiquitious distribution, their wide range of enzyme activity, and their ability to decompose organic compounds, they can compete in a wide range of environments and are often responsible for the spoilage of low E_h foods. Hence they are important food spoilage organisms, although some, such as lactobacilli, can cause desirable changes. Some facultative anaerobes, such as enterobacteria, are organisms of public health significance.

C. Obligate Anaerobes

Those clostridia most prevalent in foods can grow only at low redox potentials (ca. −300 mV) and some will grow only in the absence of molecular oxygen. The maximum E_h values at which anaerobes grow range from +30 to −250 mV for clostridia (Morris, 1976). Some anaerobes may grow at a higher redox potential in the absence than in the presence of oxygen. For example, *Clos-*

tridium acetobutylicum grew at +370 mV in the absence of O_2, although not at +100 mV when O_2 was present (O'Brien and Morris, 1971). Non-spore-forming anaerobes, such as *Bacteroides,* are often not tolerant of high redox potential, whereas many clostridia can survive for long periods at relatively high potentials (+116 mV) in the presence of oxygen and sometimes grow at similarly high potentials (+370 mV) in the absence of oxygen. Strict anaerobes grow most often as contaminantes in the inner part of unprocessed foods where free access of oxygen is limited and together with facultative anaerobes are the main spoilage organisms. Many anaerobes cannot grow below ca. 10°C and rapid chilling is used to slow their growth in post-rigor meat. Canned foods and foods in gas-impermeable wrappings are most at risk from post-processing contamination by these microbes. The addition of vitamin C (ascorbic acid) to culture media favors the growth of anaerobes such as *Clostridium perfringens,* in the presence of air, although it does not alter the oxygen tension (Kligler and Guggenheim, 1938).

III. USE OF REDOX POTENTIAL TO CONTROL MICROORGANISMS IN FOODS

Although redox potential is not generally recognized as a processing parameter during food manufacture, it undoubtedly interacts with pH and gaseous atmospheres to determine the spoilage microflora of many foods and probably influences the development of both pleasant and unpleasant flavors. The redox potential of the interior pieces of meat is sufficiently reduced to prevent the growth of aerobes (e.g., pseudomonads, bacilli, or molds), but its low redox potential (post-rigor) encourages the growth of enterobacteria and clostridia. The rate and extent of redox drop (more electronegative) in food products lacking residual enzymic activity depend on the growth rate and on the physiological type of bacteria present. Animal tissue, however, retains a proportion of its original enzyme activity for a considerable time up to 1 week after slaughter, depending on temperature and processing. Barnes and Ingram (1955, 1956) showed that the initiation of growth by clostridia in horse muscle was affected by its redox potential. Immediately after death (pre-rigor) the redox potential was +250 mV, and in muscle free from bacteria this fell to the range + 10 to −130 mV post-rigor. When bacteria were present in the sample the redox fell to −250 mV eventually. Clostridia grew in the latter only when the redox potential fell below − 36 mV, suggesting that anaerobes naturally present in meat do not grow until the onset of rigor mortis because of high redox potential in pre-rigor meat.

In this muscle the redox potential remained too high for the growth of anaerobes for about 6 hr, although the length of this period varied according to the oxygen reserves in the muscle. This period of high redox potential, when the pH also becomes acid to an inhibitory extent, allows time for cooling the carcass to a safe temperature (<15°C) before the growth of anaerobic bacteria commences (Ingram, 1962b, 1972).

During storage, the redox potential of meat may fall to ca. −200 mV; the rapid multiplication of anaerobes only becomes possible when the E_h has fallen below + 40 mV. When rigor mortis is complete the redox potential is usually about 0 mV.

In loosely packed minced fresh meats the E_h value is generally around +200 mV as the small particle size (<3–4 mm diameter) allows the rapid diffusion of oxygen (Jay, 1970). In fresh oxygen-saturated meat straight from the mincer, the core is oxygen depleted within a matter of hours; however if fresh mince is loosely packed it will remain saturated with oxygen for 1–2 days. Hence redox potential within processed meat may be controlled by packing. The redox potential of cheeses ranges from −20 to −200 mV. In dairy products, oiliness (a type of rancidity) may be prevented by the growth of bacteria which lowers the redox potential and removes oxygen so that fat oxidation is prevented. These oxidative taints do not appear when the E_h is less than +300 mV at pH 6.5 (Davis, 1932; Davies *et al.*, 1934).

Wirth and Leistner (1970) suggest that an initial redox value between −20 and −150 mV is optimum for canned meats. During long periods of storage the redox potential of canned pork and beef may drop to −350 mV, making it unfit for human consumption. Additives such as nitrite (oxidizing) or ascorbic acid (reducing) may be used to increase the buffering power of the product and to some extent increase the shelf-life by keeping the redox potential within acceptable limits.

In meats, chemicals containing -SH groups maintain reducing conditions, while in fruits and vegetables ascorbic acid and reducing sugars have the same effect. Plant foods, especially plant juices, have E_h values from +300 to +400 mV (pH 4–5) and are consequently spoiled by aerobic bacteria and molds.

IV. COMBINATION EFFECTS AND INTERACTIONS

All microbes growing in an environment without access to air reduce the redox potential to their lower limit value. When different types of microbes exist in a food, their various abilities to alter the redox potential and also tolerate altered redox values are important in microbial competition and in the establishment of microbial successions. For example, damaged microbes or those with a lag phase alter redox potential only slowly and have little influence on the course of microbial succession.

Facultative anaerobes (e.g., Enterobacteriaceae) are not usually inhibited by the growth of strict aerobes, but conversely, their growth can inhibit the growth of aerobic spoilage organisms. Species with poor tolerance of low redox potential may often reach their maximum population density before redox potential is lowered below their limit. The microflora of a food may reduce its redox potential and hence reduce spoilage by aerobes. For example, in packaged bacon, lactobacilli may prevent spoilage by micrococci.

It is possible to observe certain relationships between culture redox limits for growth and the effects of oxygen on anaerobes. Morris (1976) has suggested that the inhibitory effects of O_2 on many anaerobes stem directly from its presence and not from its effect on redox potential. Oxygen toxicity is probably caused by oxygen free radicals, and it has been suggested that many reductants, such as thiols, act as scavengers of these toxic free radicals.

Thus, obligate anaerobes cannot grow in the presence of the very toxic superoxide free radicals (formed during metabolism or by chemical reactions in food), since they lack superoxide dismutase which breaks down these free radicals to peroxides. These may then be broken down by peroxidases, catalases, or other enzymes of tissue origin (Morris, 1976). Low redox potential provides an environment that is sufficiently free of oxygen free radicals and peroxides for the growth of these organisms to occur. Microbes totally devoid of catalase/peroxidase and superoxide dismutase are killed by enzymic and nonenzymic processes which yield hydrogen peroxide, for example, the lactoperoxidase–thiocyanate–peroxide defense system in raw milk (Reiter *et al.*, 1964) and superoxide anions when oxygen is available. Hence, food components that serve as nutrients, reductants, or oxidants might themselves generate or neutralize the toxic by-products of free radical or peroxidative reactions.

Sometimes even the total exclusion of molecular oxygen is not sufficient to ensure the growth of strict anaerobes since it may reduce the redox potential only to +100 mV. Anaerobic organisms require, and may often develop and maintain, a low E_h in their cultures. Anaerobes grow best under so-called "reducing conditions" when there is minimal demand on the organism's reducing power, which they could employ more productively to yield energy or for biosynthesis. The establishment and maintenance of such desirable equilibrium conditions in many foods or culture media lacking reductants calls for reducing contributions from the organisms themselves. Anaerobes differ in their capacity to generate the excess (disposable) reducing ability and the redox buffering action or intrinsic redox couples in the food may vary; these interact to give a powerful selective factor in the foodstuff.

Those microbes with the least reducing capacity have a tightly coupled cycle of endogenous electron generation and acceptance which leaves little scope for diversion of electron flow for reduction of exogenous oxidants. Their growth in foods not prereduced will be limited, and they will be the species most threatened by addition of oxidants to a food. Conversely, their growth is benefited by gas-impermeable packaging or the addition of reductants. These species are likely to be difficult to recover from foodstuffs if the environment is disturbed and they may be unable to grow in media not pre-poised at a low E_h. Reducing agents facilitating the establishment of a low redox potential (e.g., sulfite or thioglycollate) are useful in aiding initiation of growth of anaerobes, especially from small inocula. Growth media for anaerobes are often supplemented with -SH containing compounds (e.g., cysteine), which act as reducing agents, and growth of anaerobes is probably encouraged in foods (e.g., meat), which are anaerobic and contain such compounds. The growth of *Bacillus subtilis* and *Pseudomonas fluorescens* was inhibited by $KMnO_4$, even though it increased the redox potential; growth of the *Bacillus* was stimulated by $Na_2S_2O_4$ (Olczyk, 1968). On the other hand, the growth of *Torula utilis* was independent of E_h.

Environmental redox potential is an important determinant of microbial growth, for it can act as a selective agent and as a factor influencing the metabolic products of microbes. How redox potential affects microbial metabolism or selects a certain species of microbe is at present imperfectly understood and requires further study. As this chapter has shown, it is difficult to obtain accurate estimates of redox potential and indeed to interpret such readings when they have been made. The interaction of superoxide dismutase and peroxidase with particular redox couples and toxic free radicals is barely understood; understanding of the significance of these interactions to cells could provide clues to the mode of action of many preservatives and other unrecognized inhibitors within foodstuffs.

7

Organic Acids

I. INTRODUCTION

Organic acids and their esters occur widely in nature. They are frequently found in fruits; for example, citric acid in citrus fruits, benzoic acid in cranberries and greengage plums, and sorbic acid in rowan berries (Dukes, 1969; Chichester and Tanner, 1972). Lactic acid occurs in animal tissues, methyl gallate in the leaves of several genera of plants (Carlson *et al.*, 1951), and various organic acids occur in spices (Hargreaves *et al.*, 1975). Many are produced as intermediate or end products of microbial metabolism and occur in large amounts in many fermented milk, meat, and vegetable products. In these products, the combination of desirable flavors, textural changes, and acidity brought about by the formation of organic acids were found by our forefathers to be valuable means of delaying or preventing proteolytic spoilage. This allowed many otherwise perishable foods to be stored, and thus added variety to the diet.

Today, food manufacturers make use of certain organic acids to aid the preservation of a wide variety of products. The amounts and types of organic acids permitted, however, are carefully controlled by governmental agencies responsible for food safety, and the amounts allowed are generally small in comparison to the amounts of organic acids occurring naturally in many fruits and fermented products; see FAO/WHO recommendations (FAO/WHO, 1973).

A. Properties

For reasons of solubility, taste, and low toxicity, the short chain organic acids such as acetic, benzoic, citric, propionic, and sorbic acids are most commonly used as food preservatives or acidulants. When considering the possible use of other organic aids as preservatives, it is useful to note that the antimicrobial activity of organic acids generally increases with chain length. However, aliphatic acids of chain lengths greater than C_{10} or C_{11} have very little potential applicability because of their very low solubility in water. For general information on the properties and usage of organic acids and their esters, see Table 7.1 and Chichester and Tanner (1972).

B. Methods of Measurement

Methods of analysis for organic acids in foods are based mainly on steam distillation or solvent extraction followed by estimation of the type and amount of acid(s) present by use of spectrometry or chromatography. Details of the methods will be found in most textbooks on chemical analysis, for example the AOAC *Official Methods of Analysis* (Horwitz, 1975).

II. EFFECTS OF ORGANIC ACIDS ON MICROORGANISMS

A. General Effects

It is the undissociated molecule of the organic acid or ester that is responsible for antimicrobial activity (Ingram *et al.*, 1956; Macris, 1975). In the undissociated state some organic acids are readily soluble in cell membranes (Cramer and Prestegard, 1977). Only those organic acids that are lipophilic show antimicrobial activity. One hypothesis is that they inhibit or kill microorganisms by interfering with the permeability of the microbial cell membrane, causing uncoupling of both substrate transport and oxidative phosphorylation from the electron transport system; unsaturated organic acids such as sorbic acid and the esters of *p*-hydroxybenzoic acid also inhibit the electron transport system (Freese *et al.*, 1973). This leads to acidification of the cell contents which is probably the main cause of inhibition and death (see Chapter 5).

The pK_a values (pK_a = the pH at which 50% of the total acid is undissociated) of organic acids used as food preservatives are between pH 3 and 5 (see Table 7.1). Lowering the pH of a food increases the proportion of molecules of an organic acid that are undissociated and thus increases its effectiveness as an antimicrobial agent; the relationship between pH and the proportion of undissociated acid present in a solution of an organic acid is illustrated in Fig. 7.1. These considerations limit the usefulness of organic acids mainly to foods with pH below 5.5. They are used principally as mycostatic agents. However, in high concentrations, they are very effective agents against a wide variety of microorganisms (including viruses), i.e., when their levels are greater than 1% in fruits and microbially fermented products or when acidulation drops the pH to 4.0 or less (Dakin, 1957). The effect observed in mixed cultures or in stored foods where one microorganism exerts a synergistic or antagonistic effect on another is believed to be the result in many cases of the *in situ* formation of organic acids as by-products of microbial growth.

TABLE 7.1

Data on the pH, Solubility, and the Usage of Organic Acids and Esters as Food Preservatives

		pK_a	Solubility[b] (gm/100gm)	ADI[c] (mg/kg body weight)	Typical maximum use concentration (mg/kg)	Examples of usage
Acetic acid	CH_3COOH	4.75	Very soluble	Not limited	No limit	Pickling of meat, fish, and vegetable products
Sodium diacetate	$CH_3COONaCH_3COOH\frac{1}{2}H_2O$	4.75	Very soluble	Not limited	4000	Bread and bakery products
Sodium benzoate	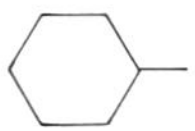 COONa	4.2	50 (25°C)	5	1–3000	Pickles, acid sauces and salads, semipreserved fish, fruit juices, soft drinks, jams, margarines
Citric acid	$(COOH)CH_2C\text{-}(OH)(COOH)\text{-}CH_2COOH$	3.1	Very soluble	Not limited	No limit	Soft drinks
Lactic acid	$CH_3CHOHCOOH$	3.1	Very soluble	Not limited	No limit	Salad creams and mayonnaise

Methyl paraben[a]		8.5	0.16 (15°C)	10	1–2000	See sodium benzoate
Ethyl paraben[a]	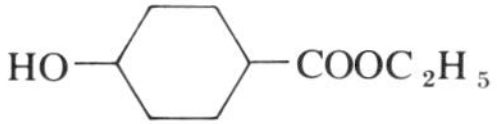	8.5	0.08 (15°C)	10	1–2000	
Propyl paraben[a]	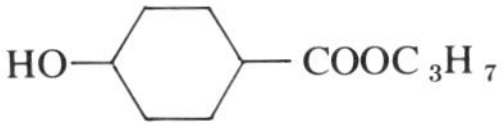	8.5	0.023 (15°C)	10	1–2000	
Sodium propionate	CH_3CH_2COONa	4.9	Very soluble	10	1–3000	Bread, bakery, and cheese products
Sorbic acid (including K salt)	$CH_3CH{=}CHCH{=}CHCOOH$	4.8	0.16 (20°C)	25	1–2000	Fresh and processed cheese, dairy products, bakery products, fruit juices, acid sauces and salads, jams, jellies, soft drinks, margarines, semi-preserved fish, and meat products

[a] Paraben = *p*-hydroxybenzoic acid.
[b] Solubility in water (temperature).
[c] Acceptable Daily Intake (FAO/WHO, 1973).
[d] Typical values; actual values will vary with product and from country to country. Food Additive Lists should be consulted for specific limits for a country.

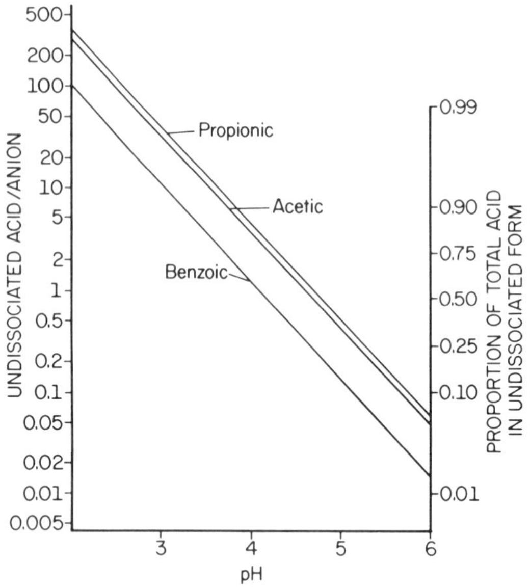

Fig. 7.1 The relationship between pH and proportion of undissociated acid in a solution of an organic acid.

Most organic acids are largely ineffective as microbial inhibitors in the pH range 5.5–6.8, within which all food poisoning bacteria and most spoilage bacteria grow. Exceptions to this are the esters of *p*-hydroxybenzoic acid, which have a pK_a value of 8.5 and show antimicrobial activities near neutrality, and propionic and sorbic acids, which show some antimicrobial activity to pH 6.0 or 6.5 (Chichester and Tanner, 1972).

There are numerous reports in the literature, and many patents, claiming broad and narrow spectrum antimicrobial activity for a wide variety of straight-chain, branched, saturated, and unsaturated organic acids. A number of the longer chain saturated organic acids are highly effective inhibitors of both gram-positive and gram-negative bacteria, although at chain lengths greater than C_8 the activity against gram-negative bacteria falls markedly (Sheu *et al.*, 1975). Chilling of cells increases the bactericidal activity of nonanoic and decanoic acids against *Escherichia coli* (Fay and Farías, 1976). Long-chain unsaturated fatty acids are particularly effective against vegetative gram-positive bacteria and they also prevent the outgrowth of spores of *Bacillus* species. Cis-isomers are more effective than trans-isomers, and activity for a particular chain length increases with the degree of unsaturation of the molecule (Kodicek, 1956).

In general, organic acids are compatible with other preservatives or preservation systems and indeed many combinations are synergistic. For example, they show greater effectiveness as microbial inhibitors as the storage temperature of the food is decreased and greater effectiveness as microbicidal agents as the temperature of a food is increased (Goepfert and Hicks, 1969); Park and Marth, 1972). Examples of specific synergistic combinations are (a) benzoate with sulfur dioxide, carbon dioxide, sodium chloride, or sucrose (Rehm, 1960; Chichester and Tanner, 1972); (b) propionate with carbon dioxide (Windsor and Thoma, 1974); (c) sorbate with sucrose, sodium chloride, or with both nisin and polyphosphate (Gooding *et al.*, 1955; Ashworth *et al.*, 1974); (d) lactic with acetic acid (Rubin, 1978); (e) propionate with sorbate (against staphylococci, Preonas *et al.*, 1969); and (f) benzoic acid and boric acid against aspergilli (Rehm, 1960). When such combinations are used, lower concentrations of each are needed for effective food preservation.

The effectiveness of organic acids as antimicrobial agents is generally improved by anions which interfere with the dissociation of the acid molecule; certain specific cations may also significantly increase the effectiveness of organic acids by increasing the solubility of the acid in the microbial cell membrane (Larsson *et al.*, 1975). Some organisms apparently possess an inducible, energy-requiring system that is able to transport organic acids from the cell, thus permitting growth in the presence of high concentrations of acids (Warth, 1977). The formation of peroxides or other products of fatty acid oxidation may increase the antimicrobial effectiveness of certain unsaturated fatty acids (Lamprecht and Elliott, 1974; Tonge, 1964). However, in common with most microbial inhibitors, organic acids are usually more effective under anaerobic than aerobic conditions (B. Freame and S. Warren, 1976 personal communication).

There are several limitations to the value of organic acids as microbial inhibitors in foods:

1. They are usually ineffective when initial levels of microorganisms are high.
2. Many microorganisms use organic acids as metabolizable carbon sources.
3. There is inherent variability in resistance of individual strains.
4. Resistant types may be selected under use conditions (Ingram, 1960; York and Vaughn, 1954).

Factors affecting the antimicrobial activity of bases are in many respects similar to those affecting organic acids. (For further information, see Simon and Beevers, 1951; Molin, 1964; Board, 1969; Kabara *et al.*, 1972.)

B. Specific Effects

The antimicrobial spectra of organic acids commonly used as food preservatives or as acidulants are listed in Table 7.2. The data listed are the levels of undissociated organic acid that have been found in the laboratory to be generally effective in inhibiting particular groups of microorganisms in microbiological media under near optimal growth conditions. This information should not be regarded as absolute but rather as a guide in comparing the relative effectiveness of particular types of organic acids. To determine the total acid required at a specific pH value to give a particular concentration of undissociated acid or ester, the percentage or proportion of total undissociated acid at different pH values (Table 7.3) can be plotted as in Fig. 7.1.

1. Acetic Acid

Acetic acid and its salts are highly effective and widely used as acidulants and preservatives. Only *Acetobacter* spp., certain lactic acid bacteria, and some yeasts and molds are appreciably resistant to acetic acid (Dakin, 1957). The presence of 1–2% undissociated acid in meat, fish, or vegetable products will usually inhibit or kill all microorganisms, although under use conditions, particularly when hygiene is poor, more acid-tolerant types may occur; this level of acid may be reduced significantly for refrigerated products or products with high salt or sugar content (Bell and Etchells, 1952). The growth of most food poisoning and spore-forming bacteria is inhibited by 0.1% and mycotoxigenic molds by 0.3% of the undissociated acid.

TABLE 7.2

Antimicrobial Spectra of Organic Acids Used in Foods[a,b]

	Concentrations of undissociated acid required to inhibit growth of most strains in microbiological media				
Organic acid	Yeasts	Molds	Enterobacteriaceae	Micrococcaceae	Bacillaceae
Acetic acid	0.5	0.1	0.05	0.05	0.1
Benzoic acid	0.05	0.1	0.01	0.01	0.02
Citric acid	>0.005[d]	>0.005	>0.005	0.001[e]	>0.005
Lactic acid	>0.01	>0.02	>0.01	>0.01	>0.03
Methyl paraben[c]	0.1	0.1	0.2	0.4	0.2
Ethyl paraben[c]	0.1	0.05	0.1	0.1	0.1
Propyl paraben[c]	0.01	0.02	0.1	0.05	0.05
Propionic acid	0.2	0.05	0.05	0.1	0.1
Sorbic acid	0.02	0.04	0.01	0.02	0.02[f]

[a] From Chichester and Tanner, 1972 and unpublished data of S. Warren and B. Freame.
[b] Values given as percent in solution.
[c] Paraben = *p*-hydroxybenzoic acid.
[d] Actual inhibitory concentrations are probably far in excess of values given here.
[e] *Staphylococcus aureus*; micrococci are much more resistant.
[f] Clostridia are generally more resistant.

TABLE 7.3

Proportion of Total Acid Undissociated at Different pH Values[a]

Organic acids	pH values 3	4	5	6	7
Acetic acid	98.5	84.5	34.9	5.1	0.54
Benzoic acid	93.5	59.3	12.8	1.44	0.144
Citric acid	53.0	18.9	0.41	0.006	<0.001
Lactic acid	86.6	39.2	6.05	0.64	0.064
Methyl, ethyl, and propyl parabens[b]	>99.99	99.99	99.96	99.66	96.72
Propionic acid	98.5	87.6	41.7	6.67	0.71
Sorbic acid	97.4	82.0	30.0	4.1	0.48

[a] Values given as percentage.
[b] Parabens = *p*-hydroxybenzoic acid.

2. *Benzoic Acid*

Benzoic acid is used primarily as an antimycotic agent and most yeasts and molds are inhibited by 0.05–0.1% of the undissociated acid. Food poisoning and spore-forming bacteria are generally inhibited by 0.01–0.02% undissociated acid, but many spoilage bacteria are very much more resistant, and benzoic acid cannot therefore be relied on to preserve foods capable of supporting bacterial growth. (For further information on the antimicrobial spectrum of benzoic acid, see Chichester and Tanner, 1972.)

3. *Citric and Lactic Acids*

These acids have in general only moderate antimicrobial activity and except at low pH values are not particularly effective as microbial inhibitors. A concentration of undissociated citric acid of 0.001%, however, will inhibit the growth of *Staphylococcus aureus* under anaerobic conditions (S. Warren, 1976 personal communication). Both lactic and citric acids apparently specifically inhibit the formation of aflatoxin and sterigmatocystin (Reiss, 1976).

4. *p-hydroxybenzoic Acid*

The esters of *p*-hydroxybenzoic acid (parabens) have broad spectrum antimicrobial activity and because of their high p*K* value are effective inhibitors of yeasts, molds, and bacteria at pH near neutral. Although they are almost completely undissociated on the acid side of neutrality (see Table 7.3), the pH of a food will have a significant effect on their spectrum of activity, because many microorganisms will be inhibited by pH per se (see Chapter 5).

The antimicrobial activity of parabens increases with increasing numbers of carbon atoms in the ester group, but the solubility decreases with increasing chain length, which limits their general use to the methyl, ethyl, and propyl esters (Chichester and Tanner, 1972). The propyl ester is particularly effective in inhibiting the growth of food poisoning and spore-forming bacteria (see Table 7.2). The propyl ester will also prevent staphylococcal enterotoxin formation at concentrations that do not inhibit growth, i.e., 0.02–0.03% undissociated acid (R. Holbrook and A. C. Baird-Parker, unpublished observations).

5. *Propionic Acid*

Propionic Acid and its salts are highly effective mold inhibitors, but at concentrations permitted in foods are virtually without effect against yeasts. They are effective inhibitors of many bacterial species at concentrations of 0.05–0.1% undissociated acid and are used to prevent mold growth and ropiness in bakery products (Chichester and Tanner, 1972).

6. *Sorbic Acid*

Sorbic acid is the only unsaturated organic acid currently permitted as a food preservative. It has an interesting antimicrobial spectrum in that it is relatively ineffective against catalase-negative bacteria such as the lactics (Emard and Vaughn, 1952). Sorbic acid has been used as a selective agent in media for the isolation of clostridia and is metabolized by a number of microorganisms including *Clostridium botulinum* (York and Vaughn, 1954). It has a wide spectrum of activity against catalase-positive microbes including yeasts, molds, and bacteria and is thus used to inhibit aerobic contaminants of fermented or acidulated foods. These latter organisms are generally inhibited by between 0.01 and 0.03% undissociated acid. Salmonellae may be killed by 0.1% undissociated acid, but the rate of inactivation depends on the acidulant used to adjust the pH, the substrate composition, and the storage temperature (Park and Marth, 1972). It is an effective antimicrobial agent at pH values up to about pH 6.0. In view of the use of sorbate to select clostridia (see above), it is surprising that recent studies have shown its value as an *inhibitor* to the growth of *C. botulinum* in cured meat products (Ivey and Robach, 1978; Ivey *et al.*, 1978). However, studies carried out by Unilever Research in England indicate that sorbate is not as effective as nitrite for this purpose (J. Crowther, personal communication).

III. USE OF ORGANIC ACIDS TO CONTROL MICROORGANISMS IN FOODS

Much can be learned about the potential antimicrobial effectiveness of organic acids by using model food systems and microbiological media. Some guidance in this respect can be obtained from Table 7.2. However, as discussed previously, the complexity of factors affecting the activity of organic acids cannot be fully covered in such model systems. It is therefore necessary to check the effectiveness of a particular choice and level of organic acid by inoculated pack studies under actual use conditions.

The effectiveness of an organic acid in a food is particularly affected by the water activity, pH value, redox potential, substrate availability, and fat content (Lubieniecki-von Schelhorn, 1964; Jarvis and Burke, 1977). Of equal importance in the selection of an organic acid is the microflora that the organic acid is intended to inhibit or destroy, e.g., numbers, types, and relative resistance of microorganisms likely to be present, together with their ability to grow under the proposed conditions of storage and use. Thus the choice of a particular organic acid depends not only on the inherent characteristics of the organic acid per se (i.e., appropriate antimicrobial activity, solubility, stability, and compatibility of organoleptic properties) but also on the microenvironmental and storage conditions of the food.

Of further consideration in the selection of organic acids are the views of regulatory authorities concerning their safety. Most countries list the maximum permitted levels in particular foods. Maximum concentrations are not usually specified for certain organic acids such as acetic acid, citric acid, and lactic acid. For information concerning the current usage of organic acids in a particular country the latest food additive lists for that country should always be consulted. As a guideline to generally accepted levels, Codex Alimentarius Commission values for Acceptable Daily Intake (ADI) values are listed in Table 7.1 together with typical maximum use concentrations and examples of use in different foods.

To be approved as a food preservative, an organic acid must be of proven direct or indirect benefit to the consumer. That is, it must maintain nutritive value, increase the food supply, improve the keeping quality in the home, substantially decrease the costs, or increase convenience to the consumer. Amounts permitted are based on toxicological considerations and are set at the lowest possible level to give the required benefit (FAO/WHO, 1973).

More detailed information on the use of specific organic acids as food preservatives will be found in Volume II, Chapters 15 to 29.

8

Curing Salts and Related Materials

I. INTRODUCTION

Curing was originally developed to preserve certain foods by the addition of sodium chloride. One natural impurity of sodium chloride, sodium nitrate, was eventually shown to be responsible for the development of a pink to red pigment in meat—the cured meat color. Subsequently it was found that nitrite rather than nitrate was the compound important in development of color, the nitrite being formed by bacterial reduction of nitrate. Today, sodium chloride and sodium and potassium nitrite or nitrate are considered curing salts.

Although curing was originally a mechanism for preservation by salting, over several millenia additional processes concommitant with curing have evolved, notably fermentation, smoking, drying, and heating. In the past half century, cured products have been developed that are not stable unless refrigerated. Indeed, most cured meat products must be refrigerated to remain safe and wholesome, and during the past two decades even the packaging of many classes of cured products has become important in extending the period during which the product remains wholesome.

In addition to the curing salts and related processes mentioned above, additives collectively known as adjuncts are used in many cured meat products. These include ascorbates, phosphates, glucono-δ-lactone, and sugars. Adjuncts are used primarily to obtain or maintain desirable changes, the ascorbates in connection with color and the others in connection with pH, texture, and in some cases, flavor. Adjuncts may also affect safety.

The term "curing" is widely used in many industries, always in the context of desirable change, e.g., in preparation of leather, manufacture of steel, hardening of concrete, as well as in the preservation of food. However, even in the food industry the appellation of cured is related only to certain meat, fish, and cheese products. Even in these foods, curing may have different connotations: in meat, salt and nitrite or nitrate are always added; in fish, salt is always added, but nitrite only rarely; and in cheese, which always contains salt, but infrequently contains nitrate, the term curing is applied to the production of desirable proteolytic and lipolytic changes. Indeed, even in the food industry, the definition of cured foods is by usage: for example, both butter and certain pickled vegetables require salt as part of their preservative system, but they are never referred to as cured products.

Curing salts, adjuncts, and related processes referred to above alter the basic food. Alterations include color, flavor, texture, and susceptibility to microbial growth. Microbial growth, depending on the product, may be desirable, cause spoilage, or cause food poisoning. These will be considered in more detail, mainly in reference to the curing of meat. See also Sections VIII, IX, and X in Chapter 15, and Section VII in Chapter 20.

The history of curing extends over several thousand years since man first began preserving food by salting. The presence of nitrate in crude salt caused the reddening of meat. However, even before the Christian era, nitrate itself was used in ancient China and India. Early Roman writing describes the curing of meat by salting, smoking, and acidulation (Binkerd and Kolari, 1975). About the end of the nineteenth century, it was known that bacteria reduced nitrate to nitrite in meat curing and that nitrite rather than nitrate was responsibile for reddening. However, it was only in 1923 that the direct addition of nitrite to meat was permitted in the United States. There is now a trend to use salt and only nitrite in most cured meat, although nitrate, or mixtures of nitrate and nitrite, are still used for some products.

Nitrate has been used in cheese making in Europe for about 150 years, almost exclusively for the manufacture of some cheeses salted by immersion in brine (see Chapter 18).

The concentration of each curing agent depends on the nature of the food products and on the technology used in individual countries. Where there is abundant refrigeration, lightly salted cured meat products that require refrigeration are the most common. Conversely, in warm climates and where refrigeration facilities are scarce, fermented and heavily salted cured products that are shelf-stable are more common. As a result, cured products reflect national economies or climates and become part of national or even regional cultures. Thus, there are vast differences among cured sausages prepared in different countries. A similar statement could be made for cured fish and cheese. However, the remainder of this chapter will be concerned with curing mainly as it pertains to meat.

Cured meats can be divided broadly into three groups: unheated, mildly heated (pasteurized to center temperature of 65°–75°C) and severely heated (shelf-stable after heating to 100°–120°C). In general, only a few unheated products (some types of ham and sausage, bacon, and lightly salted briskets) require refrigerated storage while almost all of the mildly heated products should be refrigerated after heating. Those receiving a severe heating, however, are stable indefinitely in temperate climates if in hermetically sealed containers (see also Chapter 15.)

II. EFFECT OF CURING SALTS

A. Effect on Meat

1. Fate of Nitrite

When nitrite is put into a food system a complex series of reactions ensues, the nature of which depends on the physicochemical characteristics of the system. Many of the reactions are unknown. Nitrite added to meat is converted to an equilibrium mixture of NO_3^-, NO_2^-, and NO depending on pH and E_h (Shank *et al.*, 1962; Ingram, 1973). The nitrite eventually disappears as the result of chemical reactions with components of the meat or from the metabolic activity of microbes.

Reactions with meat may involve the heme pigments (Möhler, 1973), the nonheme proteins (Woolford *et al.*, 1976), and other compounds. About 70–80% of nitrite added to meat can be recovered in the following percentages: nitrate, 1–10; nitrite, 5–20; gas, 1–5; reaction products with myoglobin, 5–15; with protein, 20–30; with sulfhydryl, 5–15; and with lipid, 1–5 (Cassens *et al.*, 1976). A few compounds in meat (cysteine and histidine) and the additives, ascorbate and isoascorbate, appear to be responsible for the major loss of nitrite: the reductants mediate the conversion of nitrite to nitric oxide, but the role of histidine is unknown (Fox and Nicholas, 1974).

Some nitrite is converted to nitrate by various chemical reactions, especially in the presence of ascorbate and prolonged curing (Herring, 1973; Fujimaki *et al.*, 1975; Möhler and Baumann, 1971) provided oxygen or some other suitable hydrogen acceptor is present. One equation for this conversion is (Lee *et al.*, 1978b):

$$2\ \text{Metmyoglobin (Met Mb)} + \text{ascorbate} \rightleftharpoons 2\ \text{Myoglobin (Mb)} + \text{dehydroascorbate}$$
$$2\ \text{Mb} + O_2 + NO_2^- \rightleftharpoons 2\ \text{Met Mb} + 2\ NO_3^- + 2\ \text{electrons}(e)$$
$$\text{Dehydroascorbate} \xrightarrow{+2e} \text{Ascorbate}$$

Nitrite in canned products, if present in grossly excessive amounts, may result in release of gaseous oxides of nitrogen during thermal processing, and swelling may occur (Morris, 1952).

The rate of disappearance of nitrite in cooked meat depends on pH and temperature; as the pH drops and the temperature rises, the rate of disappearance becomes more rapid (Olsman and Krol, 1972). The effect of pH and temperature on the half-life of nitrite can be calculated from the following equation (Nordin, 1969) in which 0.65, 0.25, and 0.35 are constants.

$$\log_{10} \text{(half-life in hours)} = 0.65 - 0.25 \text{ (temperature in °C)} + 0.35 \text{ (pH)}.$$

For example, the half-life in hours at pH 6.0 for various temperatures is 100°C, 1.8; 77°C, 6.7; 24°C, 126; and 7°C, 376. After the concentration of nitrite has decreased to about 20 ppm the above equation may no longer pertain.

Some microbes may also cause the disappearance of nitrite by using it as a hydrogen acceptor. In cured meats vacuum-packed in oxygen-impermeable plastic, yeasts (Wickerham, 1957) and bacteria (Pivnick, unpublished observations) can accomplish this. The growth of some nitrite-reducing bacteria persists in this ecosystem until the nitrite is exhausted. In canned cured meat, nitrate serves as a hydrogen acceptor for aerobic bacteria, e.g., *Bacillus* spp. and micrococci; without it they normally would not grow. Bacterial reduction of nitrate or nitrite to N_20 and N_2 may cause swelling (Eddy and Ingram, 1956).

2. *Flavor*

The principal curing agents and adjuncts that affect flavor are salt, sugar, smoke, and nitrite. Sugar is used widely in some bacons and hams, but its contribution to flavor in other products is in dispute. Smoke as a flavoring agent will be discussed later.

Nitrite reacts with components of meat, especially pork, to alter the flavor; and that altered flavor is more acceptable than pork cured only with salt (Barnett *et al.*, 1965; Mottram and Rhodes, 1973). Depending on the product, from 15 to 150 ppm of nitrite is an adequate concentration. The chemical basis of cured flavor is unknown despite numerous investigations, but one hypothesis is that part of the cured flavor is an absence of products of oxidative cleavage of unsaturated lipids, e.g., hexanal, valeraldehyde. Salt and iron accelerate oxidation, and this oxidation is more rapid in pork, which has more unsaturated lipids than beef does (Wilson *et al.*, 1976). Nitrite decreases the rate of oxidation. Thus, one difference between meat without nitrite and meat with nitrite is decreased oxidation of lipids in the latter (Cross and Ziegler, 1965; Cho and Bratzler, 1970; Mottram and Rhodes, 1973; Wasserman, 1973; Hadden *et al.*, 1975). Some support for this concept has been presented (Simon *et al.*, 1972) in pork and beef frankfurters made with and without nitrite: pork frankfurters made without nitrite, or with low levels of nitrite, were organoleptically less acceptable than beef frankfurters. Spice and smoke may improve organoleptic acceptance of products made without nitrite (MacNeil and Mast, 1973; Wasserman, 1973).

A defect known as "warmed over" flavor (WOF) occurs in meat that has been cooked and refrigerated (Wilson *et al.*, 1976; Sato and Hegarty, 1971). It results from an iron-mediated increase in oxidized lipids such as heptanal, *n*-nona-3,6-dienal, and similar compounds (Ruenger *et al.*, 1978). Nitrite prevents or decreases the production of WOF, and this defect can also be prevented by production of antioxidants during severe heating (e.g., $F_0 = 17$; Einerson and Reineccius, 1977) or by application of some antioxidants, e.g., butylated hydroxyanisole (BHA), butylated hydroxytoluene (BHT), and propyl gallate. This adds support to the hypothesis that cured flavor is at least, in part, due to inhibition of oxidation of lipids.

3. Color

The addition of nitrite to meat converts meat pigments, principally myoglobin and to a lesser extent hemoglobin, to water-insoluble pigment, the red nitric oxide myoglobin. Heating converts the nitric oxide myoglobin to pink nitrosyl hemochrome, which is stabilized by ascorbates. A few of the more common synonyms for several of the pigments in meat are presented in Table 8.1.

The cured meat pigments may result from chemical, biochemical, or enzymatic reactions depending on whether the meat–nitrite mixture is heated and the time between addition of nitrite and heating (Möhler, 1973). The curing reaction is accelerated by reducing conditions, decreased pH, and increased temperature (Fox and Ackerman, 1968); under optimum conditions, about 15 ppm of added sodium nitrite yields maximum color in comminuted or emulsified products and about 50 ppm yields maximum color in sides of bacon. Such concentrations of added nitrite have little or no antibacterial effect. Bacteria do not play a major role in production of color, except to reduce nitrate to nitrite and, in some products, to lower the E_h and pH. E_h can also be reduced by heat and ascorbate, and pH by glucono-δ-lactone and sodium acid pyrophosphate.

The main reaction appears to be formation of a nitroso-reductant intermediate between nitrite and various reducing agents, and the breakdown of the intermediate to release nitric oxide. Nitric oxide then replaces the water attached to the iron in the heme portion of myoglobin or hemoglobin. Other reactions involv-

TABLE 8.1

Synonyms for Some Important Pigments in Curing of Meat

1. Myoglobin = myoglobin Fe^{+2} = deoxymyoglobin
2. Oxymyoglobin = oxymyoglobin Fe^{+2}
3. Metmyoglobin = metmyoglobin Fe^{+3}
4. Nitric oxide myoglobin = nitrosomyoglobin = nitrosylmyoglobin = nitrogen monoxide myoglobin = pigment of cured, nonheated meat
5. Nitric oxide hemochromogen = nitrosomyochrome = nitrosyl hemochrome = nitric oxide denatured globin hemochromogen = denatured globin nitric oxide ferrohemochrome = pigment of cured, heated meat

ing formation and destruction of both nitric oxide myoglobin and nitrosyl hemochrome are shown in Fig. 8.1 (Bard and Townsend, 1971; Kemp, 1974).

4. *Ascorbic and Isoascorbic (Erythorbic) Acids*

There are many competing reactions for ascorbates in meat (Fig. 8.2). Ascorbates are important in accelerating the development of cured meat pigment and stabilizing it. Additionally, they affect anticlostridial activity of nitrite in canned pasteurized meats, increasing the activity when a proper concentration of ascorbates is present (Schack and Taylor, 1963; Tompkin *et al.*, 1978a,b,d) or decreasing it when the concentration is excessive (Tompkin *et al.*, 1979b). Their main beneficial chemical interactions in cured meats are (1) as reducing agents and (2) as chelating agents (Tompkin *et al.*, 1978e) As reducing agents ascorbates act as oxygen scavengers; decrease the E_h; participate in the reduction of metmyoglobin to myoglobin; react with nitrite to increase the yield of nitric oxide, the nitric oxide then reacting with myoglobin to form nitric oxide myoglobin. Ascorbates chelate powerful pro-oxidants such as copper and iron; chelation of iron may be instrumental in increasing the antibotulinal activity of nitrite (Schack and Taylor, 1963; Tompkin *et al.*, 1978a,b, 1979a).

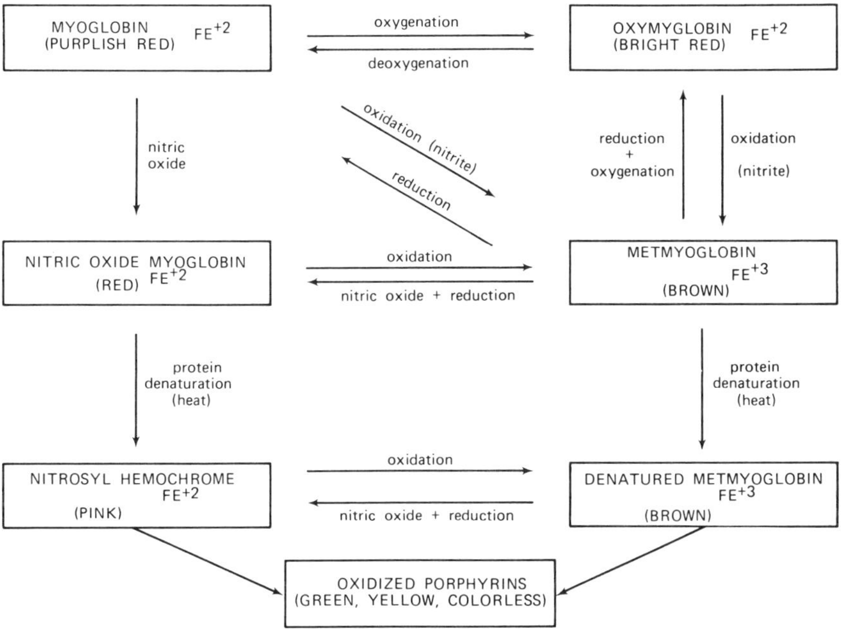

Fig. 8.1 Possible curing reactions when nitrite is added to meat. (From Bard and Townsend, 1971.)

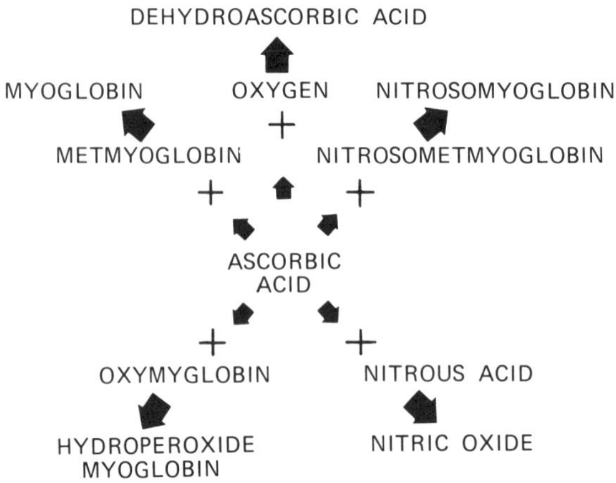

Fig. 8.2 Competing reactions for ascorbic acid in meat. (From Bauernfeind and Pinkert, 1970. *Adv. Food Res.* **18,** 219–315. Copyright by Academic Press Inc.)

The antimicrobial activity of ascorbate per se is probably nil. Neutralized ascorbate auto-oxidizes to form a sporicidal compound (Eller *et al.,* 1968), but conditions required for this to occur are unlikely to be present in meat. More important, however, is the ability of ascorbates to increase the antibotulinal activity of nitrite in canned, pasteurized meats which are not shelf-stable. In this context, reducing activity may be secondary to the ability of the ascorbates to chelate iron.

5. *Texture and pH Adjustments: Phosphates and Glucono-δ-Lactone (GDL)*

The texture of cured meat products varies from succulent to hard, depending on whether or not they are comminuted, cooked, acidified, fermented, dried, aged, etc. Salt and phosphates (Hamm, 1971) increase the water-holding capacity of meat. Phosphates may raise the pH while GDL will lower it.

Canned hams, when heated, lose liquid. The addition of alkaline polyphosphates raises the pH from about pH 6.0 to 6.5 allowing more severe heat processing while minimizing shrinkage (Hamm, 1971). The increased pH introduces two factors with respect to the contribution of nitrite to safety: more nitrite remains after processing, but that remaining is less inhibitory at the higher pH to bacteria that survive the thermal process.

The term polyphosphates includes a wide variety of polymeric phosphates (Bell, 1971). Some polyphosphates have antimicrobial activity. Although they do not prevent germination of spores, they may prevent rupture or lysis of walls of germinated spores, or prevent division of cells emerging from spores (Gould,

1964). The antimicrobial activity is a function of the particular polyphosphate and the medium in which activity is tested. In general, polyphosphates do not inhibit clostridia in cooked meat, but do inhibit nonfluorescent pseudomonads in raw meat, probably by chelating Mg^{2+} or other essential metals (Elliott *et al.*, 1964).

Lowering the pH of meat emulsions by adding acidulants decreases the likelihood of growth of spoilage- and disease-producing bacteria. But direct acidification of meat emulsions followed by an input of mechanical energy (e.g., when emulsions are stuffed into casings) decreases stability so that pockets of fat form upon subsequent heating. A delay in acidification of sausage emulsions can be obtained by adding GDL, which hydrolyzes slowly to gluconic acid. Thus the emulsions to which GDL is added are stuffed into casings before the pH drops. Sausages made with GDL can be heated and smoked at a higher temperature without separation of fat than sausages acidified by direct addition of other acidulants. The simultaneous addition of GDL and a reducing agent such as ascorbate permits much more rapid processing of emulsion to cooked sausage while obtaining adequate development of color and a decreased pH (Sair and Komarik, 1968). Although the reduction of pH is not great (0.3–0.6 units) at concentrations of GDL used commercially, the reduction is sufficiently rapid (Acton and Dick, 1977) to decrease the growth of potential pathogens while speeding up selective conditions for growth of lactic acid producing bacteria (Riemann *et al.*, 1972). This activity is probably an important contribution to the safety in some cured meats that are held at crucial temperatures permitting intense bacteriological activity, e.g., initiation of fermentation in salamis in the green room at 25°–38°C (Reuter, 1972).

A novel method for lowering the pH of cooked meats to prevent growth of pathogenic bacteria has been described. *Pediococcus cerevisiae* killed by irradiation and added to meat after cooking will produce acid from dextrose if the product is held at room temperature for a few hours; acid is not produced in refrigerated products (Riemann *et al.*, 1972). The pH may also be decreased in sausage fermentations by adding starter cultures.

6. Starter Cultures

Starter cultures are used in many types of fermented sausages to bring about a more rapid and complete fermentation than may occur from the indigenous microflora (Task Force, 1977). They may also be added to produce desirable flavor and texture (Nurmi, 1966; Deibel *et al.*, 1961; Reuter, 1972) and to reduce nitrate to nitrite. For acid production, various lactic acid producing bacteria are added and the raw mixture is held at 25°–35°C for 1–2 days to permit proliferation of the lactic acid producers. If nitrate reduction is required as a source of nitrite, *Micrococcus* spp. may be added with the acid-producing *Lac-*

tobacillus or *Pediococcus*, but the meat mixture may be held first at temperatures as low as 5°–10°C until nitrite is produced. Alternatively, nitrite may be added directly. Starter cultures are not inhibited by concentrations of salt or nitrite that are added to the mixture to inhibit spoilage and pathogenic bacteria (Zaika *et al.*, 1976) and they may even be stimulated by the spice used in some sausages (Kissinger and Zaika, 1978).

The main function of starter cultures is to decrease the pH, but the addition of the starter may not be as important as the addition of fermentable carbohydrate to effect an adequate reduction in pH. The addition of both, however, results in a rapid and marked decrease in pH during the early period of fermentation and drying. This is important in preventing growth of pathogenic bacteria [e.g., *Clostridium botulinum* (Christiansen *et al.*, 1975), *Staphylococcus aureus* (Niskanen and Nurmi, 1976), *Yersinia enterocolitica* (Raevuori *et al.*, 1978), and *Salmonella* (Smith *et al.*, 1975a; Sirviö *et al.*, 1977)] and spoilage bacteria (e.g., *Bacillus* spp. and coliforms), and in preventing spoilage, i.e., undesirable amounts of proteolytic and lipolytic activity (Nurmi, 1966).

Unfortunately one cannot rely entirely on the fermentation and subsequent drying process to bring about destruction of pathogens (Baran and Stevenson, 1975; Smith *et al.*, 1975b). Even though it will not grow, *Salmonella* may survive and cause illness (D'Aoust and Evans, 1978). Therefore, some fermented products are also heated, which eliminates this potential hazard.

On the other hand, *Staphylococcus aureus* can form enterotoxins in fermented sausages. For a discussion of this problem see Chapter 15 (especially Section VIII,B,5).

B. Effect on General Flora and on Spoilage Flora

1. Nitrite

At levels and under conditions commonly used, curing agents do not cause rapid destruction of microorganisms. Rather, they retard or prevent growth of undesirable microorganisms in unheated products and thermotolerant nonspore formers in pasteurized products; and they prevent growth of spores that survive the more severe heating given to some cured products. The relationship of salt content to a_w and the influence of a_w on inhibition of different groups of microorganisms in foods has been treated in Chapter 4. The effect of other curing agents and of adjuncts will be considered below.

Protection against spoilage of shelf-stable meat by putrefactive anaerobes has been studied thoroughly (Stumbo *et al.*, 1945; Nordin *et al.*, 1975) and expressed quantitatively for P.A.3679, a putrefactive sporeforming anaerobe commonly used in testing the efficacy of thermal processes (Nordin *et al.*, 1975). The following formula is based on percentage of spoilage occurring in tubes of cured meat processed at 115.5°C for 30 min and stored at 23°C for 150 days. Each tube contained 10^5 spores prior to processing

$$\hat{y} = -322 + 73.9\,(\text{pH}) - 0.115\,[\text{N}] - 24.7\,[\text{S}]$$

where $\hat{y}$ = estimated percent of packages spoiling within 150 days at 23°C, pH = usual definition in the range of 5.5–6.6, N = concentration of sodium nitrite in the range of 0–200 ppm, and S = concentration of salt in the range of 1.5–3.0% (2–4% in the aqueous phase).

This formula suggests that the same amount of inhibitory activity against spoilage may be obtained in shelf-stable meat by a decrease of 0.1 pH unit, an increase of 65 ppm of nitrite or an increase of 0.3% w/w of salt. The effect of each of the three factors appears to be additive.

The precise mechanism of inhibition of bacteria by nitrite is unknown. It does not prevent spore germination but prevents outgrowth (Gould, 1964; Duncan and Foster, 1968; Pivnick *et al.*, 1970). Depending on the concentration of nitrite, germinated spores and vegetative cells may grow, or they may die slowly (Pivnick, 1970; Christiansen *et al.*, 1978; Tompkin *et al.*, 1978c). If growth occurs, it may be uninhibited or it may be decreased as the concentration of nitrite is increased (Riha and Solberg, 1975b). The rate of death, however, is a function of pH and concentration of nitrite (Tarr, 1942). At pH 5.5–4.0, rapid destruction of vegetative cells and possibly spores of some species may occur (Shank *et al.*, 1962). Unexplained is the large number of viable bacteria present in nitrite-containing fermented sausages normally at pH 5.5–4.5; possibly nitrite disappears rapidly during the fermentation.

Some possible biochemical mechanisms which inhibit growth are reaction with amino groups or with sulfydryl-containing constituents of cells (O'Leary and Solberg, 1976). Some workers (Tompkin *et al.*, 1978a,b, 1979a) have proposed a reaction between nitric oxide and the iron in a ferredoxin or similar compound involved in electron transport in many clostridia. They also suggest that the reaction product is reversible, but stable, as long as a reserve of nitrite is present. Meats (e.g., heart meat) with a high content of iron in a form that binds with nitric oxide make the nitric oxide unavailable to react with the respiratory mechanism in clostridia; such meats are not normally canned and thermally processed. Respiratory pigments in aerobic bacteria do not appear to be involved (Tarr, 1941a; Castellani and Niven, 1955). From the standpoint of preservation and safety, nitrite at concentrations used commercially must be regarded primarily as a bacteriostat or as a precursor of the more stable Periogo-type factor (PTF).

Perigo-type factor is inhibitory to clostridia in shelf-stable canned meat, but Perigo factor (PF) is not. PF is an inhibitory material or materials formed when nitrite (5–20 ppm) is heated (90°–130°C) in some bacterial media (Perigo *et al.*, 1967), but not in others (Johnston and Loynes, 1971). PF is highly inhibitory to clostridia but not inhibitory to many bacilli, fecal streptococci, or salmonellae (Roberts and Garcia, 1973). Unlike nitrite, the inhibitory activity of PF is not pH dependent and is nullified by meat (Johnston *et al.*, 1969). It is highly improbable that PF is of any consequence in preventing microbial growth in meat.

Perigo-type factor (PTF) is a name given to antibacterial product(s) formed when nitrite is heated with meat. PTF is formed during the severe heating given to shelf-stable meat (Pivnick and Chang, 1973; Lee *et al.*, 1978a), but the PTF in meat has only minor antibacterial activity (Pivnick and Petrasovits, 1973) compared to PF in bacteriological media. However, PTF may play an important role in preventing growth of the small number of clostridial spores that survives thermal processing of shelf-stable products.

2. *Smoke*

Smoking was once an important component of the preservation process of many cured meat and fish products. Today, smoking is important to aid preservation of only a few products; mostly it is used mainly to contribute flavor and color (Tilgner, 1977).

Smoke contains a wide variety of organic compounds (Draudt, 1963; Gilbert and Knowles, 1975; Olsen, 1976; Hamm, 1977), including antibacterial phenolic compounds, tarry fractions, and formaldehyde. Smoke may also contain antioxidants, and oxides of nitrogen which impart a slight cured (red) color to sausages made without nitrite (Wasserman and Talley, 1972). Whether the compounds that are antimicrobial *in vitro* actually exert an antimicrobial effect in smoked meats and fish depends on numerous factors such as concentration, method of application, and interaction with other inhibitory factors.

Smoke may be applied by two essentially different methods. Natural smoke is generated from wood by combustion or friction and the particulate component is adsorbed onto the surface of the product while the soluble portion may penetrate into the product. Alternatively, natural smoke is scrubbed to remove undesirable materials and the desirable components are dissolved in water to prepare liquid smoke; the liquid smoke is applied in a chamber by spraying or aerosolization, by direct addition to comminuted material, or by dipping products into or spraying them with solutions (Hollenbeck, 1977).

Natural smoking may be carried out in two ways: hot smoking at 60°–85°C and cold smoking at 25°–35°C. Heavy smoking, even without heat, or light smoking with heat may destroy microbes (Gibbons *et al.*, 1954) and may also acidify and dry the surface (Christiansen *et al.*, 1968). Smoking may exert considerable inhibitory activity at the surface of large pieces of meat or throughout sausages of small diameter (Tatini *et al.*, 1976), but it does not affect microbes in the center of thick pieces of meat. Liquid smoke at concentrations acceptable to the consumer usually exhibits little or no antibacterial activity.

The antimicrobial effect of smoke is generally more pronounced against gram-negative rods and *Micrococcus* and *Staphylococcus* than against *Pediococcus, Leuconostoc, Streptococcus,* and *Lactobacillus* spp. Spores of molds and bacteria are still more resistant (Handford and Gibbs, 1964; Jensen, 1943; Niinivaara and Pohja, 1956). However, there are wide differences in response among strains in a single species. Depending on conditions of packaging and storage of smoked food, spoilage may be delayed as a result of a general inhibition of bacteria, or by a shift in microflora from predominantly catalase-positive bacteria to catalase-negative nonproteolytic and nonlipolytic bacteria. Smoke may also mask odors of spoilage, but does not mask the cured flavor due to nitrite (Cho and Bratzler, 1970).

3. Combination Effects and Interactions

Besides being organoleptically unique, the many hundreds of different cured products present almost as many unique ecosystems. Although the microbiological contaminants of raw meat (see Vol. II, Chapter 15) are derived mainly from the hides, feet, gut, abattoir, and chill room, and the microbiological contaminants of raw fish from the slime and, to a lesser extent, from the gut (see Vol. II, Chapter 20), the ecosystem of the cured product influences the microflora that develops. The components contributing to the ecosystem are the above-mentioned curing materials and adjuncts; the a_w, pH, and E_h of the product; the extent of thermal processing; the availability of oxygen; the opportunities for postcooking contamination by flora from the processing and packaging environment; the packaging material; and the temperature of storage. The types and proportions of microorganisms that develop in a particular ecosystem (and change it as a result of their growth) vary with the product.

Thermal processing of cured meat and fish destroys some microorganisms and damages others, but the presence of curing salts has little or no effect on that destruction. The curing salts do, however, decrease the probability of outgrowth of the surviving organisms, and increasing the severity of heat damage lowers the concentration of salt and nitrite needed to prevent outgrowth. These inhibitory levels are further decreased if other conditions, particularly a_w, pH, and temperature, are suboptimal for growth. It is assumed that E_h is also important, but direct evidence in cured products is lacking.

Although a single agent is rarely used for preservation, much information is available on the effects of single agents in sterilized bacteriological media at temperatures near optimum for growth of the microorganism. The test dose is usually at least 10^6 spores or vegetative cells. However, variation among strains of a single species is so great that published data must be used with discretion in attempting to predict inhibitory activity of a given curing agent in another context. For example, widely varied concentrations of sodium nitrite (Table 8.2) and sodium chloride (Table 8.3) are required to inhibit different strains of a single species tested under identical conditions of media, pH, temperature, etc. In cured foods, these differences are magnified by many interactions which influence preservation.

a. pH Dependence of Nitrite. The antimicrobial activity of nitrite depends on pH (Tarr, 1941a,b, 1942, 1944; Castellani and Niven, 1955; Shank *et al.*, 1962; Roberts and Ingram, 1973; Riha and Solberg, 1975a) and increases as the pH decreases from 7.0 to 5.0. The inhibitory activity of nitrite equates well with the concentration of undissociated nitrous acid (HNO_2); 3500 ppm of $NaNO_2$ at pH 6.9 and 40 ppm at pH 5.0 yield almost identical concentrations (about 1 ppm) of undissociated HNO_2 (Castellani and Niven, 1955). The proportion of undissociated HNO_2 can be calculated from

$$\log_{10} \{(HNO_2/NO_2)\} = pK - pH$$

where $pK = 3.4$. Thus, at a constant concentration of $NaNO_2$, there should be about a tenfold increase in HNO_2 per unit decrease in pH. In practice, the concentration of nitrite required to inhibit a constant number of bacteria decreases 5- to 10-fold for a decrease of one pH unit between pH 7.0 and 5.0. Some examples are given in Table 8.2.

b. E_h Dependence of Nitrite. Redox potential may also affect the antimicrobial activity of nitrite against some, but not all species (Castellani and Niven, 1955; Henry *et al.*, 1954).

TABLE 8.2

Concentrations of Sodium Nitrite Inhibitory to Foodborne Organisms[a]

Species	Number of strains tested	Inhibitory concentrations of sodium nitrite at pH 6.0 (ppm)	pH 7.0 (ppm)
Clostridium botulinum types A and B	9	10–160	640–2560
Clostridium botulinum type E	3	10–80	160–320
Clostridium perfringens type A	8	80–>240	1280–2560
Clostridium sporogenes	4	80–200	320–1280
Bacillus subtilis	4	120–800	—
Bacillus (nine species)	9	200–>800	—
Streptococcus faecalis	2	>200	—
Salmonella typhimurium	2	>160	—

[a] From Perigo and Roberts (1968); Roberts and Garcia (1973).

c. pH Dependence of Salt. The antimicrobial activity of salt, which effectively changes the a_w, has been discussed in Chapter 4 but is discussed briefly here with particular reference to curing. Many species of bacteria capable of causing spoilage in raw meat are inhibited by 3–4% salt in the water phase, while others are not. Those organisms not inhibited may be desirable (e.g., in fermentation of sausages) or may cause spoilage or disease. Even within the same species, different strains may vary greatly in inhibition by salt (Table 8.3), and these differences are magnified as the pH is decreased from 7.0 to 5.5 (Baird-Parker and Freame, 1967; Ohye and Christian, 1967; Riemann, 1967a).

d. Curing Agents, Adjuncts, Processing, and Storage Conditions. Interaction of curing salts, temperature, pH, E_h, a_w, fermentation, and other factors have been studied to assess safety against several pathogens and spoilage organisms in many foods. The complexity of these interactions is illustrated below, as is the difficulty in predicting whether or not stability or bacterial safety is maintained when one or more factors are changed.

i. Clostridium botulinum types A and B in bacon: Spore numbers, nitrite, and temperature were varied. At 20°C, 10^6 spores were unlikely to produce toxin in bacon without nitrite; but at 30°C, 10^4 spores produced toxin in bacon made with 150 ppm of nitrite (Collins-Thompson *et al.*, 1974).

ii. Clostridium botulinum types A and B in ready-to-eat ham: Nitrite and

TABLE 8.3

Concentrations of Sodium Chloride Inhibitory to Foodborne Organisms

Species	Number of strains examined	Inhibitory concentration of NaCl (%)
Clostridium botulinum types A and B	45	>6.0[a]
(proteolytic)	11	<6.0[a]
Clostridium botulinum types A and B	7	<10.5[b]
Clostridium botulinum type E and nonproteolytic type B	70	ca <4.5[a]
Clostridium perfringens	18	>4.0–<8.0[c]
Clostridium putrificum	2	4.9– 10.5[b]
Clostridium sporogenes	3	8.7– 9.5[b]
Staphylococcus aureus	—	16– 18[d]
Salmonella sp.	—	8[d]

[a] Data from Baird-Parker and Baillie (1973).
[b] Data from Tanner and Evans (1933).
[c] Data from Gough and Alford (1965).
[d] Data from Genigeorgis (1976).

temperature were varied; spore concentration was ca. 10^6 per package (Table 8.4). The meat became toxic in 2 to 8 days at 30°C, but required 15 to over 60 days at 20°C.

iii. Clostridium botulinum types A and B in fermented sausage: Spore numbers, fermentable carbohydrate, and nitrite were varied. Without fermentable carbohydrate, the pH remained at about 5.6 and most sausages became toxic despite input of up to 150 ppm of sodium nitrite; with fermentable carbohydrate the pH fell to less than 4.7 within 1 week, and no sausage made with ≥ 50 ppm of nitrite became toxic during storage up to 7 weeks (Christiansen *et al.*, 1975).

iv. Clostridium botulinum types A and B in shelf-stable canned meat: Salt and nitrite were varied. The addition of both chemicals was required to prevent toxicity, except at higher levels (Table 8.5).

v. Clostridium botulinum types A and B in meat with 3.5% salt: Nitrite and time of incubation were varied. In slurries with this salt content spoilage occurred at all levels of nitrite, but the rate of spoilage decreased with increased nitrite concentration (Fig. 8.3).

Although numerous additional examples could be cited for *C. botulinum,* and for other pathogens and important spoilage organisms, it is sufficiently obvious that curing salts per se must be considered only as components of the safety system and, for many cured products, temperatures of processing and storage are the most important factors. There is not sufficient quantitative data to permit the

TABLE 8.4

Effect of Nitrite and Temperature of Incubation on Toxin Production by *Clostridium botulinum* in Cooked, Sliced Ham in Air-Impermeable Plastic Pouches[a]

		Time to become:	
Sodium nitrite input (ppm)	Temperature (°C)	Toxic (days)	Organoleptically unacceptable (days)
0	30	2	3
	20	15	18
100	30	3	6
	20	>25	>25
200	30	8	9
	20	30	35
300	30	8	14
	20	>60	>60

[a] From Pivnick *et al.* (1967). Reprinted by permission from *Food Technol.* **21,** 204–206. Copyright © by Institute of Food Technologists.

TABLE 8.5

Effect of Salt and Nitrite Levels on Toxigenesis in Shelf-Stable Canned Luncheon Meat[a,b,c]

Percent salt in water phase	Sodium nitrite input level (ppm)		
	0	75	150
4.9	10	9	0
5.5	8	6	0
5.8	4	0	0
6.1	0	0	0

[a] From Pivnick *et al.* (1969).
[b] 10^6 spores per gram; processed to $F_0 = 0.57$; incubated 18 months at 30°C.
[c] Values given as cans toxic/10 cans incubated.

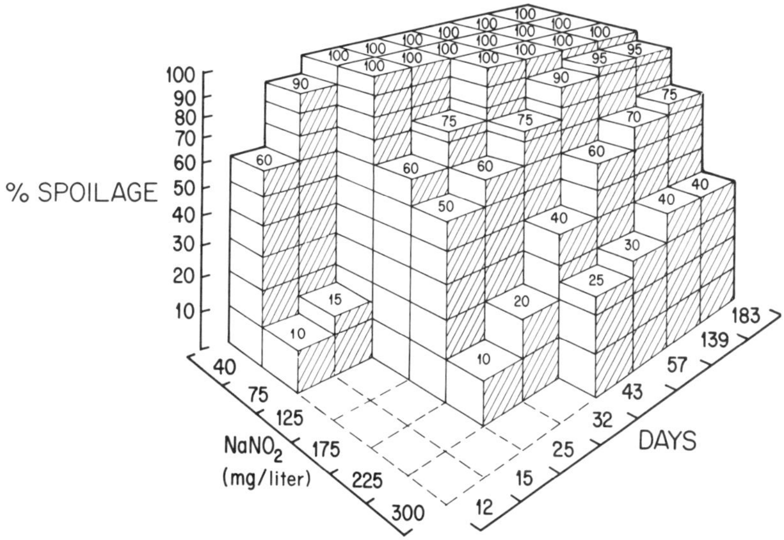

Fig. 8.3 The effect of nitrite on time to spoilage at 20°C in pork slurry containing 3.5% sodium chloride and inoculated with *Clostridium botulinum*. The percentage of containers with spoilage is given at the top of each column of blocks for nitrite concentration and incubation periods specified, e.g., 25% spoilage when slurry containing 300 mg $NaNO_2$ per liter (i.e., 300 ppm) was incubated 43 days. (From Roberts *et al.*, 1976. Copyright by Blackwell Scientific Publications, Ltd.)

prediction of stability or safety against growth of *C. botulinum* when salt, nitrite, processing temperature, storage temperature, or other factors are varied.

Because there are scientific and political pressures to reduce nitrite usage in cured meats, two types of canned products will be discussed more fully to illustrate interactions. They are pasteurized canned ham, which should be refrigerated but frequently is not, and shelf-stable canned cured meats which do not require refrigerated storage. These products contain about the same concentrations of salt and nitrite.

Pasteurization kills essentially no spores, whereas heating to obtain shelf stability kills most spores and those that survive are damaged. Many of the spores that survive either process will germinate within a few days depending on temperature of storage. In the refrigerator, germination is minimal and delayed; at room temperature, it is usually rapid with over 99.9% germination occurring within 2 weeks (Tompkin *et al.*, 1978c; Christiansen *et al.*, 1978; Pivnick *et al.*, 1969). On the other hand, some spores may remain dormant in either type of product for months; but they grow readily when inoculated into microbiological media. The reasons for this prolonged dormancy are not known.

In large pasteurized canned hams the heating is so low that it is unlikely that many clostridial spores are destroyed, or even damaged. The safety system consists of a relatively high residual nitrite level (80–100 ppm) probably potentiated by isoascorbate (Tompkin *et al.*, 1978d, 1979b), salt, and especially refrigeration. However, much pasteurized canned ham is inadvertently held without refrigeration. At room temperature there is a race between decline of nitrite and decline of spores; and when the nitrite declines below about 30 ppm, some of the germinated botulinal spores can initiate growth and produce toxin (Christiansen *et al.*, 1978). When the product is held continuously at refrigerated temperatures, both nitrite concentration and numbers of spores decline, but the nitrite declines to subinhibitory levels before the spore population has died out. If such pasteurized canned hams are then placed at ambient temperature, growth of *C. botulinum* may occur; the longer they have been stored in the refrigerator, the more rapid is growth of any viable *C. botulinum* after removal to ambient temperature (Tompkin *et al.*, 1978c). Pasteurized canned hams have an excellent safety record over many decades; but because *C. botulinum* is not uncommon in pork (Roberts and Smart, 1976, 1977), and we know that it can form toxin in unrefrigerated pasteurized canned hams, the levels of curing salts in such products should not be reduced.

Although the shelf-stable canned cured meats are similar in salt and nitrite content to the pasteurized product, the more severe heat process (to 100°–108°C) results in a completely different system for stability and safety. The heat process destroys about 75–80% of the nitrite, commonly leaving postheating concentrations of 20–40 ppm; but this more severe heating probably produces more PTF (Lee *et al.*, 1978a). Additionally, more spores are destroyed than during pasteurization, and the few remaining viable spores are heat-damaged, making them less able to initiate growth in the presence of salt (Roberts and Ingram, 1966; Pivnick, 1970; Pivnick and Thacker, 1970). Despite the more severe heat treatment, spoilage occasionally occurs. Putrefactive spoilage by clostridia is rare, because spores of putrefactive anaerobes are not numerous in raw meat. If putrefactive spoilage occurs, the nonmeat ingredients or possibly the machinery for mixing, transporting, and pumping should be investigated as sources of contamination. Strains of *Bacillus* whose spores are unusually heat resistant are most commonly the cause of spoilage of shelf-stable canned cured meats.

C. Defects in Curing

Color defects of cured meats may be caused by a variety of factors, both nonbacterial and bacterial. Nonbacterial defects are principally due to undercuring or fading. The undercuring is manifested as a brown grayish-green color due to inadequate reaction between nitrite and meat pigments prior to heating; it usually results from inadequate distribution of pickle due to pumping only partially defrosted cuts, from insufficient diffusion into large cuts, or from inadequate mixing in comminuted product. Fading of the cured meat pigment is due to oxidation. Fading is accelerated by light and inhibited by ascorbate or its derivatives.

The microbe-mediated defects are of four types: nitrite burn, surface greening, green core, and green ring. *Nitrite burn,* a brown-green discoloration, is caused by excessive nitrite especially in products of low pH. In fermented sausages which depend on acid-sensitive nitrate-reducing bacteria to reduce nitrate to nitrite and on acid-producing bacteria to terminate their reducing activity, a lag in acid production may result in production of excessive amounts of nitrite. Subsequent fermentation that reduces the pH to 4.5–5.0 completes the conditions required for nitrite burn (Deibel and Evans, 1957). The replacement of nitrate with suitable, but not excessive, concentrations of nitrite (i.e., $\leq$ 200 ppm of nitrate + nitrite) is useful in preventing bacteria-mediated nitrite burn (Bacus and Deibel, 1972). Nitrite burn may also result from direct addition of excessive nitrite: in such cases, microbes are not involved in producing the defect.

Surface greening is due to the oxidation of the red pigment of cured meat to green porphyrin ring fragments by hydrogen peroxide and does not occur unless oxygen is present. Lactic acid bacteria may grow to large numbers in cured meat under anaerobic conditions, but do not produce peroxide in the absence of oxygen. However, opening a package of product with large numbers of such bacteria results, within a few hours, in production of sufficient peroxide to cause surface greening. This type of spoilage is caused by some, but not all lactic acid-producing bacteria. Once established in a plant these ''greening'' organisms are spread by equipment used in the postcooking procedures, e.g., slicers and peelers.

Green core is also caused by specific lactic acid bacteria that produce hydrogen peroxide, but the discoloration is limited to the core. These bacteria may be somewhat more heat resistant than other lactics and survive the normal thermal process in the core. The practice of reworking, i.e., mixing in small amounts of such defective returned product, may introduce the heat-resistant, green-core producing inoculum into subsequent batches.

Green ring is a special case of green core: the greening appears as a ring between the center and the surface of the sausage. Outside the ring, the causative organisms are killed by the thermal process while inside the ring the causative organisms survive and grow, but sufficient oxygen has not diffused to permit production of hydrogen peroxide. Thus, the ring delineates an ecosystem where both living organisms and sufficient oxygen are present to produce peroxide (Evans and Niven, 1951; Shank and Lundquist, 1963).

D. Public Health Aspects

1. Toxicity of Nitrites

Nitrite consumed in large quantities may cause acute toxicity and even fatalities although small quantities may be consumed daily without effect (WHO, 1977). To prevent inadvertent addition of large quantities of nitrite (it may be mistaken for sodium chloride), some countries require that nitrite may be used only as a food additive when mixed with salt, usually to a concentration of 0.5–0.6%. Thus, addition of the mixture to obtain 2.5% salt in a meat product would furnish 125–150 ppm of sodium nitrite. This practice has one major disadvantage: a reduction in the concentration of salt would result in a reduction in the concentration of nitrite. To maintain safety against bacteria, one should be increased if the other is decreased.

The presence of nitrates, nitrites, and nitrosamines in food has been reviewed (WHO, 1977).

2. Pathogenic Microorganisms

Many pathogens which may be found in meat are capable of causing disease in both animals and man (see Chapter 15). Thus, a food that is to be cured will frequently contain some pathogens. During the curing and associated processing and handling the pathogens may either be killed, heat-damaged but not killed, reduced in number, kept at nonhazardous levels, or they may grow.

Staphylococcus aureus can grow in media containing up to 18% salt; production of enterotoxin B in the favorable pH range 4.8–8.0 is inhibited by about 10% salt (Genigeorgis, 1974), but enterotoxin A can be produced in much higher concentrations of salt (Lotter and Leistner, 1978). Most cured meats contain only 3–6% salt in the water phase, an environment suitable for toxinogenesis by *S.*

aureus if sufficient oxygen is available. Fortunately, *S. aureus* is readily killed by heat, and even in uncooked products such as fermented sausages, the anaerobic interior presents an ecosystem that is inhibitory to growth and toxinogenesis, despite the frequency of *S. aureus* in such products (Pullen and Genigeorgis, 1977; Task Force, 1977). The pathogenic fungus *Sporotrichum schenckii* has been found in frankfurters; it grows well at 5°C, but not in the ecosystem created by modern packaging (Scharding *et al.*, 1973). Salmonellae are not inhibited by the salt, nitrite, or pH of many cured products (Davidson and Webb, 1972). Although readily destroyed during cooking, they may persist in uncooked, fermented sausages or grow in canned cured meat contaminated by leakage after thermal processing (Meers and Goode, 1965). *Clostridium perfringens* may contaminate pickling brines and, when thus introduced into hams, may persist through the curing and smoking process (Gough and Alford, 1965). *Clostridium botulinum,* especially type B, is sometimes found in raw and in cured meats, and can grow in mildly cured meats that are held at temperatures above 15°C. However, commercially prepared products have an exceptionally good safety record. Although *C. botulinum* type E can grow at 3°C, it is fortunately not found in meat, and it is readily inhibited by as little as 3.5–4.0% of salt in the water phase. *Trichinella spiralis,* occasionally present in pork in some countries, more commonly in others, may persist for months in cured, uncooked products. It is destroyed by cooking or by freezing. In some countries, cured pork that is to be consumed without cooking must be made from frozen meat. Although most infectious pathogens and the toxins of *C. botulinum* are readily destroyed by moderate temperatures (60°–80°C), many cured products are not heated by the consumer. Fortunately, the safety factors introduced by the curing salts, the selective growth of lactic acid bacteria which lower pH, and the use of refrigeration minimize the probability of growth of pathogens.

Curing usually changes the food to prevent spoilage or to change the type of spoilage. Thus, the noncured foods such as fish or meat traditionally spoil with offensive odors due to growth of *Pseudomonas* and other gram-negative psychrotolerant species. In contrast, some cured products, depending on packaging and temperature, frequently spoil without repulsive odor. Indeed, growth of pathogens to dangerous levels may occur long before spoilage is sufficiently offensive to warn the consumer. It is this period between establishment of a hazardous condition and spoilage sufficient to prevent consumption that is termed "the period at risk." The period at risk ranges from less than a day to many weeks. In general, it increases in those subinhibitory situations that prevent pronounced evidence of spoilage, but still allow growth of pathogens, e.g., exclusion of air, subinhibitory conditions of pH and temperature, and concentrations of salt. Smoke and spice may increase the period at risk by masking offensive odors. Even experts in organoleptic analysis may be unable to detect spoilage in cured products that have become toxic due to *C. botulinum* (Table

8.4) (Pivnick and Bird, 1965; Pivnick *et al.*, 1967; Greenberg *et al.*, 1959). Changing conditions of curing, packaging, and distribution without concern for the period at risk is not advisable.

Salt has been considered with respect to a_w (Chapter 4). A few aspects of safety, relative to curing, will be considered here. Although salt was once the most important component of many cured foods with respect to safety, today this is true for only a few products. Moreover, its concentration may be widely different among products and even variable within a single product. This variability may result from three causes: (1) decrease in concentration of salt in a traditional product even though the salt is uniformly distributed; (2) differences in practices among countries; and (3) uneven penetration and distribution of salt in noncomminuted products. For example, shelf-stable comminuted cured meats contained 3.5% salt (about 6% in the water phase) as recently as 35 years ago (Vinton *et al.*, 1947). Today, the concentration is usually 2.5% (about 4.5% in the water phase), which is much less inhibitory to heat-damaged spores that may survive the thermal process (Roberts and Ingram, 1966; Pivnick and Thacker, 1970). Salt concentrations in the same type of product vary from country to country. In pasteurized canned hams, 3.5% salt (5.0% in the water phase) is recommended in the United Kingdom (Department of Health and Social Security, undated), but in North America, the average salt content of this product can range among manufacturers from 2.0 to 2.5%. Moreover, there is uneven distribution of salt in noncomminuted meats: the range between containers in a single lot may vary from 2.0 to 3.0%. Thus, attempts to estimate the contributions of salt to safety, except in comminuted products that have an even distribution of both salt and nitrite, become almost useless.

Regardless of the intricacies of the safety system, it is useful to obtain an estimate of the safety of the cured product in relationship to the most probable or the most serious hazard that may result from consumption of the product. And to do this, the probable abuse of the product, the potential for offensive spoilage, and the manner of use by the consumer must be considered. Such estimates are especially important when any one of the components of the safety system is to be altered.

The consequences of changing the safety system of bacon and pasteurized canned ham are considered in the context of, for example, drastically decreasing nitrite. In North America bacon is rarely stored without refrigeration and rarely eaten without cooking because of the prevalence of *Trichinella spiralis* in raw pork. Cooking quickly destroys botulinum toxin (Collins-Thompson *et al.*, 1974; Losikoff, 1978). For these reasons, it is unlikely that more than one package of nitrite-free bacon in 10^{12} would cause botulism (Hauschild, 1979). In contrast, pasteurized canned ham is frequently stored without refrigeration and rarely heated by the consumer. Thus, the danger of botulism from bacon made with a decreased concentration of nitrite is probably miniscule compared with the danger from pasteurized canned ham made with decreased nitrite.

Additional examples such as the prevalence of *Salmonella* in poultry and pork that may be used for fermented, unheated sausages make it imperative to consider the probability of adverse consequences. Such considerations are important when new products, materials, methods of preparation, and changes in concentrations of curing salts are to be introduced. The probabilities are best estimated from studies of inoculated product, processed in the usual way, and stored at abusive temperatures.

Concepts of safety for shelf-stable canned products are based on destruction of 10^{12} spores of *Clostridium botulinum* (12*D* process; see Chapter 1, Section IV,B) in products with pH greater than 4.6, or inhibition of an infinite number of such spores in products with pH at or below 4.6. However, most canned shelf-stable cured meats do not receive a 12*D* thermal process (F_0 ca. 3.0); in fact, most receive a process of $F_0 = 0.2–0.6$ (ca. 1.0–2.5*D*) (Greenberg *et al.*, 1965; Riemann, 1966) and some shelf-stable products are processed with even lower heat treatments (Silliker *et al.*, 1958; Riemann, 1967b). Despite the fact that they are subjected to processes of about $F_0 = 0.2–0.6$, shelf-stable comminuted cured meats have about the same degree of protection against *C. botulinum* as products without curing salts that receive a thermal process of about $F_0 = 1.5$ (Pivnick *et al.*, 1969). The additional safety is due to heat damage of spores that are not killed and the inhibitory activity of salt, nitrite, and PTF against surviving damaged and undamaged spores. The inhibitory activity has been measured in pseudo-*D* values with one pseudo-*D* value equivalent to prevention of outgrowth of 1 $\log_{10}$ of surviving spores (Ingram and Roberts, 1971).

A formula for quantitating the safety of shelf-stable cured meat has been proposed (Pivnick and Petrasovits, 1973):

$$\mathrm{Pr} = \mathrm{Ds} + \mathrm{In}$$

where Pr = Protection, Ds = destruction due to the heat process, and In = inhibition by curing salts of the spores that survive the heat process. Pr, Ds, and In are measured in units of $\log_{10}$ of the number of spores of *C. botulinum*. In canned, cured shelf-stable comminuted meats made with 150 ppm sodium nitrite, 5% salt in the water phase, and heated to $F_0 = 0.6$, the protection against *C. botulinum* spores probably approaches the following: Pr = 9, Ds = 4, and In = 5. Nitrite plus PTF contribute about 1.5 units of In, and salt, about 3.5 units (Pivnick and Petrasovits, 1973). In an unheated product with 10% salt in the water phase or with a pH < 4.6, the protection against *C. botulinum* could be expressed by Pr = ∞, Ds = 0, and In = ∞. The formula Pr = Ds + In may also be used generally in food microbiology for estimating protection against sporeforming bacteria other than *C. botulinum,* and even against nonsporeformers.

Formulas for predicting the number of *Staphylococcus aureus* required to initiate growth with inhibiting agents pH and salt interacting in laboratory media have been developed (Genigeorgis *et al.*, 1971a), but the number of cells required to initiate growth in meat under the same conditions of pH and salt is considerably less than in media (Genigeorgis *et al.*, 1971b).

Increasing the number of cells increases the probability that one or more cells in the population will have the physiological attributes required to initiate growth in less than ideal conditions. In many cases, the changes in the safety system required to reduce the probability of outgrowth from 1 to 1×10^{-6} may be very small. Many examples of this concept as applied to major foodborne disease hazards have been reported for bacteriological media (e.g., Riemann, 1966). However, without thoughtful limitation it is risky to use studies in bacteriological media to predict the efficacy of curing salts in food (Anonymous, 1973).

III. CONTROLS

A. The Safety System—Interactions of Bacterial Numbers, Curing Agents, and Processing Conditions

The preservation and safety of a cured product is the end result of interaction of many factors in a system designed to prepare and deliver an attractive, safe product. The system is delicately balanced and any change in one or more of the factors may decrease or increase the probability that the product will remain organoleptically acceptable and safe for consumption.

The main components of the system are the curing salts; the microbial content of the raw product; the type, temperature, and duration of processing; the types and numbers of microbes that survive the process or that gain entry after processing; and, most important, the conditions that determine whether or not these microorganisms can grow in the processed product (i.e., the pH and E_h of the product, the continuously decreasing concentration of nitrite, the type of packaging, and the temperature and duration of storage). Even adequate labeling may contribute to the safety system. However, labeling that states a product is best if purchased by the consumer before a specific date may create potentially hazardous situations for some cured foods unless the required temperature of storage is also boldly and clearly stated.

B. Storage Temperature

Because they are less prone to odoriferous spoilage than uncured products, cured meat products are more likely to be held at unsafe temperatures by vendors and purchasers. Occasionally, even the manufacturer may not realize the implications of improper temperatures. Packages containing pasteurized products are usually labeled "keep refrigerated" if the pH or a_w of the product are within the range permitting microbial growth. However, there remains a need for an established system that relates the interacting influences of pH and a_w to the need for different degrees of refrigeration. Table 4.4 illustrates one such system.

C. Control of Nitrite Levels

1. Regulations for Nitrites and Nitrates

The use of curing salts, particularly nitrite and nitrate, is usually regulated. In general, about 150–200 ppm of $NaNO_2$ may be added to most products and about 200–500 ppm of $NaNO_3$ (or equivalent amounts of K salts) may be added to some products. However, some countries do not regulate levels and others permit considerably less than those indicated above (Meester, 1973). Because of intense current interest in limiting additives in foods, the permitted levels of nitrite and nitrate are continually reviewed and have been reduced in many countries.

2. Products without Nitrite

Both frankfurters (MacNeil and Mast, 1973) and bacon (Wasserman and Kimoto, 1976) made without nitrite have been considered organoleptically acceptable in limited tests; and a ban on the use of nitrate and nitrite in some cured meat products in Norway has not led to sustained reduction in sales (Høyem, 1976). However, in view of the demonstrated antibotulinal activity of adequate levels of nitrite, decreases in nitrite are best preceded by adequate experimentation to determine the probability of food poisoning resulting from such decrease.

D. Laboratory and Production Controls

Some products, especially shelf-stable canned meats should be routinely tested for levels of nitrite and salt; fermented salamis that do not require refrigeration should be tested for pH, and questionable lots tested for heat-stable desoxyribonuclease or for salmonellae. They should also be tested during fermentation for viable cells of *S. aureus* (Task Force, 1977), because the cells die out during drying while any toxin that is formed will remain (see also Chapter 15, Section VIII,B,5).

9

Antibiotics in Foods

I. INTRODUCTION

Antibiotics, produced either by microbes or synthetically, inhibit microbes at high dilution. Antibiotics as "wonder drugs" to combat disease in man and animals need no discussion here. For a time, they were used in foods for inhibiting spoilage bacteria, but for several reasons discussed in this chapter, food control agencies now prohibit their addition to foods. Even small residues in foods carried over from treating animals are now considered unacceptable.

Organisms present in food can become resistant to antibiotics and colonize the gut of animals and man. Residues can also create a selective pressure favoring the resistant flora already present in the bowel. Antibiotics used therapeutically may then become ineffective. One of the most important reasons for the prevalence of resistant bacteria is the indiscriminate use of antimicrobial drugs in man and livestock (WHO, 1976). Antibiotic resistant microbes are unimportant in heat-processed foods because in general they are likely to be killed by the processing. However, they can be present in raw foods and they can cross-contaminate other foods.

II. ANTIBIOTICS AS FOOD PRESERVATIVES

A. General

Following the great therapeutic success of antibiotics in bacterial diseases, it was natural that antibiotics should also be tried as preservatives against microbial food decay. This development occurred in the period between 1945 and 1960, and most antibiotics were tested on many perishable foods in every imaginable way. The enormous literature which resulted from this will not be reviewed (see

Marth, 1966) mainly because few of the original applications are still in use. As fear of antibiotic-resistant microorganisms increased (see Section III), the application of antibiotics as food preservatives diminished. In addition, Ng *et al.* (1957) and Walker and Ayres (1958) observed the development of bacteria resistant to those antibiotics regularly used in plants treating poultry meat. Highly antibiotic-resistant spoilage organisms were selected readily (Vaughn *et al.* 1960). It thus became apparent quite early that regular use of antibiotics might render them ineffective food preservatives because antibiotic-resistant spoilage floras were being selected.

The two principal antibiotics presently used in many countries for food preservation are natamycin and nisin, neither antibiotic having therapeutic application. Tylosin (a macrolid antibiotic), listed in Section III,B among the low risk antibiotics, is still used as a feed-stuff additive. This antibiotic occupies an intermediate position because it had some application, chiefly in Japan, as a food preservative (Amano *et al.*, 1968). Its use appears to have been discontinued.

B. Natamycin*

Natamycin is a macrolide† antibiotic produced by *Streptomyces natalensis*. Its effect is primarily antifungal, and its use is permitted in some countries (WHO, 1969, 1976). *In vitro* and in raw ground peanuts, natamycin inhibits the development of fungi known to produce mycotoxins (Gelda *et al.*, 1974). This is an important advantage and one regarded by some experts as sufficient to offset any misgivings about the use of this antibiotic in food. There are several reports on the use of natamycin to inhibit mold development on sausages. For example, natamycin, at a level above 1000 ppm successfully preserved Dutch sausages against mold. The antibiotic could be added with the brine, or by dips or sprays, and the sausage could be stuffed into different types of casings. This treatment led to a surface concentration of 2 ppm/cm^2 which was considered to have no toxicological significance (Moerman, 1972). Probably the most important application of natamycin is in the treatment of the rind of soft or hard cheese, which may be dipped in or sprayed with suspensions of 500 ppm of natamycin. The antibiotic can be detected in the rind; its penetration varies according to the type of cheese (Gripon and Bergère, 1973). Natamycin may also be used on film-packaged cheese. Its preservative effect can be lost if the antibiotic is applied after mycelium has developed. Some molds (e.g., *Aspergillus flavus*) are strong producers of enzymes which inactivate natamycin.

*Natamycin is the international nonproprietary name for the antibiotic previously called pimaricin.

†Macrolide antibiotics contain a large lactone ring, few double bonds, no nitrogen atoms, and the ring is substituted with one or more sugar residues.

C. Nisin

Nisin is an antibiotic produced by strains of the common milk-souring organism *Streptococcus lactis*. It occurs naturally in soured milk and farmhouse cheese, so it is likely that small amounts have been ingested since cows have been domesticated. The application of nisin as a food preservative has been extensively studied and much of the pertinent literature has been surveyed (Hurst, 1981; Jarvis and Morisetti, 1969). Nisin has application neither in feeds nor in human and veterinary medicine. Organisms that become resistant to it are not known to become resistant to other antibiotics.

Nisin is a polypeptide, readily inactivated by proteolytic enzymes at pH 8, making it likely to be digested. Extensive studies with rats, fed for 2 years with up to 8 mg/kg body weight failed to reveal toxicity. The use of nisin as a food preservative "should be considered acceptable, the unconditional Average Daily Intake being 0–33,000 U/kg of body weight" (WHO, 1969). (There are approximately 40,000,000 units in 1g of pure nisin; approximately 100 to 400 units per gram of food (or 2.5–10 ppm) is recommended for food preservation). Nisin is at present manufactured in the U.S.S.R., Poland, and United Kingdom (Gudkov *et al.*, 1973), and its use is permitted in some foods in about 20 countries (Fowler and McCann, 1972).

Nisin has a narrow spectrum affecting only gram-positive bacteria. Because gram- negative bacteria are not affected, use of nisin as a food preservative cannot cover up unsatisfactory hygiene. Its narrow spectrum and acid stability lead to rather special conditions of application. It can be used effectively only when the spoilage organism, which is nisin sensitive, is virtually the only one in the food. This is the case, for instance, in the clostridial "blowing" of swiss-type cheese. However, for other reasons, this process is seldom used in practice. The antibiotic is also heat-stable, especially under acid conditions. Hence, nisin can be used as an adjunct to heat processing. The heat used can be reduced to that required to destroy *Clostridium botulinum,* which is the most nisin-resistant sporing anaerobe. This reduced heating improves the quality of the food product. In addition, products may have better storage properties (especially at high ambient temperatures). In the combined heat–nisin process, nisin prevents the outgrowth of bacterial spores surviving the heat process. At the concentrations commonly used (100 U/gm), however, nisin does not appear to be sporicidal. Therefore, the process must be so devised that sufficient nisin remains after the heat treatment and during storage to ensure the continued inhibition of spore outgrowth.

When the law permits, nisin is used not only to prevent "blowing" of cans, but also to preserve processed cheese and chocolate milk. Other food industries use it in the preservation of green peas, beans in sauces, and catsups.

III. ANTIBIOTICS FOR GROWTH ENHANCEMENT AND DISEASE CONTROL IN FARM ANIMALS

A. Animals and Man as a Single Ecological System

The common practice of including antibiotics in the diet of farm animals followed the publication of a paper by Stokstad and Jukes (1950), who noted that aureomycin fed at low levels increased the growth rate. The resulting increase in food production averaging about 4–10% depends on a number of factors. Feeding of antibiotics has also permitted the development of intensive and crowded rearing conditions, a characteristic of modern farming practice (Kiser *et al.*, 1971).

Soon concern was expressed about the development of resistance by the microflora to the antibiotics being fed. Jukes (1973) reviewed the effect of antibiotics 23 years after the introduction of widescale feeding and found no evidence for antibiotic toxicity either to farm animals or to man. The live weight gain attributable to low level feeding of antibiotics for extended periods has remained unchanged.

Nevertheless, the increasing incidence of antibiotic-resistant microbes has caused concern. Resistant microbes have been known since the pioneering work of Paul Ehrlich at the turn of the century. These bacteria were thought to become resistant by spontaneous chromosomal mutation or by selection in the presence of the chemotherapeutic agent of the resistant individuals already in the population. The chromosomal type of resistance is specific to one group of related antibiotics.

A new development in the story of bacterial resistance was the discovery of multiple resistance in the late 1950s by Japanese workers; bacteria can become resistant simultaneously to several and unrelated groups of antibiotics (for reviews, see Watanabe, 1963; Falkow, 1975). Multiple resistance is acquired by the transfer, from a resistant microbe to a sensitive one, of a genetic element that carries the resistance information. This genetic element, called a plasmid, exists independently of the bacterial chromosome, and one or several different kinds of plasmids may be present in one microorganism. The type of plasmid concerned with antibiotic resistance has also been called resistance factor (R factor) and a microbe possessing it can be denoted as R^+.

Like the chromosome, the plasmid is composed of deoxyribonucleic acid (DNA) molecules. The larger R plasmids represent 1–2% of the total genetic information of a bacterium. Transfer can occur between members of different species; hence R factors are especially important in enterobacteria where the possibility exists for tranferring antibiotic resistance from a nonpathogenic *Escherichia coli* to a pathogenic *Salmonella*. All genera of enterobacteria can contain R factors. Resistance plasmids also occur in pseudomonads, staphylococci, and streptococci. The possibility exists that such transfer of R factors occurs under natural conditions between bacteria belonging to the same family. It is now generally accepted that the large majority of "resistances" in wild-type bacteria are due to R factors and are not of the chromosomal type. The concept thus emerging is of an ecological system common to man and animals. Recent work by Levy *et al.* (1976a,b) amply confirms these suppositions.

Concern over the transfer of R factor from nonpathogenic to pathogenic organisms and the consequent fear of the failure of antibiotic therapy in man, especially in the case of multiple antibiotic resistance, has led to the establishment of commissions of enquiry. Perhaps the most important ones were the Swann Committee (Anonymous, 1969) in the United Kingdom and the Task Force of the Food and Drug Administration in the United States (U.S. Department of Health, Education and Welfare, 1972). In general the Task Force and the Swann Committee considered the problem from the same points of view, but the recommendations and the resulting legislation in the two countries differed appreciably.

B. Proposals to Break the Ecological Cycle between Animals and Man

The Swann Committee and the Task Force found no evidence that antibiotic residues in products from animals fed subtherapeutic levels of antibiotics are toxic to the consumer. Furthermore, they found no direct evidence that such products contribute to allergic reactions in persons who had been previously sensitized to them (but see Section IV,D below). On the other hand, concern was expressed by both bodies that antibiotic-resistant forms of bacteria harmful to man might appear in animals and pass from them to man.

A WHO report (World Health Organization, 1973) emphasizes that the benefits from feeding low level antibiotics are too great to contemplate their withdrawal from agricultural use. Nevertheless, because antibiotics are indispensable to human and veterinary therapy, this report arranges antibiotics in an order of increasing risk. Those with the lowest risk are suitable for feeding, while those with high risk should not have agricultural application. Following is the classification of antibiotics used for therapy, prophylaxis, or feeding purposes in order of increasing risk with respect to the development of bacterial resistance.

1. Bacitracin, flavophospholipo,* virginiamycin
2. Polymyxin, furans, lincomycin, tylosin, and related macrolides

*This is the nonproprietary name for an antibiotic with the trade name Flavomycin.

3. Erythromycin, spiramycin, oleandomycin
4. Penicillins, tetracyclines
5. Ampicillin, cephalosporins
6. Sulphonamides, streptomycin, neomycin
7. Chloramphenicol

The European Economic Community (EEC) in the 1960s recognized the risk of feeding antibiotics, which may select resistant organisms carrying plasmids, and the use of tetracyclines, streptomycin, and related antibiotics as additives to feed was discouraged. The use of penicillin was forbidden because of the importance of this drug in human and animal medicine. Chloramphenicol, polymyxin, ampicillin, cephalosporins, and sulphonamides have never been used in the EEC countries for nutritional purposes. Certain sulfonamides, however, continue to be used as coccidiostats in poultry. Discussions are still continuing concerning whether the use of macrolides, including lincomycin, should be forbidden in animal disease therapy or in feeds; some antibiotics may be totally excluded for animal use.

C. Significance of Multiple Resistance

1. Occurrence of R factors in Nature, Especially in Escherichia coli

Several surveys show that there is a high incidence of R^+ *E. coli* in healthy humans. Datta (1969) found 52% of fecal specimens positive; Moorhouse (1969) found 71% positive in specimens from healthy infants under 2 years old. Up to 10% of the *E. coli* isolated from sewage in the United Kingdom were R^+ (Smith, 1971). Surveys for multiply resistant organisms (e.g., *Salmonella*) in countries where antibiotic feeding is rarely, if ever, practiced, also show the natural occurrence of such microbes (e.g., Lafaix *et al.*, 1969, in Senegal). A comparison of the R^+ *E. coli* content of meat eaters and vegetarians revealed no difference between the two groups (Guinée *et al.*, 1970). *Escherichia coli* is present and apparently stable in many persons who have never been treated with antibiotics (Anderson *et al.*, 1973). Anderson *et al.* (1975) reported that the autotransferable plasmids isolated from enterobacteria of man and animals had a high degree of DNA homology, showing that these plasmids were in all important respects identical.

2. Occurrence of R^+ Escherichia coli in Food

Babcock *et al.* (1973) in the United States isolated R^+ *E. coli* from 80% of the beef samples they examined. Similar studies on chickens by Kim and Stephens (1972) showed that the majority of the poultry products are contaminated with R^+ *E. coli*.

3. R Factor Transfer and the Persistence of Such Organisms in the Gut

Anderson *et al.* (1973) were unable to demonstrate *in vivo* transfer, and similar results in chickens were reported by Smith (1970). Both teams, however, reported that simultaneous feeding of the appropriate antibiotics prolonged the persistence of the R^+ *E. coli*. The evidence favors the view that R factor transfer does not readily occur *in vivo* and that "foreign" strains of *E. coli* do not readily colonize the intestinal tract (Smith, 1975). In one study, when very large numbers of organisms of an R^- *E. coli* were fed to humans, the organism persisted for 120 days in the absence of antibiotic pressure (Cooke *et al.*, 1972).

Nevertheless, there is more recent and quite conclusive evidence that R factor transfer occurs in animals (Guinée, 1970; Levy *et al.*, 1976a,b). The transfer occurs when the animals receive antibiotic supplemented diet; the antibiotic exerts a selective pressure that favors the establishment of R^+ organisms.

4. Virulence of R^+ Microorganisms

The transfer of R^+ factor occurs readily *in vitro* and can also occur *in vivo*. The transfer occurs more readily from, for instance, *E. coli* to a rough recipient, which is lacking the 0 specific polysaccharide (Wiedemann and Schmidt, 1972) than to the "complete" bacteria. Jarolmen and Kemp (1969) observed that *in vivo* the R factor of *E. coli* was only accepted by rough mutants of the recipient strain of *S. cholera-suis* var. *kuzendorf*. There are many other similar reports which show that the possession of R^+ does not in itself enhance the virulence of an organism. However, the contrary should be emphasized also, namely, the possession of R^+ does not diminish virulence.

5. Conclusions

Organisms possessing multiple resistance to antibiotics occur ubiquitously in nature. R^+ factor transmission does not frequently occur *in vivo*, and there is no evidence that such organisms are more virulent than the R^- strains; nor is it easy to implant "foreign" strains into the gut of an animal or man. To get a significant degree of transfer *in vivo* (*without the use of antibiotics*) requires bacterial strains that donate and receive R factors at very high rates and that colonize the alimentary tract well. These requirements are seldom met under natural conditions (Smith, 1974).

Antibiotic feeding changes this situation. Selective pressure due to antibiotic feeding then favors R^+ organisms, and they could then form an important part of the gut flora. For this reason, it is advisable to restrict the agricultural use of antibiotics to those that do not readily form cross-resistance with therapeutically important antibiotics (see, for example, Section III,B). R factors are discussed in detail by Falkow (1975) and Hahn (1976).

IV. ANTIBIOTIC RESIDUES

Antibiotic residues should not occur in foods. However, because antibiotics are widely used for growth promotion and for disease control in animals, residues sometimes carry over into meats and milk.

A. Sources of Residues

Antibiotics added to feeds to promote growth are used at low levels and thus are not important sources of residues. However, under modern conditions of intensive farming antibiotics are also fed to farm animals for prophylaxis. The World Health Organization (WHO) has judged that a concentration of 20 ppm is adequate for growth promotion, whereas 100–200 ppm is considered a prophylactic level (WHO, 1963, 1969). These higher prophylactic levels may leave residues in meat, milk, or eggs. Because a given antibiotic may be used for both purposes and because growth effects may in part be due to prophylaxis, it is difficult to distinguish between these two uses. The carcasses of animals treated therapeutically with antibiotics are especially likely to contain residues. From the point of view of the food microbiologist the most important consideration arises from antibiotic residues resulting from mastitis treatment in cows. These residues can effect the flora of raw milk and the manufacture of fermented foods, particularly yogurt (see Section IV,C), and can be a cause of allergies in man (see Section IV,D).

B. Tests for Detection

These undesirable effects can be largely avoided if the residue levels are below that of the minimal inhibitory concentration (MIC) for the most sensitive test organism. Microbiological tolerances have been established, therefore, in relation to the MIC values (van Schothorst and Peelen-Knol, 1970). In the United States (Mercer, 1975) and in the WHO report (WHO, 1969) tolerance levels are based on limits of detection. In milk, maximal sensitivity is attained by using lactic starter cultures or *Bacillus subtilis, B. stearothermophilus* or *Micrococcus luteus* (ATCC 9346) as test organisms for penicillin detection. The International Dairy Federation (1970) has published an official method for detecting penicillin in milk. For the control of residues in meat, carcasses are examined. Maximum sensitivity is attained by examining organs in which antibiotics are known to accumulate, e.g., kidney (Huber, 1971).

Detection tests are designed first to find a ''general inhibitor'' and for this, microbiological assays are the most sensitive. Such tests may use cups or disks impregnated with food extract placed on agar seeded with standardized spore suspensions, or the food extract or milk may be placed in cups on the agar

(Kavanagh, 1963; Hausler, 1972; U.S. Department of Health, Education and Welfare, 1974). Zones of inhibition of growth are attributed to antibiotics or other antimicrobial substances. Alternatively, lactic starter cultures may be used for penicillin detection. They are strong reducers of dyes such as methylene blue or resazurin; failure of the culture to reduce the dye implies the presence of an inhibitor.

An inhibitory sustance so detected can then be identified in several ways. Inactivation by a specific enzyme may give a clue, e.g., penicillinase inactivates penicillin (Jarvis and Berridge, 1969). Antibiotics can also be distinguished by using a variety of test organisms known to be sensitive or resistant to certain antibiotics (van Schothorst and Peelen-Knol, 1970). More often, chemical separation techniques need to be used. For example, antibiotics are separated by thin-layer chromatography or electrophoresis. The chromatograms or electrophoresis slabs are not stained as would usually be done in chemical laboratories, but are overlayered with agar medium seeded with a test organism. The antibiotics are then identified by the position of the zones of inhibition (Lightbown and de Rossi, 1965; Langner *et al.,* 1973).

C. Levels in Foods and Effect on Fermented Products

The antibiotic content of milk samples has been examined by many workers, and recently there appears to be a general reduction. A significant decrease in the percentage of raw milk samples containing penicillin occurred in Ontario between 1950 and 1976 (Schiemann, 1976). In Germany about 1% of samples of raw milk may contain inhibitory substances. In several European and North American countries meat inspection procedures now include examination for antibiotic residues. Carcasses are condemned when residues are found. Residues due to feeding antibiotics for growth promotion are negligible and therefore, the occurrence of residues in meat is infrequent and unpredictable (van Schothorst *et al.,* 1973; Takács and Kovacs, 1969; Levetzow and Weise, 1974).

Antibiotic residues cause abnormal fermentations and starter failure in salami and other fermented meat products and in fermented milks and cheeses. Starter failure can permit the development of staphylococci and the production of staphylococcal enterotoxins. Yogurt is probably the most sensitive product; as little as 0.02 IU of penicillin per milliliter of milk can allow penicillin-resistant organisms, which are normally inhibited by the starter culture, to gain dominance, whereupon the flavor and appearance become unacceptable.

D. Allergies

Antibiotics, with the exception of penicillin, in foods are not known to cause allergic responses (Jukes, 1973). In the United States, 1–10% of the population may be allergic to penicillin (Stewart, 1973). Penicillin residues, which may or may not have antibiotic activity, are strongly suspected to be allergenic. The most common vehicle is milk from cows that have been treated for mastitis. Using undiluted milk from a single farm has been reported an important cause of allergics. Serious effects can also be obtained from the bulked milk supply in which the penicillin-containing milk is diluted with uncontaminated milk. The only safe course of action is to withhold such milk from commerce. It is proposed that raw milk should not contain >0.01 IU of penicillin per ml (Olson and Sanders, 1975).

There is little conclusive evidence that milk and milk products are the cause of penicillin allergies and the literature dealing with this subject is sparse. Some episodes are discussed by Olson and Sanders (1975). Others were reported in the United Kingdom (Borrie and Barrett, 1961), in the United States (Wicher *et al.*, 1969), and in France (Gounelle and Szakvary, 1966).

10

Gases as Preservatives

I. INTRODUCTION

For many years various naturally occurring and manufactured gases and vapors have been known to kill or inhibit the growth of microorganisms. A number of these have been studied to determine their potential for increasing the storage life of foods. Of these, only those which have been used in commerical practice—namely, carbon dioxide, ethylene oxide, propylene oxide, sulfur dioxide, and ozone—will be discussed in detail. Table 10.1 lists some of the essential characteristics of these gases with respect to their action on microorganisms and use with foods.

Nitrogen and oxygen are frequently used in the packaging and storage of foods but not directly to inhibit the growth of microorganisms. Liquid nitrogen, when used as a cryogenic agent with foods, indirectly inhibits or kills microorganisms by freezing or chilling or by excluding oxygen but has no direct effect on microbial cells (Huffman, 1974; Scott and Smith, 1971). Oxygen is used in controlled atmosphere storage to maintain the physiological health of fruits and vegetables (Smith, 1959, 1963) and the desired red color of fresh meat (Clark and Lentz, 1973; Georgala and Davidson, 1970), but is not applied to inhibit spoilage microorganisms.

A number of powerful biocidal gases have been found useful for disinfecting hospital rooms, animal housing, and ships holds or as soil fumigants, but these have not been applied to foods. Among these are β-propiolactone, chloropicrin, glycidaldehyde, glutaraldehyde, and peracetic acid. Methyl bromide, while used as an effective fumigant for insect-infested fruits and grains, has not been used specifically to control microorganisms in foods. Formaldehyde is used as a funigant for the decontamination of poultry houses (Ministry of Agriculture, Fisheries and Food, 1970) and eggs for hatching (Graves and MacLaury, 1962; Ministry of Agriculture, Fisheries and Food, 1971), but again not directly to control microorganisms in foods. Formaldehyde unquestionably does have a detrimental effect on microorganisms on meat and fish during smoking, but since heat and drying are also involved, its actual contribution to decontamination is not known.

Recent research shows that carbon monoxide and acetaldehyde have a potential use with foods. One percent carbon monoxide prolongs the color shelf-life of fresh meat and has an inhibitory effect similar to that of carbon dioxide on psychrotrophic bacteria (Clark *et al.*, 1976). Acetadehyde vapors at 0.5% concentration in the atmosphere markedly inhibit yeasts and molds causing postharvest decay in fruits and vegetables (Aharoni and Stadelbacher, 1973; Barkai-Golan and Aharoni, 1976).

II. CARBON DIOXIDE

A. Properties

Carbon dioxide, a gas at all normal temperatures of storage, is colorless, odorless, and noncombustible. It is nontoxic to humans at concentrations below about 10%, but above this level unconsciousness occurs with prolonged exposure, an important consideration where the gas is used in high concentrations in walk-in storage chambers and transportation vehicles. It leaves no toxic residues and presents no serious mechanical problems in application. The gas is usually marketed as a liquid in steel cylinders under about 58 kg/cm^2 (830 lb/in^2) pressure, or in the solid form as Dry Ice. At atmospheric pressure the solid form changes into the gaseous phase without liquefaction, subliming at −78.5°C at standard pressure. Gloves should be used when handling Dry Ice since at −78.5°C momentary skin contact can cause serious frostbite and blisters.

Carbon dioxide dissolves readily in water (1.71 ml CO_2/ml H_2O at 760 mm of pressure and 0°C) and hence in the juices of foods, but the absorption appears to be wholly a physical phenomenon involving no more tightly bound chemicals than carbonic acid. When absorbed in foods the pH is lowered in proportion to the amount of carbonic acid formed and the buffering capacity of the food, but rises again when the CO_2 is driven off by exposure to air or slight warming (Killeffer, 1930).

B. Methods of Measurement

Carbon dioxide in a gas mixture can be measured gravimetrically by passing the gas through a barium hydroxide scrubber and drying and weighing the barium carbonate formed. The gas can be determined in foods or beverages by distilling weighed samples, collecting the distillate in standard hydroxide, and titrating the excess hydroxide with standard acid (Horwitz, 1975). Fairly accurate percent-

TABLE 10.1

Factors Relevant to the Use of Various Gases for the Preservation of Foods

Gas	Action on microorganisms	Usual range of levels used	Main factors affecting antimicrobial activity	Main organisms killed or inhibited in order of susceptibility	Main foods treated	Safety considerations
Carbon dioxide	Mainly biostatic	10–40%	Age of cells Concentration	Molds Gram-negative bacteria Yeasts	Chilled fresh red meats and poultry, fresh fruits and vegetables Mineral water Soft drinks	Concentrations of 10% or higher can cause unconsciousness
Sulfur dioxide	Both biocidal and biostatic	70–3000 ppm	pH <4 Concentration	Molds Yeasts Gram-negative bacteria	Fruit juices Wines Sausages Dried fruits Starch	Suffocating odor Possibly mutagenic at high concentrations

Ethylene oxide	Biocidal	400–1000 ppm	25–50% RH Concentration Temperature	Molds Yeasts Bacterial vegetative cells Bacterial spores	Whole or ground spices	Flammable Explosive Vesicant Eye and lung irritant Toxicity of degradation products (glycol and chlorohydrin)
Propylene oxide	Biocidal	800–2000 ppm	25–50% RH Concentration Temperature	Molds Yeasts Bacterial vegetative cells Bacterial spores	Cocoa Gums Spices Starch Nutmeats	Flammable Less toxic than ethylene oxide but same precaution should be used
Ozone	Biocidal	1–200 ppm	Age of cells Low pH Low temperature 60–80% RH Relative absence of organic matter	Bacteria Molds and yeasts Bacterial spores	Water Surface of stored foods Eggs	Eye, nose, and throat irritant Explosive

ages of CO_2 in gas mixtures in fermentation tubes or in environmental samples can be determined differentially by absorbing the CO_2 from a known volume of the sample with 20–40% sodium or potassium hydroxide and determining the remaining volume of gas. Gas chromatographic analysis is the most accurate, convenient, and modern method of monitoring the concentration in a supply stream or closed environment. One milliliter or even less injected into a gas chromatograph containing a molecular sieve column in conjunction with a thermal conductivity detector will provide a result in about 1 min.

C. Effects of Carbon Dioxide on Microorganisms

1. General Effects

Carbon dioxide kills (Coyne, 1932; Killeffer, 1930), stimulates (Valley and Rettger, 1927; Gladstone *et et al.*, 1935), inhibits (Fränkel, 1889; Valley, 1928; Clark and Lentz, 1969), or has no noticeable effect on the growth of microorganisms (Parekh and Solberg, 1970), depending on the organisms, the concentration of the carbon dioxide, the temperature of incubation, the age of the cells when CO_2 is applied, and the water activity of the food or medium. Of these effects, killing and inhibition are naturally the ones of greatest importance to food processing and preservation.

There is considerable variation in CO_2 sensitivity among genera, species, or strains. For example, some cultures of *Bacillus, Enterobacter, Flavobacterium*, and *Microococcus* are killed by exposure to 100% CO_2 for 4 days at room temperature (Coyne, 1932), while *Proteus* spp. (Haines, 1933), *Clostridium perfringens* (Parekh and Solberg, 1970), and *Lactobacillus* types (Ogilvy and Ayres, 1953) are not noticeably affected, or affected only slightly. Thus, while there is a selective action of CO_2 on microorganisms, most yeasts, molds, and bacteria are inhibited by concentrations from 5 to 50% (v/v gas phase), but are not killed or entirely prevented from growing; lower concentrations are usually without effect or actually stimulate growth (Valley and Rettger, 1927). Inhibition increases nearly linearly with increasing concentration from 5% to about 25–50% depending on the food and microflora involved (Ogilvy and Ayres, 1951a); higher concentrations generally show little or no increased effect (Ogilvy and Ayres, 1951a, 1953; Clark and Lentz, 1969). While the degree of inhibition varies with the organisms and food involved, 10% usually gives about a 50% inhibition on the basis of total counts after a given incubation time (Ledward *et al.*, 1971; Clark and Lentz, 1969).

The inhibitory effect of CO_2 on microorganisms in foods appears to increase with decreasing temperature of storage (Coyne, 1933), as indicated by the relatively large extension of shelf-life of the food as the temperature is lowered. However, when the temperature and CO_2 effects are isolated, the effect of CO_2 is shown to increase as the temperature increases (Fig. 10.1; Clark and Lentz, 1969). The inhibitory effect of CO_2 is manifested by an increased lag phase and an increased generation time during the log phase for both bacteria and fungi (Brown, 1922; Tomkins, 1932). However at concentrations between 5 and 20% the greatest effect is on the lag phase. Application of the gas once the cells become adjusted to the medium, or are in the log phase, results in a substantially reduced effect. For example, with mixed cultures of psychrotrophic strains of *Pseudomonas* and the *Acinetobacter–Moraxella* group, delaying the application of 20% CO_2 until 1 or 2 days after inoculation of meat surfaces resulted in the loss of most of the inhibitory effect (Fig. 10.2; Clark and Lentz, 1969). However, when the CO_2 concentration was increased to 40%, time of application had much less influence. The growth rate under CO_2 does not increase with time, that is, the slope of the gorwth curve does not change (King and Nagel, 1967). This indicates that the metabolic system does not adapt during growth under CO_2 nor is there selection for a more CO_2 resistant organism. Reduction of the water activity of the growth medium increases the inhibitory effect of CO_2 on organisms (Scott, 1957).

The manner in which CO_2 inhibits cell growth is not yet clearly understood. As

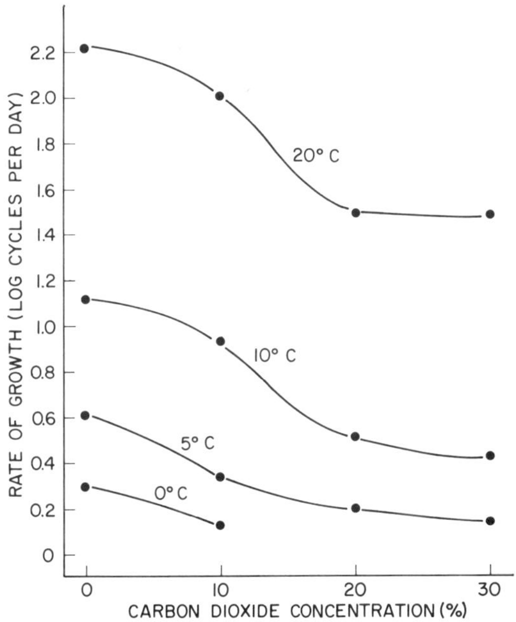

Fig. 10.1 Effect of concentration of carbon dioxide (% v/v) on rate of growth of psychrotrophic bacteria on meat slices during period between inoculation (10^4 cells/cm^2) and development of off-odor (count of 10^8/cm^2 of surface) (From Clark and Lentz, 1969.)

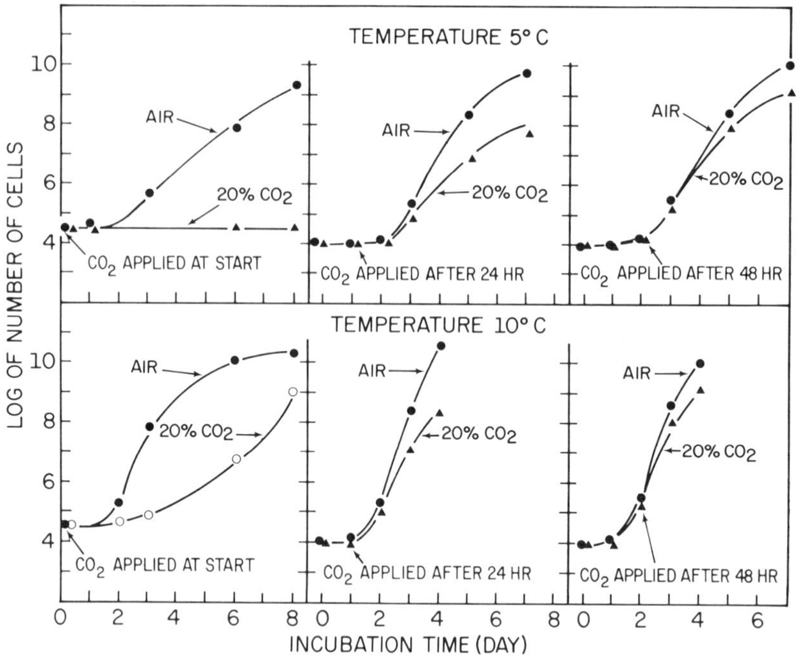

Fig. 10.2 Effect of delaying application of carbon dioxide 24 and 48 hr after inoculation on growth of psychrotrophic bacteria on meat slices. (From Clark and Lentz, 1969.)

early as 1889, however, it was found that inhibition is due to the actual presence of gaseous carbon dioxide and not to the mere absence of oxygen, and that the effect is reversible since the treated organisms will eventually resume normal growth rates when the CO_2 is removed (Fränkel, 1889). Nevertheless, for meat spoilage organisms residual inhibitory effects after removal of the gas have been reported (Silliker *et al.*, 1977).

The direct effect of the gas on cells was confirmed many years later in exacting experiments with *Pseudomonas aeruginosa* (King and Nagel, 1967). Reduction of pH due to formation of carbonic acid in the medium would have a detrimental effect on some organisms, because an atmosphere of 20% CO_2, for example, above unbuffered medium can lower the pH as much as 1 pH unit (6.9 to 5.8; Tomkins, 1932). But experiments with buffered media (King and Nagel, 1967) and naturally buffered foods like meat (Haines, 1933) have shown that pH external to the cell does not explain fully the adverse effect of CO_2. A plausible explanation is interference with certain enzymes involved in terminal metabolism. High concentrations of CO_2 (above 20%) inhibit the metabolism of succinate in *Pseudomonas aeruginosa* (King and Nagel, 1967), *Rhizopus nigricans* (Foster and Davis, 1949), and higher plants (Ranson *et al.*, 1957). Since plant cytochrome oxidase is inhibited by bicarbonate (Miller and Evans, 1956), this may be the component of succinic oxidase which is inhibited by CO_2. Another possible explanation is that CO_2 dehydrates the cell membrane and impedes the passage of water-soluble food materials into the cell (Sears and Eisenberg, 1961). Yet another possible explanation is a mass action effect which interferes with decarboxylation (King and Nagel, 1975). Further work is needed in this area.

2. *Spoilage Organisms*

Concentrations of CO_2 above about 5% inhibit the growth of a broad spectrum of food spoilage bacteria, yeasts and molds, a fact which makes the gas effective in the preservation of food. It is particularly inhibitory to psychrotrophic spoilage organisms that grow on refrigerated poultry (Ogilvy and Ayres, 1951a), beef (Empey and Vickery, 1933; Clark and Lentz, 1969, 1972), pork (Huffman, 1974; Silliker *et al.*, 1977), lamb (Newton *et al.*, 1977), frankfurters (Ogilvy and Ayres, 1953), bacon (Callow, 1934; Hansen, 1960; Cavett, 1962), fish (Makashev, 1959; Killeffer, 1930), eggs (Moran, 1939), and fruits and vegetables (Brown, 1922; Brooks *et al.*, 1932; Montgomery, 1958; Smith, 1959, 1963; Yackel *et al.*, 1971). The degree of inhibition varies widely because the effect is distinctly selective (Coyne, 1933; Ingram, 1962a), but within broad groups certain generalizations can be made. Molds are sensitive (Tomkins, 1932), yeasts comparatively resistant (Ingram, 1958) except for some of the nonfermentative types (Scott, 1938; Ogilvy and Ayres, 1953) and bacteria are highly variable. Among the bacteria gram-negative organisms are generally more sensitive than gram-positive types. Fortunately, the bacteria most affected are the pseudomonads and achromobacteria (*Acinetobacter-Moraxella* group) which grow rapidly and produce off-odors in raw meats. Equally fortunate is the fact that the Lactobacillaceae generally are among the most resistant forms; they do not readily spoil foods and tend to predominate in the microflora of vacuum packaged, cured and raw meat products (Ingram, 1962a; Roth and Clark, 1972; Sutherland *et al.*, 1975a; Patterson and Gibbs, 1977). Micrococci (Coyne, 1933; Ogilvy and Ayres, 1953) and *Bacillus* spp. (Coyne, 1933) are also spoilage types that are strongly inhibited by CO_2.

3. *Pathogens*

The effect of carbon dioxide on foodborne pathogens has not been studied thoroughly, probably because these organisms do not grow, or grow slowly, at normal storage and distribution temperatures for perishable foods (0°–10°C). For example, the minimum temperature for growth in foods of *Salmonella* is about 5.3°C, of *Staphylococcus aureus* 6.7°C, and of *Clostridium perfringens* 12°C (Angelotti *et al.*, 1961; Table 13.1). Also, in perishable foods, like meats, the normal presence of rapid-growing psychrotrophic organisms reduces by competitive inhibition the growth rate of pathogens when temperature abuse does occur. However, the little evidence available shows that pathogens in general survive the combination of low temperature and high CO_2 concentrations. When minimum growth temperatures are exceeded, some pathogens grow normally in the presence of high levels of CO_2 while others are inhibited. For example, *Salmonella* (Shaw and Nicol, 1969), *C. perfringens* (Parekh and Solberg, 1970), and *Brucella abortus* (Valley, 1928) are not affected, while *S. aureus* is inhibited (Hays *et al.*, 1959).

Of potential public health concern is the finding that high numbers of various strains of *Yersinia enterocolitica* develop on vacuum-packed beef and lamb (Hanna *et al.*, 1976). The organism grows on both raw and cooked meats at temperatures as low as 0–1°C (Hanna *et al.*, 1977). Certain serotypes (mainly 0:3, 0:9, 0:17, and 0:5,27) cause acute gastroenteritis or enterocolitis and an assortment of other clinical symptoms. The significance of this CO_2–resistant psychrotrophic organism in vacuum-packaged meats has not been determined.

D. Use of Carbon Dioxide to Control Microorganisms in Foods

Solid CO_2 (Dry Ice) is used extensively in many countries as a refrigerant in the storage and transport of unfrozen foods such as eggs, meats, and poultry, and of frozen foods such as ice cream (Reich, 1945). As Dry Ice sublimes, the gas helps to prevent the growth of psychrotrophic spoilage organisms. Of more direct application, CO_2 gas allowed to accumulate naturally in contained environments of respiring fresh plant and animal produce or added directly to a select environment helps to extend the storage life of millions of tons of fresh meats, fruits, and vegetables and soft drinks throughout the world.

1. Vacuum-Packaged Meats

Refrigerated fresh meats vacuum-packaged in gas-impermeable plastic films keep many times longer than meat stored in air. The reason is the rapid increase in CO_2 that occurs in the gas phase of the package, to 10–20% in 4 hr, reaching a maximum of about 30% (Ledward *et al.*, 1971; Taylor, 1971). There is a concomitant reduction in the oxygen content (down to 1–3%), as a result of intrinsic enzyme activity in the meat. This altered gaseous environment discourages the development of the fast-growing aerobic spoilage bacteria (mainly pseudomonads and the *Acinetobacter–Moraxella* group (Clark and Lentz, 1972) and encourages the development of the slower-growing CO_2-resistant lactobacilli (Roth and Clark, 1972; Ledward *et al.*, 1971; Tändler and Heinz, 1970a,b) which do not normally produce off-flavors or off-odors until after several weeks of storage at 0–5°C.

The practical application of this method of handling has been remarkable. Primal or subprimal cuts of fresh meats are being stored and transported in increasing amounts in vacuum packages (Marriott *et al.*, 1977). For example, over half of the fresh meat in North America is handled in this fashion. In addition to the shelf-life advantages, the method improves sanitation, prevents shrinkage, and reduces retail and handling costs. Aging (tenderization) is not affected (Minks and Stringer, 1972). The method applies generally to all types of fresh meat.

2. *Controlled Atmosphere Storage*

In controlled atmosphere (CA) storage the gaseous environment in food storage rooms, transport vehicles, or gas-impermeable packages is maintained at a desired composition; usually the concentration of CO_2 is increased and that of oxygen decreased relative to air. Large volumes of fresh meats, vegetables, and fruits are stored and transported in such environments in many countries. The specific gas mixtures vary with the product and in some cases are trade secrets.

With refrigerated fresh meats, CO_2 prolongs shelf-life by inhibiting the growth of aerobic spoilage bacteria, mainly pseudomonads and the *Acinetobacter–Moraxella* group. The best concentration is about 20% (Clark and Lentz, 1969); higher concentrations give marginally better results, but at 40% and above cause off-color (Silliker *et al.*, 1977). The use of 50–80% oxygen in conjunction with 20% CO_2 in CA storage increases the color shelf-life and to a slight extent odor shelf-life (Clark and Lentz, 1972); Georgala and Davidson, 1970). This gas mixture is being used extensively in gas-impermeable bags to prolong the shelf-life of whole or ground fresh meat in Denmark and England. The Transfresh Corporation in the United States holds a patent on the Tectrol atmosphere—an atmosphere of 35–75% CO_2, 21–28% oxygen, and the remainder nitrogen (Karmas, 1975). The Techtrol system has been used for the past several years for shipments of fresh meats from the United States to Japan and from the continental United States to Hawaii and Alaska. Recently, carbon monoxide has been shown to give better results than oxygen as a color stabilizer (Clark *et al.*, 1976; El-Badawi *et al.*, 1964).

Carbon dioxide is used to delay senescence and to inhibit the development of decay-causing molds and yeasts in refrigerated fruits and vegetables. It also controls ripening in fruits. The optimum combination of CO_2 concentration and temperature varies greatly with the product and must be worked out for each variety under the prevailing conditions of climate and cultivation (Smith, 1963). For example, some varieties of apples will tolerate only 5% CO_2 while others will tolerate 10%. The first sign of CO_2 damage is off-flavor, followed by discoloration and breakdown of tissues. Generally speaking, concentrations of 5–10% are used.

The microbiological advantage of CO_2 is due in part directly to the inhibitory effect of the gas on microbial growth, and in part indirectly to the maintenance of the physiological health of the plant tissues (Brown, 1922). Fully hydrated tissue is more resistant to invasion by fungus and bacteria than partially dehydrated tissue (Van den Berg and Yang, 1969). Published literature indicates that the indirect effect has the greater influence. Culture studies with the common fruit-rotting organisms *Botrytis, Rhizopus, Cladosporium, Mucor,* and *Penicillium* show that 10% CO_2 is often without effect, while concentrations between 20 and 50% apply a strong check to fungal growth at all normal storage temperatures. Excellent reviews on the use of CO_2 in CA storage of fruits and vegetables have been written by Brooks *et al.* (1932) and Smith (1963).

3. *Carbonated Beverages*

Carbonation of soft drinks (soda and fruit drinks) and mineral waters to levels of 3–5 atmospheres of CO_2 kills or strongly inhibits the growth of spoilage and pathogenic bacteria (Donald *et al.*, 1924; Insalata, 1952; Koser and Skinner, 1922; Sharf, 1960). The microbial content of carbonated beverages normally decreases with the age of the drink, frequently until sterility is attained (Sharf, 1960). Vegetative cells of spoilage and food-borne pathogens are killed within 2 to 6 days depending on the CO_2 pressure and the sugar content (Insalata, 1952). The higher the CO_2 pressure and the lower the sugar content the faster the death rate. Sugar protects the cells marginally, particularly at relatively low pressures of CO_2 (3 atmospheres or less). (See also Chapter 22).

Studies have indicated that carbon dioxide per se causes inhibition and death (Hays *et al.*, 1959; Insalata, 1952) while other work has demonstrated that reduced pH (4–5 atmospheres of CO_2 will reduce pH of a neutral unbuffered liquid to near 3) is the more important factor (Koser and Skinner, 1922). The combined effect of high CO_2 pressure and low pH is the most plausible explanation, because CO_2 is more effective on organisms outside their normal range of growth, increased atmospheric pressure itself is not a factor (Hays *et al.*, 1959).

III. SULFUR DIOXIDE

A. Properties

Sulfur dioxide (SO_2) is a colorless, nonflammable gas with a suffocating odor. It condenses to a colorless liquid at −10°C, solidifies at −72°C, and is soluble in water (85% at 25°C) and organic solvents. It is applied to foods and beverages as the liquefied gas or, more commonly, in the form of sulfite, bisulfite, or metabisulfite salts. Sulfur dioxide and these salts set up a pH-dependent equilibrium mixture when dissolved in water (see Fig. 10.3). As the pH falls the proportion of sulfite ions in the mixture is decreased and the proportion of SO_2 increases at the expense of bisulfite ions. This is important in relation to the antimicrobial activity of SO_2 (see Section C).

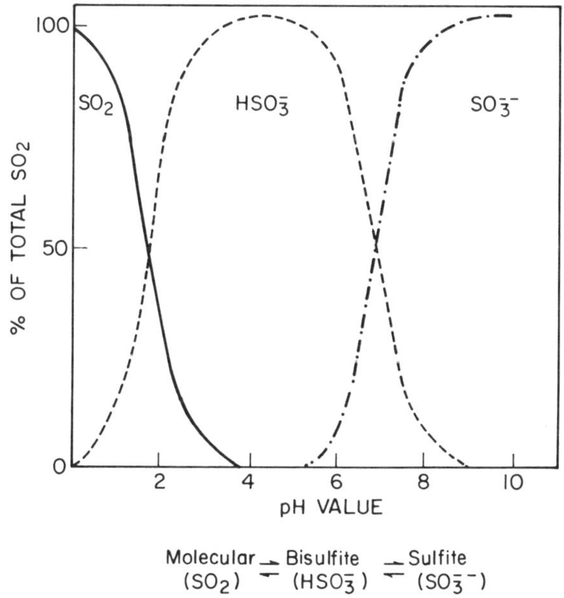

Fig. 10.3 Percentage distribution of sulfite, bisulfite, and molecular SO_2 as a function of pH, aqueous solution. (From Hammond and Carr, 1976.)

Sulfur dioxide reacts with many organic compounds in aqueous solution to form so-called sulfite addition compounds. Reactions take place with aldehydes, ketones, olefins, sugars, organic acids, thiol groups, enzymes, cofactors, vitamins, amino acids, nucleic acids, and lipids. The involvement of some of these reactions in destroying microorganisms will be dealt with in Section C. Hammond and Carr (1976) have reviewed the chemical activity of SO_2 in fruit beverages and in microbial cells.

B. Methods of Measurement

Free SO_2 can be determined by acidifying with sulfuric acid, expelling air with sodium carbonate, and titrating the sulfite ion with standard iodine (Horwitz, 1975). For most foods and beverages the reference method for total sulfurous acid is usually the Monier–Williams method (Horwitz, 1975) where the acidified sample is refluxed and steam-distilled under nitrogen. Gaseous SO_2 in the distillate is converted to sulfuric acid by H_2O_2 and the acid titrated with standard alkali. Alternatively the SO_2 can be determined gravimetrically by adding barium hydroxide to the acidified distillate and drying and weighing the barium sulfate formed. In fruits, SO_2 can be measured colorimetrically using mercurate and rosaniline reagents (Horwitz, 1975).

C. Effects of Sulfur Dioxide on Microorganisms

The antimicrobial activity of SO_2 is related largely to the unbound nonionized molecular form. Therefore, treatments to kill or inhibit microorganisms are most effective at pH <4 (see Fig. 10.4). Sulfur dioxide is >1000 times as active as bisulfite or sulfite ions against *E. coli*, 500 times more effective against yeasts and 100 times more effective against *Aspergillus niger* (Rehm and Wittman, 1962). Like undissociated antiseptic acid molecules in general, SO_2 probably

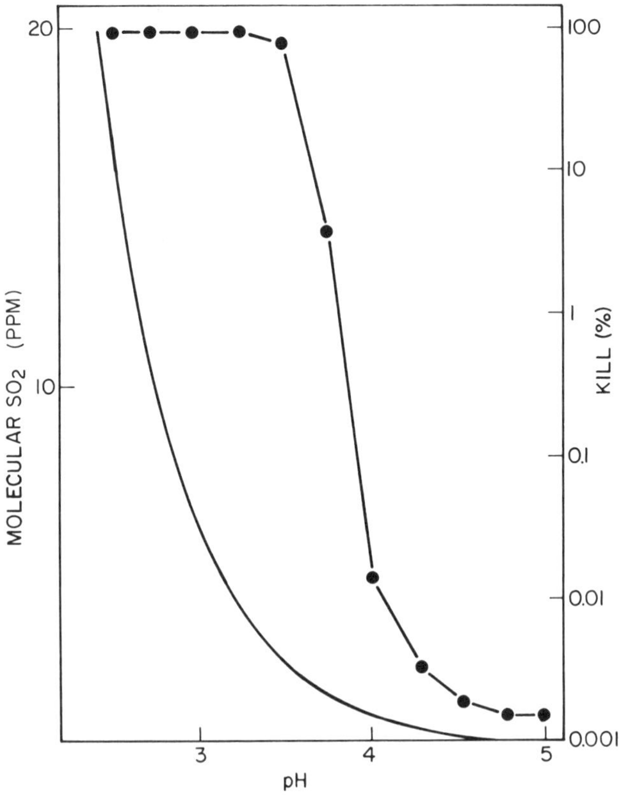

Fig. 10.4 Effect of 100 ppm sodium metabisulfite on the viability of a yeast suspension (8×10^6 organisms/ml) at different pH values (•). Solid line shows the molecular SO_2 concentration. (From Hammond and Carr, 1976.)

penetrates the microbial cell more readily than ionic species (Ingram *et al.*, 1956; Rahn and Conn, 1944).

Because SO_2 is very reactive and probably interacts with many cell components, the precise cause of inhibition or death is not known. Possible cell damage results from interaction with thiol groups in structural proteins, and interaction with enzymes, cofactors, vitamins, nucleic acids, and lipids. Sulfur dioxide cleaves interchain disulfide bonds in proteins and induces changes in the molecular conformation of enzymes, thus modifying the enzyme active site or destroying the coenzyme; it destroys by cleavage the activity of thiamine (vitamin B_1) and thiamine-dependent enzymes and produces possible cytotoxic effects by inducing cross-linking between individual nucleic acid residues or between nucleic acid residues and proteins. It also damages cell metabolism and membrane function by peroxidizing lipids. Death or inhibition could be a combination of these factors or could also result from removal of an essential metabolite from the growth medium (Hammond and Carr, 1976).

The toxic action of SO_2 on microorganisms is selective. Against yeast and fungi, SO_2 is fungicidal even in low concentrations; less than 25 ppm of free SO_2, for example, will kill 10^5 yeast cells in a few hours (Hammond and Carr, 1976). Bacteria are much more resistant; low concentrations (1–2 ppm) appear to be bacteriostatic, and while higher concentrations cause death, the rate of death is relatively slow. The death rate increases with decrease in pH. This selective effect could be due to differential uptake of SO_2 by fungi and bacteria. For bacteria, SO_2 is more inhibitory against gram-negative rods such as *E. coli* and *Pseudomonas* than against gram-positive rods such as lactobacilli (Roberts and McWeeny, 1972).

D. Use of Sulfur Dioxide to Control Microorganisms in Foods

Sulfur dioxide is used extensively for controlling the growth of undesirable microorganisms in soft fruits, fruit juices, wines, sausages, fresh shrimp, acid pickles, and during the extraction of starches. With soft fruits (grapes, raspberries, gooseberries, strawberries, and cherries) it controls the growth of *Botrytis, Cladosporium,* and other molds enabling jam production to be spread over a longer period than would otherwise be possible (Roberts and McWeeny, 1972). It is added to expressed grape juices used for making wines to inhibit molds, bacteria and undesirable yeasts; the concentration added depends on the cleanliness, stage of maturity and general condition of the grapes (Amerine and Joslyn, 1940). Sulfur dioxide at appropriate concentrations interferes neither with the wine yeasts nor with the flavor of the wine. For fruit juices in general, and for starch extraction processes, the chemical is used to prevent fermentation during processing and/or storage. Added as sulfite or meta-bisulfite in sausages, it delays the growth, but does not necessarily reduce the eventual total cell yield, of molds, yeasts, and gram-negative mesophilic bacteria. This is particularly true during chilled storage where sulfite favors the development of acid-producing *Lactobacillus* spp. and *Microbacterium thermosphactum** (Brown, 1977). In pickle manufacture it confers a high resistence to mold growth. For fresh shrimp, sodium bisulfite is used routinely to control "black spot," a discoloration associated with decay (Idyll, 1956; U.S. Department of Health, Education and Welfare, 1978b) (see Chapter 20).

Besides being used to control microbial growth, sulfur dioxide is added to a great variety of foods (Roberts and McWeeny, 1972) as an antioxidant or a reducing agent and to inhibit (a) various enzyme-catalyzed reactions, notably enzymic browning, and (b) nonenzymic browning. For a comprehensive review of the uses of sulfur dioxide and related substances in foods, see a report of the FAO/WHO Expert Committee on Food Additives (FAO/WHO, 1971).

The use of sulfur dioxide in food material is subject to government regulation in many countries, and, therefore, the extent of commerical usage varies. Usually it is not permitted in foods that are recognized as sources of thiamine. Demonstration that SO_2 produces mutations in bacteria and phages (Mukai *et al.*, 1970) has raised safety questions, but animal tests have shown that the "no-effect level" is greater than 70 mg of SO_2 per kg per day. Applying the usual 100-fold safety factor to this value, gives the 0.70 mg/kg/day figure established by FAO/WHO as the Acceptable Daily Intake (about 50 mg/day for a 70 kg body weight). In developed countries, the average daily intake is less than 20 mg (Institute of Food Technologists, 1975).

IV. ETHYLENE OXIDE

A. Properties

Ethylene oxide is a colorless, noncorrosive gas at ordinary temperature, which liquefies at 10.9°C and freezes at −111.3°C. The liquid is readily miscible with water and organic solvents. The gas form has an ether-like odor with a detectable concentration of about 700 ppm. Concentrations over 100 ppm are generally regarded as unsafe for humans; the gas irritates the lungs and eyes and prolonged exposure can cause nausea and mental disorientation. Breathing high concentrations can cause pulmonary edema. It has an additional toxic effect in that it is a vesicant and will cause blisters if its vapors come in contact with the skin for a short period of time. The liquid form, if spilled on the skin, evaporates too

*This organism is considered *incertae sedis* in the eighth Edition of Bergey's Manual (Buchanan and Gibbons, 1974).

quickly to cause harm, but if it is confined by clothing, gloves, or shoes, blistering will occur. Rubber articles in particular should be aired for 24 hr following treatment before being worn.

The main hazard in handling ethylene oxide, however, is not its toxicity but rather its extreme flammability; as little as 3% in air is flammable with explosive violence if confined. Mixtures of 10% ethylene oxide, 90% carbon dioxide (Jones and Kennedy, 1930), or somewhat higher concentration of ethylene oxide in a fluorinated hydrocarbon (Haenni *et al.,* 1959) are nonflammable when mixed with air in any proportion, and are commerically available. Neither the carbon dioxide nor the fluorinated hydrocarbons add or detract from the biocidal activity of the gas. Mixtures of about 90% ethylene oxide and 10% of inert gas can be used safely in vacuum chambers. A mixture of 50% ethylene oxide and 50% methyl formate (commerically available) has several advantages: it lowers the smell detection limit to below 50 ppm, leads to less than half the residual ethylene oxide in the product after treatment and avoids pH and color changes in some treated spices (Mayr and Suhr, 1973).

The gas will penetrate most organic materials including plastics, rubbers, paper, textiles, soil, and powders, and causes little or no damage. It is, therefore, excellent for sterilizing heat-, moisture-, or radiation-sensitive objects like bedding, clothing, books, medical instruments, and supplies and powders such as spices. Sufficient aeration time must be allowed after treatment to permit the gas and its degradation products (ethylene glycol and ethylene chlorohydrin) to dissipate to acceptable levels (Bruch, 1972).

B. Methods of Measurement

Ethylene oxide can be measured by passing the gas through a salt solution containing standardized acid to form ethylene glycol and metallic hydroxides; the amount of acid consumed as determined by titration with standard base is a measure of the initial ethylene oxide content (Phillips and Kay, 1949). It can also be determined colorimetrically by first converting it to formaldehyde (Critchfield and Johnston, 1957). In the past 15 years there has been a shift away from wet chemical determination to the more sensitive and reproducible gas chromatography or gas-liquid chromatography methods (Adler, 1965; Ben-Yehoshua and Krinsky, 1968; Spitz and Weinberger, 1971).

C. Effects of Ethylene Oxide on Microorganisms

In sterilization by ethylene oxide, microbial death follows first-order kinetics (see Fig. 10.5). The time required for sterilization depends on concentration of the gas, temperature, relative humidity (RH), and resistance of the material to penetration. Doubling the concentration, or increasing the temperature 10°C (Figure 10.5) will approximately halve the sterilization time. However, at higher temperatures and concentrations where sterilization times are short these factors become less important and time required to penetrate becomes the limiting factor (Ernst and Shull, 1962). Evacuating the chamber containing the product before introducing the ethylene dioxide mixture greatly speeds penetration. The effect of moisture is more complicated than temperature and concentration and has frequently been a source of trouble (Phillips, 1968). The high relative humidities needed for other gases (essentially 100% for formaldehyde and 60–80% for ozone) are not required for this compound. Ethylene oxide sterilization is most rapid at about 30% RH and becomes progressively slower as the RH increases

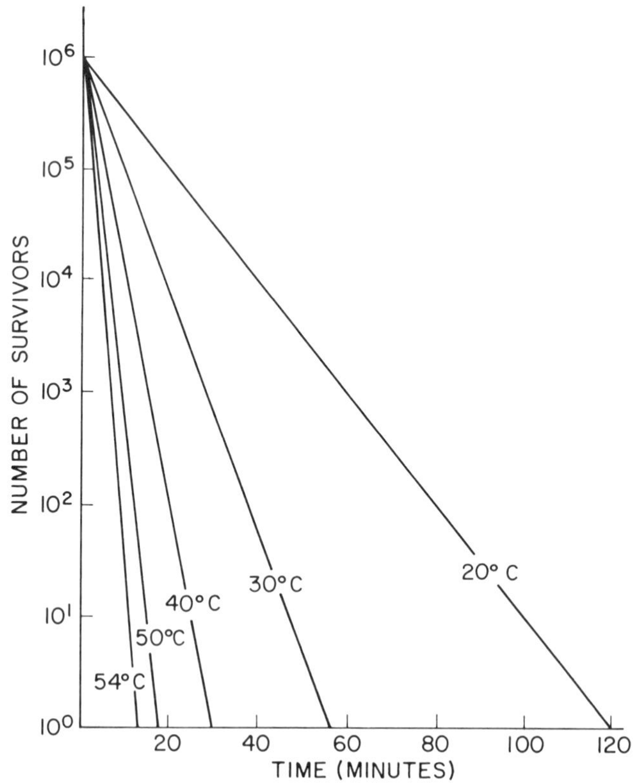

Fig. 10.5 Inactivation rates at various temperatures for *Bacillus subtilis* var. *niger* spores on paper strips in gaseous ethylene oxide at 1200 mg/liter and 40% relative humidity. (From Ernst, 1972.)

to 100% or decreases toward 0%, but there is high activity through the range 25–50%. Some microorganisms, once they have been highly desiccated, acquire a resistance that is not completely overcome when the relative humidity is again raised to 30%. The phenomenon is not well understood but is a problem when one attempts to sterilize lyophylized material or material that has been previously subjected to high vacuum or heat (Phillips, 1968).

The mechanism by which ethylene oxide kills microorganisms has been linked to its chemical activity as an alkylating agent (Phillips, 1968). During alkylation, ethylene oxide replaces labile hydrogen atoms with hydroxy ethyl ($-CH_2CH_2OH$) groups thus blocking many reactive groups needed in essential metabolic reactions. Carboxyl, amino, sulfydryl, and hydroxyl groups can be attacked in this way (Fraenkel-Conrat, 1944). Work with *Salmonella senftenberg, Escherichia coli* (Michael and Stumbo, 1970), and *Clostridium botulinum* (Winarno and Stumbo, 1971) indicates that the lethal action of ethylene oxide on vegetative cells or spores is through alkylation of the guanine and/or adenine components of DNA.

Microorganisms can be grouped into three categories on the basis of their susceptability to ethylene oxide.

1. Molds and yeasts are the least resistant, requiring an exposure of about 1 min to 400 mg of ethylene oxide per liter, at 30% relative humidity and 22°C.
2. Nonspore-forming bacteria and vegetative cells of spore formers are about twice as resistant as yeasts and molds.
3. Bacterial spores are about 10 times as resistant as vegetative cells (Toth, 1959).

Bacillus subtilis var. *niger* is commonly used to monitor the effectiveness of an ethylene oxide sterilization process (Whitbourne and West, 1975). No general distinction is apparent between pathogenic and nonpathogenic microorganisms within these categories.

D. Use of Ethylene Oxide to Control Microorganisms in Foods

Gaseous ethylene oxide has been used widely to reduce microbial contamination and kill insects in various dried foods. Many of these applications have been based on the patents of Griffith and Hall (1938, 1940) who presented extensive data on the treatment of gums and spices. Among the foods that have been treated with ethylene oxide are dried fruits (Whelton *et al.*, 1946), corn, wheat, barley, and potato flour (Pappas and Hall, 1952), corn starch (Pappas and Hall, 1952), dried egg, and gelatin (Mayr and Kaemmerer, 1959; Rauscher *et al.*, 1957). However, when concern arose over the toxicity of the residues of ethylene oxide hydrolysis (ethylene glycol and ethylene chlorohydrin), regulations on the use of the gas were introduced. In the early 1960s it was still permitted as a fumigant for dried fruits, ground spices, dried mushrooms, and edible gums in the United States, but at the present time it is permitted only for whole or ground spices except mixtures containing salt; the residue of the gas in the product must not exceed 50 ppm. Canada and other countries have similar specifications.

It is very unlikely that ethylene oxide will be used to sterilize protein foods, even if the toxic residue problem could be solved, because certain vitamins and essential amino acids are destroyed (Windmueller *et al.*, 1959).

V. PROPYLENE OXIDE

A. Properties

The epoxide propylene oxide is similar in properties to ethylene oxide but is less volatile and less active biologically. It is a very flammable colorless ether-like gas at ordinary temperatures, liquefies at 34.23°C, and freezes at −112.13°C. Even though it is less biologically active, it should be handled as carefully as ethylene oxide. It is less penetrating than ethylene oxide but nonetheless can be used on a similar range of materials without damage and has the advantage that its breakdown product, propylene glycol, is nontoxic.

B. Methods of Measurement

Propylene oxide can be determined chemically and by gas or gas–liquid chromatography as described for ethylene oxide above.

C. Effects of Propylene Oxide on Microorganisms

Propylene oxide has been much less studied than ethylene oxide, but it appears that the general effects of concentration, relative humidity and temperature are similar (Bruch and Koesterer, 1961; Phillips, 1968; Skinner and Hugo, 1976). Bacteria are more resistant than yeasts and molds; bacterial spores are more resistant than vegetative cells. *Staphylococcus aureus* and *Streptococcus faecalis* are among the more resistant nonspore-forming organisms (Himmelfarb *et al.*, 1962). Since the biological activity of propylene oxide is less than that of ethylene oxide, about double the concentration (800–2000 mg/liter compared to 400–1000 mg/liter for ethylene oxide) is used in sterilizing treatments (Bruch, 1961). Since propylene oxide is an alkylating agent, the mechanism by which it kills microorganisms is thought to be the same as that of ethylene oxide (see Section IV,C).

D. Use of Propylene Oxide to Control Microorganisms in Foods

Its use in foods was overshadowed by the more biologically active ethylene oxide until the late 1950s when the use of the latter compound was discouraged because of the toxicity of derivatives. The breakdown product of propylene oxide, propylene glycol, is harmless and is itself a permitted food additive (humectant). Some countries like Canada still do not allow the use of propylene oxide for foods, but others like the United States permit its use for a number of dried foods. In the United States it is used as a fumigant for the control of microorganisms and insects in bulk quantities of cocoa, gums, processed spices, starch, and processed nutmeats (except peanuts) when these foods are to be further processed. Only one treatment at a temperature of 51.7°C or below is permitted and the residue limit is 300 ppm of propylene oxide.

VI. OZONE

A. Properties

Ozone is an unstable, bluish-colored, water-soluble gas with a pleasant, characteristic odor at concentrations below 2 ppm. It is produced continuously in the outer layers of the atmosphere by the action of solar ultraviolet irradiation on oxygen. Ultraviolet sterilizing lamps operate on the same principle. About 0.04 ppm is the upper limit tolerable by humans; in spite of its pleasant odor, 0.1 ppm is objectionable to all normal people becasue of irritation in the nose, throat, and eyes (Witheridge and Yaglou, 1939). The gas is a powerful oxidizing agent which decomposes spontaneously in air and water leaving only oxygen, and therefore is usually generated at the point of application. Solutions containing ozone will explode on warming. The rate of decomposition increases with increase in relative humidity (twice as rapid at 80% RH as at 50% RH) and organic matter. This instability makes it difficult to maintain a fixed concentration of ozone for controlled experiments.

Decomposition is so rapid in the water phase of foods that its antimicrobial action takes place mainly at the surface. It rapidly oxidizes unsaturated compounds leading to rancid flavors in edible fats, oxidizes aldehydes to acids thus lowering pH, oxidizes sulfhydryl groups and amino compounds, coagulates proteins, and inactivates enzymes, particularly catalase, peroxidase, and dehydrogenases.

Ozone is prepared by passing dry air between two plate electrodes connected to an alternating current of several thousand volts. It is also produced by uv lamps at wavelengths of 1750–2100 Å (Nagy, 1959) and is considered to be the cause of part of the germicidal effect of uv-treatment of air and surfaces.

B Methods of Measurement

Ozone in a gas mixture can be measured by bubbling a given volume of the gas through a potassium iodide solution and titrating the freed iodine with standard reducing agent, according to the standard procedure (Rand *et al.*, 1976). The liberated iodine can also be measured electrophotometrically (Scott and Lesher, 1963).

C. Effects of Ozone on Microorganisms

It is difficult to draw conclusions on the relative sensitivities of different microorganisms from the often discrepant results of separate investigators. The problem is that susceptibilitiy is influenced by factors that are not generally standardized, *viz.*, the stage of growth at which ozone is applied, the pH of the medium, the temperature, the humidity (when cells are treated in air), and the amount of organic matter surrounding the cells. Rapidly growing cells are more resistant than resting cells (Kefford, 1948; Ingram and Barnes, 1954). The biocidal effect of ozone appears to increase with reduced pH in the acid range but few organisms have been tested in this regard. Decreasing the temperature increases the effect of ozone, so that treatments that are practical at chill temperatures are ineffectual at room temperature. Ozone has no appreciable disinfecting action on airborne bacteria at humidities of 45% or less but has strong bactericidal action at 60–80%. Organic matter surrounding cells protects them from the destructive action of ozone; thus unwashed cells are more resistant than washed cells, and cells suspended in broth more resistant than airborne cells.

In spite of the above, certain general relationships are evident from the literature. Bacteria are more susceptible than yeasts and molds (Ingram and Barnes, 1954) and bacterial spores are 10 to 15 times more resistant than vegetative cells (Broadwater *et al.*, 1973). Gram-positive cocci are more sensitive than gram-negative bacteria and less so than gram-positive bacilli, but the apparent activity of ozone is highly dependent on the test methods (Baird-Parker and Holbrook, 1971).

Both vegetative cells and spores of bacteria exhibit an all-or-none response Broadwater *et al.*, 1973; Fetner and Ingols, 1959), a phenomenon in which a threshold dose for killing must be attained before any cells die and at which dose the total population is killed in time (see Fig. 10.6). The threshold dose for killing and the rate of death depend in large measure on the amount of organic matter present. Comparatively small concentrations (below 10 ppm) suffice to destroy bacteria suspended in clean water; much larger concentrations (several hundred ppm) are required for nutrient broth or food surfaces. Inhibition of cell multiplication has been noted at levels below those required for killing; the cause of inhibition is thought to be a complex process depending in part on pH changes

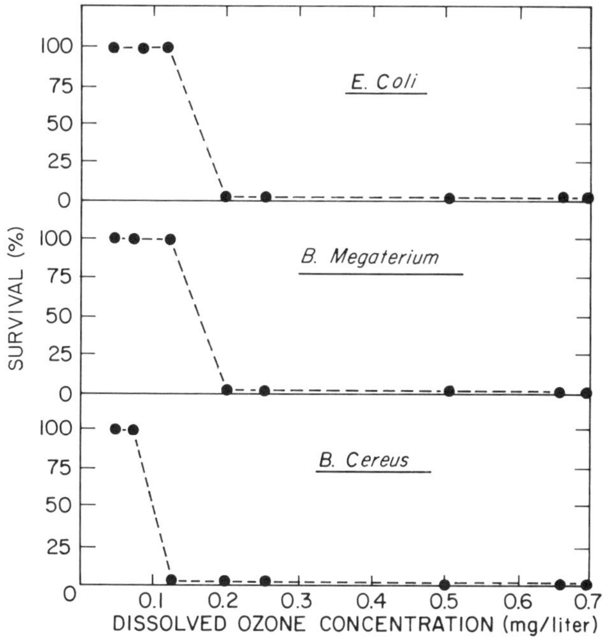

Fig. 10.6 Effect of ozone on washed, vegetative cells of *Escherichia coli, Bacillus megaterium*, and *Bacillus cereus* in deionized water. Points on the figure are averages of three replicate plate counts for a single experiment at each dosage species. (From Broadwater *et al.*, 1973.)

and the secondary effects of decomposition of the medium (Ingram and Barnes, 1954).

Ozone can attack many vital constituents of microbial cells but the precise cause of death is still not clear. General destruction of dehydrogenases could lead to death by interference with cell respiration (Ingram and Haines, 1949), as could oxidation of the sulfhydryl groups, or leakage of vital cellular constituents as a result of oxidation of unsaturated lipid material in the cell wall or membrane (Scott and Lesher, 1963). Possibly several of these destructive oxidation processes may be involved (Ingram and Barnes, 1954).

D. Use of Ozone to Control Microorganisms in Water and Foods

Commerically, ozone has been used mainly for the treatment of water; for example, for sterilization purposes in France (Guinvarch, 1959) and the United States (Bean, 1959) and for removal of iron, manganese, color, or tastes and odors (O'Donovan, 1965). It has also been used as a maturing agent in ciders and wine and to sterilize the interior of soft drink and mineral water bottles prior to filling (Torricelli, 1959). It can be used to preserve eggs at high relative humidity (Frazier, 1958), to sterilize and to destroy botulinal toxins in raw sewage using a 30 min exposure at 100–200 ppm (Miller *et al.*, 1959) and to retard the growth of spoilage organisms on the surfaces of stored foods (Ewell, 1950; Kaess and Weidmann, 1968). Ozone produced by uv lamps provides part of the killing effect of uv treatment of air and surfaces in cold storage rooms.

11

Packaging

I. INTRODUCTION

Packaging preserves the quality of foods and protects against damage during storage, transport, and retail handling. Protection may be classified as follows.

1. *Chemical*. The package may prevent the passage of water vapor, oxygen, or other gases, or may act selectively, permitting only some gases to pass.
2. *Physical*. The package may protect against light, dust and dirt, weight loss, and mechanical damage.
3. *Biological*. The package may prevent entry of microorganisms or insects, or it may affect the mode or rate of spoilage or the survival and growth of pathogens in the food.

Only microbiological aspects of packaging are considered in this chapter.

II. PACKAGING MATERIALS

Containers are either rigid (can, paper, cardboard, glass, plastics) or flexible (plastics, foil). Plastics are increasingly popular (Effenberger and Schotte, 1971; Briston, 1971, 1976). By various combinations of material and processing techniques, it is possible to produce packages having almost any desired functional property.

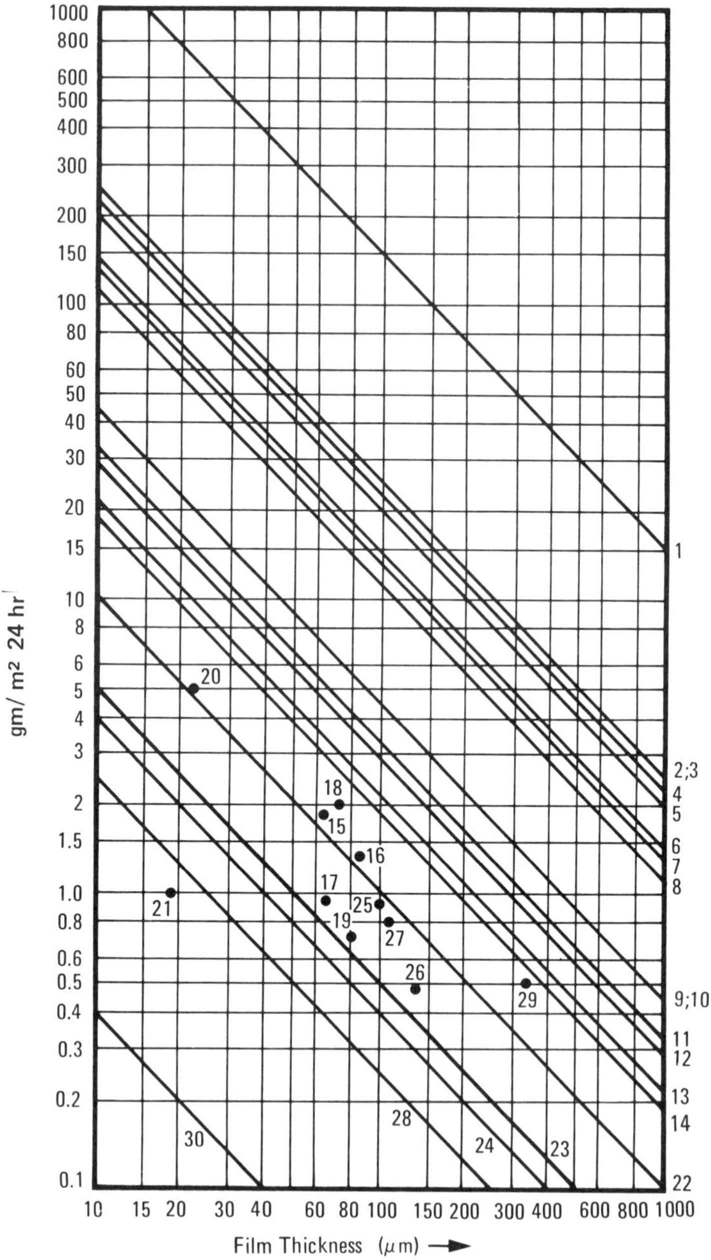

Fig. 11.1 Water vapor permeability of various packaging films at 23°C and 85% relative humidity difference, measured according to Deutsche Industrie Norm (DIN) 53 122. Based on Neitzert

A. Chemical Safety

The components and impurities of packaging material and their adhesives must not endanger human health (Bergner, 1962) and must not react with or adulterate the food.

B. Microbial Safety

Packaging material must not contain any pathogenic microorganisms that may introduce a hazard to the consumer. For example, salmonellae would be unacceptable in packaging material used for a precooked frozen dinner. On the other hand, a few spores of *Clostridium perfringens* in packaging material used for dry spices would be of little concern because such organisms are often found in dry spices anyway. Packaging material should also introduce very few spoilage organisms.

A simple steam jet apparatus has been developed for the decontamination of cans to be used for cooked cured hams (Lerche *et al.*, 1957). Many countries impose bacterial standards for surface contamination of milk bottles and similar containers intended for re-use (Cousins, 1976). A surface contamination with 10 organisms per 100 cm^2 is very clean (von Bockelmann, 1975). In one study on raw paper, *Bacillus* spp. and occasionally micrococci predominated, usually at levels below 200 per gram, but sometimes up to 2800 per gram (von Bockelmann and von Bockelmann, 1974). In another study, molds and yeasts were present at levels up to 10 per 100 cm^2 (Achtzehn, 1964).

In the United States, paper for food containers should have no more than 250 organisms per gram, and containers and seals for milk not more than one per square centimeter (U.S. Department of Health, Education and Welfare, 1966). A standard method has been proposed (Anonymous, 1974a) for the determination of counts of bacteria, molds, yeasts, and coliform organisms on nonabsorbent packaging materials.

(1974 and 1979). These data were taken from the pertinent literature, and are provided as a first comparison to aid the selection of a suitable film material. 1. 3-Acetate. 2. Plasticized polyvinyl chloride (PVC) stretch film (ca. 25% special plasticizer). 3. Acrylonitrile butadiene styrene copolymer (ABS). 4. Styrene acrylonitrile copolymer (SAN). 5. Polyamide (PA) (technical grade copolymer). 6. 6-Polyamide. 7. Polystyrene (PS) (standard). 8. Polystyrene (high impact). 9. Ethylene vinyl acetate copolymer (EVA). 10. Plasticized PVC (22% Plasticizer).* 11. Copolymer vinyl chloride–vinyl acetate. 12. Emulsion PVC (rigid). 13. Suspension PVC (rigid). 14. Polyester. 15. Polyester/polyethylene (PE) 12/50.† 16. Polyester/PE 12/75.† 17. Polyester–Polyvinylidene Chloride–polyethylene (Polyester Pex) 15150.† 18. Cellophane/PE D 35/40 (Cellophane decrease-coated on one side only). 19. Cellophane PE XX 35/60. 20. Cellophane ms 330. 21. Cellophane XS 280. 22. Low-density polyethylene. 23. Polypropylene (PP) (nonoriented). 24. High-density polyethylene. 25. Polyamide/PE 40/60.† 26. Polyamide/PEX 60/75.† 27. Polyamide/PP 40/75. 28. Polypropylene (biaxially oriented). 29. Suspension PVC/PE 250/75.† 30. Polyvinylidene chloride (PVDC).

*Largely dependent on amount and type of plasticizer. †Dependent on density of PE.

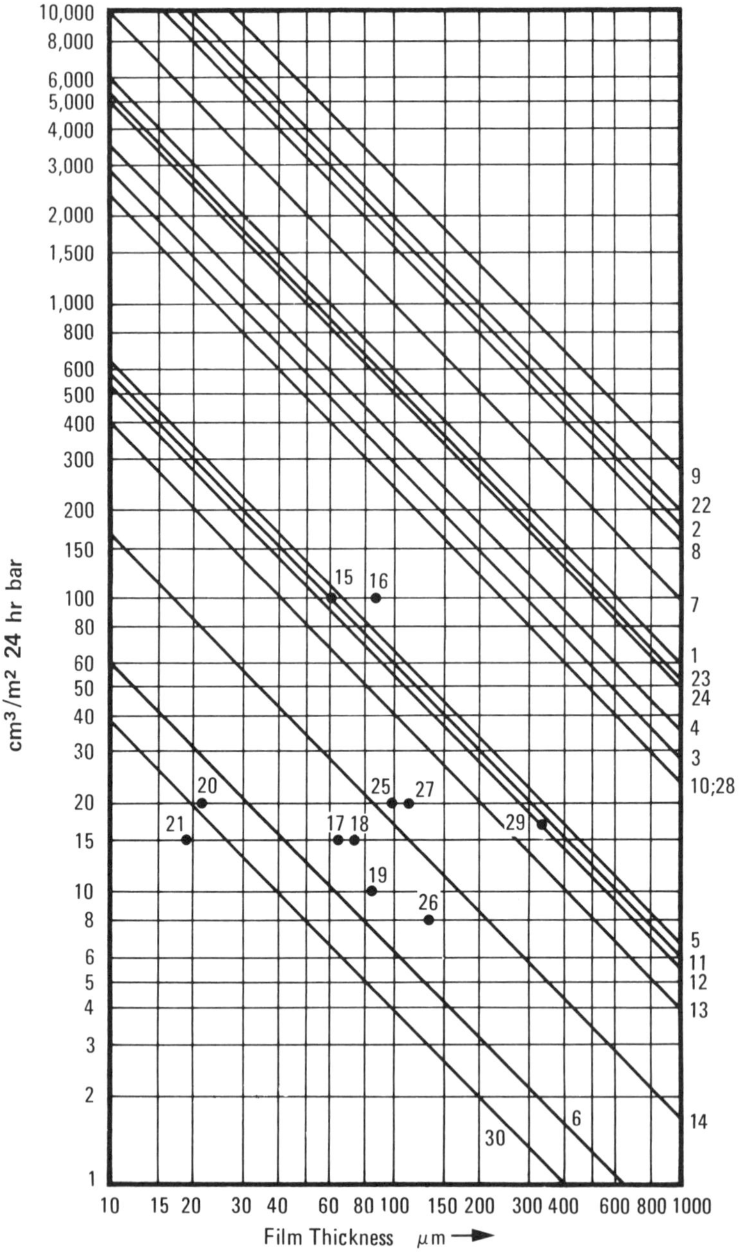

Fig. 11.2 Oxygen permeability of various packaging films at 23°C, measured according to the American Society of Testing Materials (ASTM) 1434–63. Based on Neitzert (1974 and 1979). These

The bacterial levels on the surface of plastic foils used for the packaging of foods are even lower. Plastic tubes and foils contained 1–20 organisms per 1000 cm^2 and usually fewer than 10. Similar values are found on plastic beakers (Hartman *et al.*, 1963). A few organisms remain alive even on plastics extruded at 220°C (Placzek and Witter, 1972; Voss and Moltzen, 1973). The significance of microbial contamination of the packaging material is frequently overemphasized. In general the levels on the product exceed that of the packaging material by several orders of magnitude (Yanai *et al.*, 1969).

Because they are inert to microbial activity, polystyrene or other plastic materials are more hygienic than pressed paper or wood products for packaging eggs (Pfeiffer, 1972), meat (Böhme, 1971), or berries (Ayres and Denisen, 1958).

Certain plastics have antibacterial properties, e.g., those containing alkyd varnishes, phenol resins, polyvinyl chloride, or polyacetal (Gundermann and Glück, 1971). However, before choosing a packaging material for its antimicrobial properties, one should ascertain that the antimicrobial material will not adulterate the food.

C. Permeability

Packaging material should prevent the entry of microorganisms; bottles, cans, and most of the plastic films on the market do so (Ronsivalli *et. al.*, 1966). Penetration takes place if the seal fails, or upon puncture, thus the packaging material must also have adequate mechanical strength to resist damage during processing and subsequent handling. The contents itself can damage the pack: sharp bones in meat and poultry; muscle fibers or pieces of skin in highly dried and smoked products.

A "bio-test" is the best means to determine whether a film will resist bacterial penetration. A sterile sealed package filled with a nutrient medium is immersed into a bath containing a test organism (e.g., *Enterobacter*) or a mixture of

---

data were taken from the pertinent literature, and are provided as a first comparison to aid the selection of a suitable film material. 1. 3-Acetate. 2. Plasticized polyvinyl chloride (PVC) stretch film (ca. 25% special plasticizer). 3. Acrylonitrile butadiene styrene copolymer (ABS). 4. Styrene acrylonitrile copolymer (SAN). 5. Polyamide (PA) (technical grade copolymer). 6. 6-Polyamide (PA6). 7. Polystyrene (standard). 8. Polystyrene, high impact. 9. Ethylene vinyl acetate copolymer (EVA). 10. Plasticized PVC, 22% Plasticizer. 11. Copolymer vinyl chloride–vinyl acetate. 12. Emulsion PVC (rigid). 13. Suspension PVC (rigid). 14. Polyester. 15. Polyester/polyethylene (PE) 12/50. 16. Polyester/PE 12/75. 17. Polyester–Polyvinylidene chloride–polyethylene (Polyester Pex) 15/50. 18. Cellophane/PE D 35/50 (Cellophane decrease-coated on one side only). 19. Cellophane/PE XX 35/60. 20. Cellophane MX 330. 21. Cellophane XS 280. 22. Low-density polyethylene. 23. Polypropylene (PP) (nonoriented). 24. High-density polyethylene. 25. Polyamide/PE 40/60. 26. Polyamide/PEX 60/75. 27. Polyamide/PP 40/75. 28. Polypropylene (biaxially oriented). 29. Suspension PVC/PE 250/75. 30. Polyvinylidene chloride (PVDC).

TABLE 11.1

Plastic Material for Packaging[a]

Compound	Brand	Comments	Thickness (μm)	Permeability at 20°C: Water vapor (gm/m^2 24 hr) at 85% relative humidity	Permeability at 20°C: Oxygen (cm^3/m^2 24 hr bar) at 43% relative humidity
Cellulose hydrate[b]	Cellophane	Casings for sausages, wrapping for ham and raw meat cuts ready for sale	25–100	1–1000[c]	5–500[c]
Cellulose hydrate coated or laminated with polyethylene	Cellophane	Vacuum packs	25–50	1–3	8–15
Polyethylene, low density	Suprathene	Sealable and shrinkable packages for short-term storage	75	1.3	2000
Polypropylene	Trespaphane	Heat-stable containers for sterilized preserves; "stretched shrinking foils" for prepacking of frozen meat	15	1.6	1600
Polyester	Nalophane	Stretched shrinking casings;	20	11	80
	Nalophane NS	casings not stretched; no shrinking	15	6	85

Polyamide					
Nylon 6			25	45	35
Nylon 11	Rilsan; Nalo	Casings for packaging	40	8–11	200
Nylon 12	OK; Tripan		40	8–11	300
Rigid polystyrene		Like rigid polyvinyl chloride, risk of breakage	200	7.5	700
Polyvinylidene chloride copolymer	Saran	Heat shrinkable bags for vacuum packing meat	25	1	5
Vinylidene chloride	Cryovac BB1	Heat shrinkable bags for vacuum packing meat	61	1.2–1.4	20–25
Vinyl chloride copolymer	Cryovac BK	Films for packaging cheese	51	2.8	185
Aluminum/Polyethylene		Sealable vacuum packages; heat stable; for ready-to-eat dishes, freeze-dried products			When >20 μm, without pores, impermeable to vapor, gas, and light
Polyester/ Polyvinylidene chloride/	Hostaphane	Extremely tear-resistant, sealable	12–75	0.7	15
Polyethylene		Vacuum package; frozen products			
Rigid polyvinyl chloride	Genotherm	Molded trays for menu packs	150	2.1	24

[a] From Böhme (1969); Cavett (1968).
[b] Usually called "regenerated cellulose."
[c] Depending on the type of coating (nitrocellulose or, e.g., polyvinylidene chloride).

organisms. The presence of gas or turbidity in the medium indicates penetration took place (Kleniewski, 1970; Maunder *et al.*, 1968; Ronsivalli *et al.*, 1966). For heat-resistant plastic-aluminum foil laminates, Schmidt-Lorenz (1973) recommended an ''agar cooking test'' in which the packages are boiled for 45 min in 2% agar. Before the agar cools, spores of *Bacillus stearothermophilus* are added, and during cooling the inoculated medium is drawn through leakage sites.

A number of authors have shown that microbes can attack synthetic packaging materials, and under favorable conditions, the organisms may penetrate intact film (Schwartz, 1964; Booth *et al.*, 1968; Hartman *et al.*, 1963; Ronsivalli *et al.*, 1966; Toepfer and Kanz, 1976). In some instances prolonged exposure to bacterial enzymes is necessary before entry. For example, cellulase-forming organisms, especially molds, penetrate sausage skins of cellulose hydrate only after a long period at relatively high temperatures (Leistner, 1956).

Plastic films vary considerably in their permeability to gases. Exclusion of oxygen diminishes the oxidation rate of the product and slows the growth of many types of bacteria and yeasts or prevents growth of strict aerobes (e.g., molds). The high oxygen permeability of polystyrene and polyolefins can be reduced by combining them with other material, i.e., by varnishing, lining, layering, gluing, or co-extruding. Likewise, some films are permeable to water vapor. The high permeability of cellulose hydrate can be reduced by varnishing.

Data on the permeability of various films are given in Figs 11.1 and 11.2 and Table 11.1. From these data one can choose a suitable film for any packaging problem.

III. THE MICROBIOLOGY OF PACKAGING

The factor most important to the microbiology of packaged foods is the relative permeability of the packaging material to oxygen, carbon dioxide, and water vapor, particularly if the air spaces within the package are evacuated or filled with preservative gases at time of closure, and particularly for perishable foods like meat, poultry, and fish (Cavett, 1968).

A. Permeable Packages

Films highly permeable to water vapor and gases, or more permeable to oxygen than to CO_2, or films that are only loosely applied, may protect against contaminating organisms but do not affect the growth of organisms already on the food. The intrinsic conditions in a food packaged in highly permeable material are similar to those in the unpackaged product (McDougall, 1971). For example, cellulose films permeable to oxygen permit pseudomonads to grow in packaged ground beef while gas-impermeable films inhibit pseudomonads (Jaye *et al.*, 1962). Films like polythene that are moisture impermeable and oxygen permeable, protect against contamination and loss of moisture, but serve only to encourage surface growth of the normal spoilage bacteria (McDougall, 1971).

B. Impermeable Hermetically Sealed Packages

The growth and activities of microorganisms within a package are affected by (a) the adequacy of the food as a microbial medium, (b) the temperature, (c) the a_w, (d) the pH, (e) the nature of entrapped gases, and (f) competition among microorganisms.

In gas-impermeable, hermetically sealed but unevacuated packages, respiration of fresh meat tissue and the accompanying bacterial flora soon uses the oxygen and increases the carbon dioxide content of the entrapped atmosphere. At the same time, the pH gradually drops due to the activity of lactic organisms. These changes slow the development of the typical aerobic spoilage organisms, so that shelf-life increases by 50% (Kraft and Ayres, 1952). Evacuation of hermetically sealed impermeable packages accentuates these effects.

The growth of anaerobes is rarely a problem in fresh meats, regardless of packaging (McDougall, 1971). In fact, toxin formation by *Clostridium botulinum* has not been reported in fresh meat, whether packed under vacuum or not. In cooked or cured meats or fish, where there are few microbial competitors, clostridia (including *C. perfringens* or C. *botulinum*) can sometimes grow regardless of the presence or absence of gaseous oxygen and therefore are unaffected by packaging (Pivnick and Bird, 1965; Pivnick and Barnett, 1965; Tabatabai and Walker, 1970; Thatcher *et al.*, 1962; Kautter, 1964; see also Chapter 15). Botulinum toxin was formed in slices of inoculated bologna that had been packed both with and without evacuation (Christiansen and Foster, 1965). Thus under these special circumstances, evacuation inhibits spoilage but not toxin formation, so that the consumer is not alerted to the danger (Thatcher *et al.*, 1962; Kautter, 1964; Solberg, 1967).

Carbon dioxide within the package not only slows the spoilage rate of meats, but also alters the spoilage flora (Cavett, 1968; Reuter, 1970; Leistner, 1974; Sutherland *et al.*, 1975a,b; Patterson and Gibbs, 1978; see also Chapters 10 and 15). The gram-negative bacteria such as *Pseudomonas* are inhibited, whereas lactic acid bacteria such as *Lactobacillus, Leuconostoc,* and *Pediococcus* are favored. Salted and cured products packed in evacuated packages support the growth of fecal streptococci and lactics. Unevacuated packages sometimes permit the growth of yeasts (Shank and Lundquist, 1963). Members of the Enterobacteriaceae, including *Salmonella,* survive on the surface of packaged fresh meats and sometimes multiply despite the presence of millions of lactic acid bacteria (Leistner, 1974).

Staphylococcus aureus survives in evacuated packages, but usually its growth

(Eddy and Ingram, 1962) and especially its ability to form enterotoxin (Genigeorgis and Prucha, 1971; Crowther *et al.*, 1976) are inhibited by the reduced access to oxygen. However, enterotoxin was formed in vacuum-packaged bacon which had been severely abused by being held at 37°C (Thatcher *et al.*, 1962).

Carbon dioxide can be applied pure or mixed with air or nitrogen to replace the atmosphere within the hermetically sealed package. The microbial level on red meat or poultry meat packaged with an atmosphere of 2 parts of air and 1 part of CO_2 will remain at about one-tenth that of meat not so packaged, and spoilage time will be increased by 2- or 3-fold (Thomson, 1970; Clark and Lentz, 1972; Huffman, 1974; Huffman *et al.*, 1975; Partmann *et al.*, 1976; Böhme, 1971). The inhibition is most marked at low temperatures (Ogilvy and Ayres, 1951a,b). Gas packing is used also to maintain acceptable meat color (Heiss and Eichner, 1969; Partmann *et al.*, 1975, 1976; Clark *et al.*, 1976; see also Chapter 10).

C. Aseptic Filling

Aseptic filling is the process of filling a sterile food into a sterile container under aseptic conditions. Ultrahigh temperature (UHT) milk is a good example of such a product.

There are three types of aseptically filled packages:

1. Tin cans, sterilized by steam and filled and closed in a sterile chamber.
2. Flexible containers, decontaminated by ultraviolet light, alcohol, peroxide, or a combination of these, and filled by injection and sealed in a sterile atmosphere. In a related system, the forming, "sterilizing," filling, and sealing are carried out in a single operation in a closed chamber. This latter process might better be called "ultraclean" rather than aseptic (Shaw, 1977).
3. Extruded polyethylene or polystyrene containers, sterilized in the heat of the extrusion or forming process, and filled in a sterile room (Berry, 1975).

The shelf-life of chilled foods increases with a decrease in the initial load of spoilage organisms (see also Fig. 15.2). For example, the shelf-life of packaged frankfurters can be dramatically extended by wrapping them in sterile film while still hot from cooking, in a closed chamber whose air is continuously treated with uv light. Simply sterilizing the packaging material before filling (a) kept the bacterial count of bottled water at <1/ml after 4 weeks storage, whereas that of water in conventional nonsterile containers was 1000/ml (von Bockelmann, 1975); (b) increased the shelf-life of orange juice to 4 months, compared to 4 weeks in nonsterile containers (Anonymous, 1971); and (c) doubled the shelf-life of pasteurized milk and related products (summarized in von Bockelmann, 1975).

D. Pressurized Containers

Pressurized foods are packaged with a nontoxic compressed gas which propels the food through a valved closure. The gas may change the nature of the food as it is propelled (whipped cream), or simply eject it unchanged from the container (chocolate syrup) (Graham, 1950).

Pressure packing with an inert gas such as nitrogen has no inhibitory effect on microbial growth (Riester *et al.*, 1958), although nitrous oxide, once used for cream, did show considerable antimicrobial activity (Graham, 1950; Hays *et al.*, 1959).

Some foods, such as oils, do not support microbial growth, and thus can be filled and gassed cold (Hardy *et al.*, 1961). High acid or low a_w products which cannot support the growth of heat-resistant organisms can be filled and gassed hot. Foods whose neutral pH and high a_w would permit the growth of such organisms must either be retorted after closure or sterilized separately and aseptically filled into presterilized containers; such procedures are feasible, though difficult (Riester *et al.*, 1958; Giggard and Gottschall, 1958). In this respect, pressurized foods are no different from other canned foods.

IV. PRODUCTION CONTROL

A. Canned Foods

For assurance of safety of canned foods it is necessary to control and monitor the total process. Incubation tests on end-product samples in lieu of such monitoring is not recommended (see Chapters 1 and 15).

Semipreserved canned cured hams may contain large numbers of fecal streptococci as well as bacterial spores. Micrococci may be present, but the Enterobacteriaceae do not survive the heating process.

Control of the integrity of the can or jar is very important because any pathogens that enter may grow without competition from other microorganisms. The side and end seams of cans should be examined for leakage points (Bee and Denny, 1958) and for adequacy of seam construction (National Canners Association, 1975). After the cans have been retorted they should be protected from contamination and cooled in chlorinated water to prevent microbial entry during the period when the mastic which seals the ends to the body is still semiliquid.

B. Aseptically Filled Products

The only satisfactory way to control the sterility of aseptically filled products is to design the equipment and process so that: (1) the specified thermal process used to sterilize the product can be controlled and monitored; (2) the filling equipment can be effectively cleaned, sterilized, and maintained in a sterile condition; (3) procedures used to sterilize containers are effective; and (4) the containers are sealed to prevent microbial contamination.

If these steps cannot be adequately controlled and monitored, a number of containers of the product must be examined microbiologically. However, only lots with a high incidence of contaminated units would be detected by any reasonable sampling plan. A low incidence of such units would generally go undetected [See Microorganisms in Foods (ICMSF, 1974); see also Volume II, Chapter 15].

Tests for the adequacy of the final closed container to protect against contamination from external sources would be the same as those for conventionally canned foods (Section III, A).

C. Packed Nonsterile Products Not Subsequently Heated

This category includes retail packages of meat or meat products, cheese, and a variety of sliced products usually evacuated under a plastic film or in a plastic bag, and sometimes with a protective gas. For such products, control of the hygiene of processing before filling is of primary importance. Examination of sample units of the finished product is much less effective, except for tests for container leakage.

12

Injury and Its Effect on Survival and Recovery

I. INTRODUCTION

Treatments used in food processing often kill or damage microbes. If the treatment is not severe enough to kill, the surviving microbes may be injured or stressed. Some treatments that damage microbes are heat, cold, drying, freeze-drying, freezing, low water activity, and irradiation. Adverse chemical treatments (starvation, low pH, food preservatives, and disinfectants) similarly kill and injure microbes. Organisms that are injured include the gram-positive cocci (*Staphylococcus aureus* and the fecal streptococci), gram-negative rods (*Escherichia coli, Salmonella typhimurium, Vibrio fischeri* [*marinus*]), the spore-forming bacteria (*Bacillus* and *Clostridium*), as well as yeasts. Injury is thus a general phenomenon in microbiology and probably of greater significance than hitherto realized.

Bacteria grown in pure culture under optimal conditions are tested in the laboratory for their response to selective agents. These laboratory methods are then used with food samples in which the microbes are neither in pure culture nor have they been grown under optimal conditions. Their response to selective media may be different from the anticipated one because these organisms are in a different physiological state from that in laboratory cultures. It is the purpose of this chapter to describe these physiological changes briefly and to discuss the practical steps that can be taken to help ensure that selective enumeration is carried out satisfactorily. This chapter is thus an amplification of the description of injury and stress described on page 99 in the second edition of Book 1 of this series (ICMSF, 1978).

The effect of a number of physical treatments on survival of microbes is the subject of two recent books (Gray and Postgate, 1976; Skinner and Hugo, 1976).

The borderline between injury and death is difficult to define, and there are many and complex biochemical changes that may start as injury and then end in the death of the organism. Most forms of injury involve membrane damage with loss of internal solutes and increased sensitivity to inhibitors; for example, in gram-negative bacteria, the outer membrane as well as the protoplasmic membrane may be damaged. There is too little information to draw definite conclusions about damage to other microbial organelles, but the subject has been recently reviewed by Hurst (1977) and Mossel and Corry (1977). Other useful review articles are by Busta (1976) and van Schothorst (1976).

II. CHARACTERISTICS OF INJURED MICROBES

A. General

One of the important characteristics of injured organisms is loss of resistance to selective chemical agents. As shown in Fig. 12.1 when *S. aureus* cells are heated in phosphate buffer at 52°C some cells die and others become injured. Death is assessed by the absence of colonies in an optimal (complex) medium, and injury is assessed by counting the colonies that could grow in the optimal medium, but could not in the same medium containing 7.5% NaCl. Because salt tolerance is a characteristic of *S. aureus,* the high salt medium grows only uninjured organisms while the optimal medium grows both injured and uninjured organisms (Iandolo and Ordal, 1966; Hurst *et al.*, 1976). Salt is often included in selective media for staphylococci.

The second feature of Fig. 12.1 concerns the repair period. There is no growth during the first 5 hr of incubation, but salt tolerance is gradually regained and then growth starts. This prolonged lag phase is another characteristic feature of injury and repair. After repair, enterotoxigenic staphylococci can synthesize enterotoxin B (Collins-Thompson *et al.*, 1973).

A similar effect of the selective agent on heat injured *Salmonella typhimurium* has been reported by Clark and Ordal (1969). Heat-injured cells were unable to recover fully and grow on brilliant green agar, Levine's eosin methylene blue (EMB) agar, or *Salmonella-Shigella* agar, and the bacteria found both EMB agar with 2% NaCl and desoxycholate citrate agar especially poor as recovery media.

Freeze-drying also renders *S. anatum* sensitive to deoxycholate (Ray *et al.*, 1971). The toxicity of the selective medium to freeze-dried cells of Enterobacteriaceae was found to be related to its cholate content (Northolt, 1972), and taurocholate media are preferable to media containing oxbile of unknown compositon (van Schothorst and van Leusden, 1975). These experiments were carried out with cells injured by freeze-drying rather than by heating, suggesting that injury caused by different physical treatments can manifest itself in similar ways.

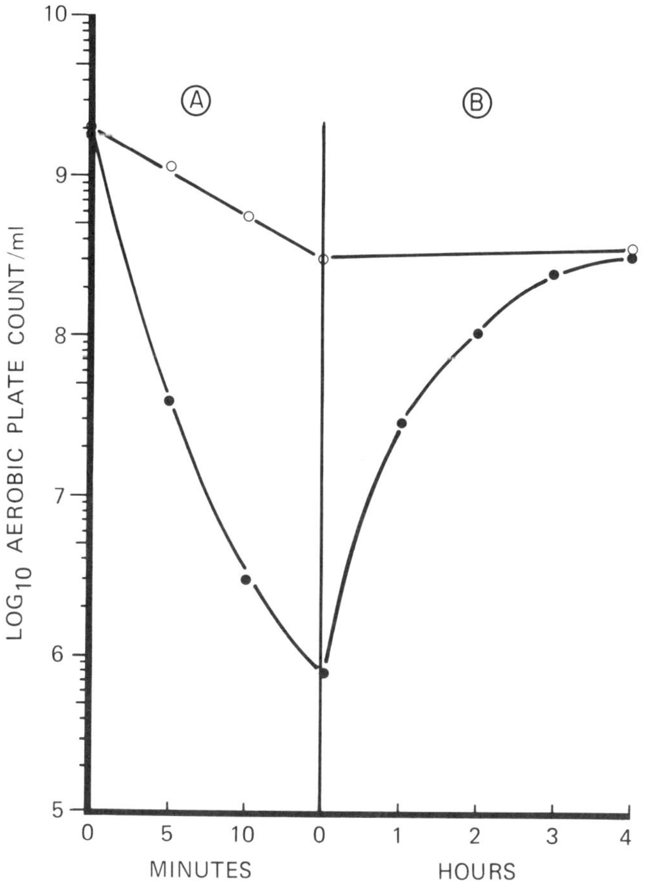

Fig. 12.1 Survival and recovery curve for *Staphylococcus aureus*. (A) Heat injury at 52°C for 15 min in 100 m*M* potassium phosphate buffer, pH 7.2. (B) Recovery from heat injury in nutrient broth at 37°C. Symbols: (○) samples plated on tryptone-soy agar to give a total viable count; (●) samples plated on tryptone-soy agar containing 7.5% NcC1 to give an estimate of the uninjured population. (Slightly modified from Iandolo and Ordal, 1966, by permission of the authors.)

B. Chill Injury

The rapid cooling of mesophilic bacteria from their normal growth temperature to about 0°C can kill or injure a proportion of the culture. Gram-negative bacteria, including *Escherichia coli*, *Pseudomonas aeruginosa*, *P. fluorescens*, *Salmonella* spp., and *Enterobacter aerogenes* appear to be more susceptible to cold than are gram-positive organisms, although similar effects have been noted with *Bacillus subtilis* and *Clostridium perfringens*.

The susceptibility of a microorganism to cold injury depends on where in the microbial growth curve the cells are harvested for test. We thus recognize three categories of injury: (1) a steady decline in viability upon chilling stationary phase cells, or upon slowly cooling exponential phase cells; (2) "cold shock" (loss of viability and some loss of internal solutes) on rapidly cooling exponential phase cells; (3) "cold osmotic shock" (leaking of minerals, pool substances, and proteins) on resuspending exponential phase cells in cold distilled water (Rose, 1976).

An example of the first effect is provided by the work of Jackson (1974). A culture of *S. aureus* in trypticase soy broth at 5°C became increasingly sensitive to mannitol salt agar, in a manner similar to heat-injured cells.

To exhibit cold shock, exponential phase cells have to be cooled rapidly. Many investigators have cooled cells from 37°C to about 0°C but there are reports that it is the temperature difference that matters, rather than the actual temperature to which the suspension is cooled (Sato and Takahashi, 1968, 1969). The lethal effect of cold depends on the growth medium, the growth phase, the time of exposure to low temperatures, the concentration of bacteria and the composition of the suspending medium (Strange and Dark, 1962). Cells grown in a minimal medium are generally more sensitive to cold shock than those grown in a rich medium (Strange and Ness, 1963). For a more detailed discussion of this subject see Ingram and Mackey (1976). Enough has been said, however, to emphasize the possibility that refrigeration of foods will injure susceptible organisms (e.g., *E. coli* and *Salmonella*) so that they may not recover on selective laboratory media, yet may continue to be viable.

C. Freeze Injury

Freezing also kills or injures microorganisms. With "metabolic" injury, there is a requirement for certain nutritional supplements before growth can occur. The organism develops on rich but not minimal media. In another type of freezing injury, the cells become highly sensitive to many surface-active agents. For example, injured *E. coli* may fail to grow on selective media containing bile salts (Ray and Speck, 1973a). Freezing and thawing can also cause (a) leakage of low and high molecular weight cellular components into the suspending medium, (b) an extended lag before initiation of growth after thawing, and (c) increased sensitivity to uv irradiation. However, cells in the frozen state are less affected by high energy irradiation than are cells irradiated at normal ambient temperatures.

From the public health point of view the occurrence of freeze injury is important, because organisms may escape detection when sought in the frozen product. They are capable, however, of repair and development if the product is allowed to thaw.

Many hypotheses have been proposed to explain cryo injury including (a) temperature shock, (b) effect of concentration of extracellular or intracellular solutes, (c) dehydration, (d) injury from internal ice formation, and (e) mechanical injury through the attainment of a minimum cell volume (Merryman, 1974). This complex subject has been recently reviewed by MacLeod and Calcott (1976).

A number of compounds protect bacterial cells against death from freezing, but they are less effective in protecting against reversible injury (Ray and Speck, 1973b). Glycerol, diethylene glycol, mono- and disaccharides, and amino acids are examples of low molecular weight compounds. More complex materials include milk proteins, serum, yeast extract, and various broths. The mechanism by which these compounds provide protection is not known, but it has been suggested that in some way they act on the cell surface (Ray and Speck, 1973a). As a consequence of the presence of protective substances in many foodstuffs, freezing often has much less effect on the microbial population in a food than would be expected from the behavior of the particular organism in a nonprotective medium (Georgala and Hurst, 1963).

D. Other Effects

Injured cells appear to suffer damage to their permeability barrier and become sensitive to antibiotics to which the uninjured cells are resistant. This point may be important for the cultivation of injured microbes because antibiotics are frequently included in selective media. For example, freeze-dried *E. coli* becomes sensitive to chloramphenicol, streptomycin, and actinomycin D (Sinskey and Silverman, 1970); cold shock renders *C. perfringens* sensitive to neomycin (Traci and Duncan, 1974); and heating of *C. perfringens* and *S. aureus* makes them polymixin sensitive (Flowers and Adams, 1976; Hurst *et al.*, 1976).

Shigella flexneri grown in mixed culture with coliforms was injured, probably due to the production of formic and acetic acids by the coliforms (Hentges and Maier, 1970). Freezing also appears to injure *Shigella* (Nakamura and Dawson, 1962).

E. Nutritional Requirements of Injured Cells

Another manifestation of injury is the apparent increase in nutritional requirements for the development of stressed cells. The original, classical investigation of such injury (Curran and Evans, 1937) showed that nutrient agar enriched with glucose gave up to 3000 times higher counts with heated spores of *Bacillus* than did nutrient agar alone. A number of subsequent studies with a great variety of microbes and with different methods of injury confirmed these early findings

(Straka and Stokes, 1959; Freeman and Bridges, 1960; Nakamura and Dawson, 1962; Moss and Speck, 1966; MacLeod *et al.*, 1966; Ray and Speck, 1972).

Organisms repair sublethal damage more readily in simple minimal media than in complex media. Thus heat-damaged *S. aureus* recovered its salt tolerance in a 10-fold diluted and dialyzed complex medium (Hurst *et al.*, 1973), and cells damaged by drying in milk recovered better when the milk was reconstituted to 1% rather than to 10% milk solids (Hurst *et al.*, 1976). Heat-stressed *Salmonella typhimurium* LT2 gave higher counts on minimal medium agar than on trypticase soy agar with yeast extract (Gomez *et al.*, 1973). Enhanced recoveries of heated *Salmonella senftenberg* were reported from foods using this minimal medium technique (Wilson and Davies, 1976). However, the consensus of opinion is that this technique has only limited application in food microbiology.

III. ENUMERATION OF INJURED MICROBES

A. *Staphylococcus aureus*

Several surveys show that Baird–Parker agar is suitable for the direct enumeration of stressed staphylococci (Baird-Parker and Davenport, 1965). This medium contains neither salt nor polymyxin, which are known to inhibit injured organisms, and yet it is highly selective. The success of the Baird–Parker agar may be due to its pyruvate content because pyruvate degrades H_2O_2. Flowers *et al.* (1977) showed that stressed cells are more sensitive to H_2O_2 than are uninjured cells and that media normally not suitable for counting injured cells can be much improved by the inclusion of catalase.

For enrichment of staphylococci, a complex liquid medium containing 10% NaCl (Baer *et al.*, 1971) cannot be recommended because injured cells will not recover in it. Giolitti and Cantoni (1966) described a liquid version of the Baird–Parker agar for most probable number (MPN) counting. Some resuscitation probably takes place in the medium while the tubes are incubating but the procedure has not been an unqualified success. Pre-enrichment in a nonselective medium has been used successfully (van Schothorst, personal communication). For isolation of staphylococci from dried milk, resuscitation in 1% milk containing 1% sodium pyruvate and penicillin at 10 μg/ml is useful (Hurst *et al.*, 1976). Penicillin does not effect repair but prevents multiplication of uninjured staphylococci so that quantitation is maintained. After 4 hr at 37°C the suspension is plated for a colony count in Baird–Parker agar containing penicillinase. As few as 10 organisms per gram of food can be estimated. The rationale of the method is that in powdered milk the great majority of the organisms present are injured, and there is need for a preliminary resuscitation step during which growth is inhibited by the penicillin. Penicillinase is a necessary ingredient in the

plating medium so that the penicillin-sensitive bacteria can grow. However, when only some of the organisms are injured, as they might be in raw meat, this procedure should not be used and a direct plating on a good selective medium is preferable.

B. Spore-Forming Organisms

Recovery media for stressed spore-forming organisms are discussed in a detailed and careful review by Roberts (1970). Only the more recent findings are discussed here.

1. Clostridium perfringens

The medium of choice for counting uninjured cells of *C. perfringens* appears to be trypticase–sulfite–cycloserine agar without egg yolk (Hauschild and Hilscheimer, 1974b). This yields a presumptive *C. perfringens* count, which is confirmed by examination of suspect colonies (Hauschild and Hilscheimer, 1974a). However, injured cells of *C. perfringens* are better counted on media containing egg yolk emulsion or lysozyme (Barach *et al.*, 1974). The suitability of the selective media is greatly affected by the presence of certain foods.

2. Clostridium botulinum type E

This organism differs from other *C. botulinum* types (A–F) in its heat sensitivity. The usual procedure of activating spores and killing vegetative cells by heating to about 80°C can be used only if lysozyme at 5 μg/ml is included in the counting medium (Sebald and Ionesco, 1972). Resuscitation methods for this organism have not been published.

C. *Escherichia coli*, Coliform Group, and Other Enterobacteriaceae

Escherichia coli is most commonly enumerated by most probable number techniques but if the liquid medium contains bile salts and certain dyes, injured cells may be inhibited. A rapid, direct method of enumerating *E. coli* has been modified to take account of injured cells, particularly in frozen and dried foods (Anderson and Baird-Parker, 1975). In the modified technique, the food is first inoculated onto a membrane which is then incubated on nutrient agar at 37°C for 4 hr. The membrane is then transferred to the selective medium at 44°C for 24 hr. The resuscitation step increases the recovery of *E. coli*.

Techniques similar to those described above for *E. coli* may be used to enumerate the coliform group. A lactose–glutamic acid medium was shown to be less inhibitory than other media for enumerating the coliform group from frozen or dehydrated foods (Moussa *et al.*, 1973).

The toxicity of selective agents to injured coliform organisms has been considered by many authors. Lauryl sulfate is less inhibitory than bile salts (Clark and

Ordal, 1969; Mossel *et al.*, 1974). Several techniques have been described to resuscitate the cells prior to selective plating: (1) a short resuscitation period in a liquid medium before plating out, suitable when the organisms are not severely injured (Mossel and Ratto, 1970; Warseck *et al.*, 1973); (2) a replica technique in which the microorganisms are first grown on a nutrient agar and then transferred by the method of Lederberg to the selective agar (Mossel *et al.*, 1965); (3) a double agar technique in which the microorganisms are grown on a nutrient agar for a short period of time after which the selective agar is carefully poured over the nonselective one (Speck *et al.*, 1975).

D. *Salmonella*

Liquid media are used in the primary steps to determine the presence of salmonellae. If the first medium is nonselective, the injured cells recover and are able to grow on transfer to selective media; if it is selective, many of the injured cells will not grow (Byrne *et al.*, 1955; North, 1961). Lactose broth is most commonly used, though lactose is not necessary for resuscitation and trypticase soy broth (TSB), nutrient broth, or buffered peptone water can also be used (Clark and Ordal, 1969; Ray *et al.*, 1972; van Schothorst and van Leusden, 1972). For complete recovery and sufficient multiplication of recovered cells, pre-enrichment should be carried out for more than 6 hr, especially when tetrathionate bile brilliant green is used as the selective medium and when this medium is incubated at 43°C (Pietzsch *et al.*, 1975; van Schothorst and van Leusden, 1975; van Schothorst, 1976). For details of the recommended procedure, see pages 160–162 of the second edition of Book 1 of this series (ICMSF, 1978). Of the solid media used in *Salmonella* isolation procedures, brilliant green phenol red agar and Wilson–Blair agar have frequently been demonstrated to be less inhibitory toward injured salmonellae than other commonly used selective media (Clark and Ordal, 1969; Corry *et al.*, 1969; Janssen and Busta, 1973; Licciardello *et al.*, 1970).

E. *Shigella*

After freezing injury *Shigella sonnei* is unable to grow in a synthetic medium. The addition of meat extract, peptone, or casamino acids improve the ability of the synthetic medium to recover injured *Shigella* (Nakamura and Dawson, 1962). Park *et al.*, (1972, 1976) studied the selective enrichment of *Shigella* in the presence of coliforms using the phenolic galactoside 4-chloro-2-cyclopentylphenyl β-D-galactopyranose (CPPG). The system makes use of the β-galactosidase activity of *E. coli* which hydrolyzes the phenolic derivative of β-galactoside to yield galactose and a phenolic compound, toxic to *E. coli* but tolerated by *Shigella*.

IV. THE SIGNIFICANCE OF INJURED MICROBES TO FOOD PROCESSING

The practical significance of injured microbes concerns all aspects of food microbiology. In food safety, we are concerned with the effect of selective agents and inadequate detection of pathogens. In food processing and technology, we are often concerned with the avoidance of injury and the maintenance of starter activity to ensure uniformity of fermentations. In food preservation the practical interest in injured microbes is exactly the opposite. The aim is to ensure that any surviving bacteria have extremely long lag phases and are inhibited by reagents which may not affect uninjured microbes. This helps to ensure product stability.

A. Effects of Injury on Starter Activity

In the manufacture of fermented foods, especially cultured milks, cheese, and dry sausages, starters are used extensively. The food manufacturing process relies on high and uniform starter activity. Injury might result in a prolonged lag phase when the starter is inoculated into the new food substrate it is supposed to ferment. Cultures handled by mail generally have a long lag phase and must be resuscitated before use.

New cheese-making techniques furnish lactic cultures that require no resuscitation and begin growing immediately, by (a) growing the starter with pH control to prevent damage from excess acid, (b) concentrating the culture by centrifuging, and (c) preserving it with carbon dioxide ice or liquid nitrogen.

B. Effects of Injury on Spoilage and Hazard

Spores surviving heat treatments are considerably more sensitive to the curing salts commonly used in meats than are unheated spores (Roberts and Ingram, 1966). Heating at 85°, 90°, and 95°C sensitized the spores of *Clostridium botulinum* type B to 2.5–3.5% NaCl, but heating at lower temperatures did not. Heated spores also became more nitrite sensitive (Jarvis *et al.*, 1976). At low incubation temperatures (below 15°C) pronounced salt–nitrite interaction was observed, but this was not evident at above 20°C (Roberts *et al.*, 1976).

D values can be used for estimating the lethality of heating (Chapter 1, Section IV,B), and D' values (pseudo *D*) can be used for estimating the effect of heat when a selective rather than an optimal medium is used for counting surviving organisms. When these two values are considered, canned cured meats are as safe as uncured meats, which have received a $12D$ heat process, after heat treatments approximately one-fifth as great (Ingram and Roberts, 1971). In considering that the ratio of D'/D quantitates injury of *S. aureus,* the loss of cellular Mg during heating causes injury rather than lethality (Hurst *et al.*, 1974a). Also, the same consideration showed that *S. aureus* heated in milk at 50°–60°C was more rapidly injured than killed. At >60°C death became faster than injury and at 70°C the two processes coincided (Firstenberg-Eden *et al.*, 1977).

Heat-injured microbes appear to contribute very little to spoilage, for example, of pasteurized milk (Claydon, 1975; Mourges and Auclair, 1973). Pasteurized milk aseptically packaged to prevent recontamination kept on the average 57 days at 4°C before organoleptic deterioration. Similarly stored good quality commercially pasteurized milk (not aseptically packaged) kept only 9–13 days (Mourges and Auclair, 1973). If recontamination is prevented, the surviving injured psychrotrophs of milk develop only slowly at 4°C.

Similarly, when spores of an organism originally isolated from canned green peas were reinoculated into sterile green peas at a level of 1–3 per gram, they spoiled the product. However, the spore level had to be 10^4 or more per gram to cause spoilage if they were first heated in the product (Vas, 1970).

V. EFFECTS OF INJURY AND RESUSCITATION ON MICROBIOLOGICAL CRITERIA

It is almost self-evident that the number of microbes a food apparently contains depends on the techniques and media used for the examination. The trend is toward ever-increasing sensitivity, and from time to time any existing microbiological criteria need to be revised in this light.

The state of the art of estimating the major food pathogens and indicator organisms after injury varies from organism to organism. Baird–Parker agar is probably adequate for direct plating of injured staphylococci in most situations. Salmonellae, also, are probably adequately detected at present because almost invariably the method contains a pre-enrichment step in which injured organisms have a chance to repair. The situation with other organisms is less certain. However, as a general principle, the same method should be used to evaluate conformity to a microbiological criterion as the investigators used when they collected data to establish the criterion. Otherwise, a lot of food that is in compliance may be rejected while another that is not in compliance may be accepted.

13

Interacting Factors Affecting Mixed Populations

I. INTRODUCTION

A food that can support microbial growth will have a constant microbial composition only after imposition of an external factor (e.g., freezing or drying), or after all the available nutrients are exhausted (i.e., complete spoilage), or after there is a sufficient accumulation of the end products of metabolism to halt further growth (e.g., fermentation or pickling). After stabilization, a steady population decrease takes place. The microbiology student will recognize what we have just described as a variation of the bacterial growth curve.

Until the steady state is reached, the quantitative and qualitative composition of the microflora is in constant change. At any given time, each food has its characteristic microbial profile, starting with the "initial flora" at time of harvest or slaughter, then progressing to that at time of transportation, processing, and storage. The specific microenvironment of the particular food and the stimulatory or inhibitory actions that microorganisms exert on each other will affect the final result.

This chapter reviews the principles that influence the development of a microflora on or in foods and the way microbes influence the survival or growth of other organisms in mixed populations.

II. ENVIRONMENTAL FACTORS

Certain conditions favor the development of specific types of microflora and inhibit the development of others; these may be classified as intrinsic, processing, and extrinsic factors.

A. Intrinsic Factors

Intrinsic factors are qualities inherent to the food that influence microbial growth. These include chemical, physical, and biochemical characteristics (nutrients, natural antimicrobial factors, pH, a_w, E_h). Their combined effects will determine the selection of that part of the initial flora capable of growth or survival; organisms unable to compete in a particular environment are gradually eliminated. For instance, most gram-negative bacteria originally present in large numbers in the orchard fail to survive on citrus fruits because of the unfavorable pH. Therefore, spoilage is caused by molds, yeasts, or gram-positive organisms that comprise only a minor element of the initial flora. Nutrients must be available in a form that is utilizable or degradable by microoranisms. Some structures are resistant to microbial attack and can be degraded only by organisms that have specific enzymes. Most spoilage organisms are not nutritionally fastidious. The spoilage of proteinaceous foods was once believed to be the result of attack by proteolytic bacterial enzymes, making available low molecular weight amines and peptides; but microbial growth on aerobically stored meat does not result in significant protein breakdown until an advanced stage of spoilage is reached (Dainty *et al.,* 1975). Initial bacterial growth is at the expense of glucose and ribose, then of lactate and amino acids (Gill, 1976; see also Chapter 15). Usually only particular groups of microorganisms have the enzymatic potential to attack complicated biological structures and to induce spoilage. For example, the initial stages of degradation of many plant tissues may often be brought about only by microorganisms that secrete enzymes such as cellulases or pectinases. These enzymes depolymerize cell wall polysaccharides to release metabolizable low molecular mass components, which can be utilized by other microorganisms.

Certain proteins such as keratin or elastin are very resistant to degradation by microbial enzymes. Another protein, collagen, is broken down by collagenase, which is formed only by a very few organisms, e.g., *Pseudomonas* spp., *Clostridium perfringens,* and other pathogenic clostridia. Soluble proteins and peptides are are much more readily broken down. As a result of the degradation of proteins, a particular pattern of free amino acids can be recognized (Miller and Kandler, 1967). Once proteins have been broken down by the proteolytic flora the low molecular mass fragments can then be readily utilized by the remaining flora. Some species produce amino acids that can be utilized by other organisms so that dependent associations can develop. Others produce peptides that are stimulatory. For example, in yogurt, *Streptococcus thermophilus* produces peptides utilized by *Lactobacillus bulgaricus.* The fat-splitting lipolytic enzymes are formed by many spoilage organisms, e.g., *Pseudomonas* and *Bacillus* spp., most molds and some yeasts and enterobacteria (Mossel and Harrewijn, 1970; Mossel, 1971). While proteolysis and glycolysis are essential metabolic activities of many spoilage organisms, lipolysis is less common. There are two reasons for this. First, proteolysis and glycolysis usually proceed much more rapidly, because the substrate proteins and carbohydrates are often soluble and therefore readily accessible to the extracellular and intracellular microbial enzymes. Second, fats are relatively insoluble, so that the amounts directly accessible in foods to the extracellular enzymes of spoilage microorganisms are generally very small, being restricted initially to the surface of monolayers of fat particles. In some water-in-oil emulsion foods, the distribution of microorganisms within the isolated aqueous phase droplets is such that most of the lipid is simply not available to microbial enzymes. In fact, oxidative fat deterioration is mainly caused by physical and chemical factors, whereas microbially induced rancidity is rare. Slow lipolysis brought about by microorganisms, however, does contribute to the development of the typical flavors of several kinds of cheese (e.g., Roquefort, Gorgonzola) as well as that of some naturally fermented and air-dried meat products (e.g., salami). Lipase activity by itself is important in flavor changes only as a first step in a series of reactions. There is no correlation between lipase production and the oxidative activities of microorganisms (Alford *et al.*, 1971). The vitamin content of a food is not normally an important factor influencing the development of spoilage, so lack of vitamins in a food seldom prevents or delays spoilage. But there are exceptions. For instance, the antimicrobial defense mechanisms in the avian egg are believed to include the tight binding of biotin by the protein avidin, thus restricting its availability to potential spoilage microorganisms (Board, 1969).

All foods contain minerals and generally in amounts well in excess of those required for the growth of microorganisms. In substrates completely free of minerals growth of microorganisms is not possible, and fruit juices extensively demineralized by ion exchange are more resistant to attack by yeasts than are nontreated products.

Some foods contain antimicrobial constituents; this area has been reviewed by Mossel and Ingram (1955) and Mossel (1971, 1975). Such constituents have been found in some plant tissues such as spices (Fuchs, 1958), onions, garlic (Wei *et al.*, 1967; Drechsel, 1965), cress and parsley, and fruit (e.g., grapes). Raspberries contain salicylic acid, rowanberries have sorbic acid, and citrus fruits contain antibacterial oils. There are antimicrobial activities in foods of animal origin, particularly in blood (e.g., complement, properdin, lysozyme, histones, protamines, and hematin). Also colostrum, milk, saliva, leukocytes, and some other body fluids and tissues contain an effective antimicrobial system in which peroxidase enzymes (''lactoperoxidase'' in milk) catalyse the peroxidation of low molecular mass substances, such as thiocyanate, to form products that inactivate microorganisms. Hens' eggs contain lysozyme, which lyses certain types of bacteria.

Growth and development of microorganisms are influenced by the physical structure or state of the food, e.g., the liquid or frozen state of tissue water, the distribution of the aqueous phase in emulsions, and the presence of biological barriers (the shells and membranes of eggs, scales of fish, and skin of poultry).

The antimicrobial activity of tissue enzymes is diminished or even destroyed by heat treatment. Freezing can liberate large enough amounts of lysozome from animal tissues to inhibit growth of the test organisms used during meat inspection for detecting residues of antibiotics or other inhibitory substances in meat. As a consequence, false positive results can occur (Heinert *et al.*, 1976).

The relevance of many antimicrobial activities in foods is often overestimated. Mostly their effects are confined to limited groups of microorganisms. Several components may in fact nullify the effects of each other. In this way, certain divalent cations counteract the inhibitory effects of long chain fatty acids (Galbraith *et al.*, 1971). Intentional control of factors such as pH and a_w is basic to food technology.

B. Process Factors

All factors influencing microbial colonization, survival, and growth during food processing are termed *process factors* (see Volume II, Chaps. 15–29). These factors can inhibit or even destroy a part of a population or the entire population. Individual components of the flora, on the other hand, are favored in certain instances so that the microbiological profile changes.

C. Extrinsic Factors

Extrinsic factors are the environmental factors such as storage temperature, water vapor pressure, and partial pressures of storage gases that select a particular flora. They are discussed in several chapters in this Volume.

D. Combined Effects

The potential for microbial growth resulting in either spoilage or foodborne illness is a consequence of the combination of intrinsic, processing, and extrinsic factors that are interdependent. Table 13.1 lists the lowest levels of water activity, temperature, and pH that allow the growth of principal foodborne disease organisms. If two of these factors are optimum, growth will occur at extremes of the third; but the growth range in the third is more limited when the other two factors are not optimum. Thus at pH 7 and 37°C, growth of *Clostridium botulinum* type A occurs at a_w 0.94 whereas at pH 5.3, the limiting a_w for

TABLE 13.1

Lowest Reported Levels of Water Activity (a_W), Temperature, and pH That Allow the Growth of Principal Foodborne Disease Organisms

	Minimum permitting growth		
Organism	a_w	°C	pH
Staphylococcus aureus	0.86	6.7	4.5
	0.90[a]	10.0[a]	—
Salmonella	0.945	5.3	4.0
Clostridium perfringens	0.93	12	5.0
Clostridium botulinum			
Type A	0.94	10.0	4.7
Type B	0.945	10.0	4.7
Type E	0.975	3.0	—
Vibrio parahaemolyticus	0.937	5.0	4.8
Yersinia enterocolitica	—	0.0	—
Bacillus cereus	0.95	7.0	4.35

[a] Toxin formation.

growth is 0.99 (Baird-Parker, 1971). Salmonellae will grow at pH 5.8 at an a_w value of 0.971, but at pH 5.0 only if the a_w is 0.986 or above. Similar interdependence for *Staphylococcus aureus* has been reported (Genigeorgis and Sadler, 1966; Genigeorgis *et al.*, 1971a).

Aspergillus glaucus at pH 5 multiplies at a_w values down to 0.73, yet at pH 3.0 only at a_w values above 0.77 (Lubieniecki-von Schelhorn, 1972-1974). Generally, unfavorable effects combine to cause inhibition.

The combination of pH, NaCl (a_w), and $NaNO_2$ that prevents growth of *S. aureus* (enterotoxin A-producing strain) at 35° and 15°C, and the effect of heating at 65°C to 0.1% survival (Bean and Roberts, 1974) illustrate the additional inhibitory effect of heating when coupled with incubation at reduced temperature. *Staphylococcus aureus*, *Salmonella typhimurium*, and *E. coli* are all more sensitive to $NaNO_2$ in the presence of NaCl at 10°C than at 15°-35°C (Albalas and Roberts, 1977). A similar increased sensitivity to $NaNO_2$ in the presence of NaCl is evident with *C. botulinum* types A and B (spore inocula) incubated for 6 months (Roberts *et al.*, 1976). Whereas similar growth and toxin production occurred at 20°-25°C, less was evident at 17.5°C and at 15°C growth was often prevented for 2-3 months. In all combinations of NaCl and $NaNO_2$ growth continued to increase up to 6 months.

The combined effects of reducing pH and a_w values and applying a heat

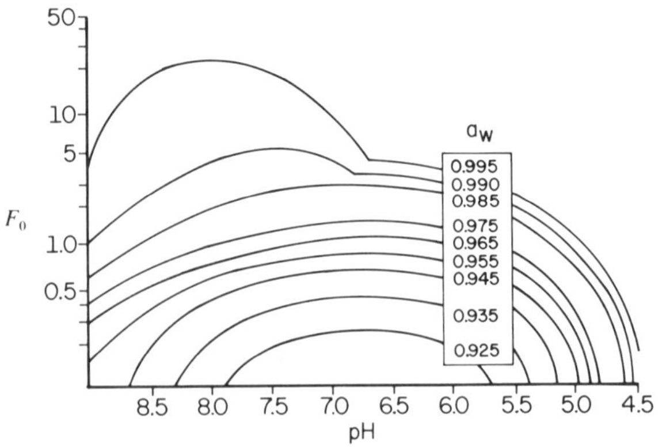

Fig. 13.1 Combinations of pH and a_w which result in various limiting F_0 values for mixed inocula of spores of 151 *Bacillus* strains. (From Braithwaite and Perigo, 1971.)

process are particularly dramatic. Braithwaite and Perigo (1971), using an inoculum containing spores of 151 strains of *Bacillus* spp., determined that the necessary F_0 values for sterilization at pH 7 varied from more than 10 at a_w values above 0.99 to less than 0.1 at a_w values below 0.91. At pH 5.0 this low F_0 value of 0.1 was sufficient with a_w 0.955, and an F_0 of 0.5 was sufficient at a_w 0.975. The complex interaction of the three parameters is illustrated in Fig. 13.1.

TABLE 13.2

Effect of Sodium Nitrite at Various pH Values on Recovery of Heated Spores of *Clostridium sporogenes* PA 3679[a]

Sodium nitrite added (ppm)	No. ($\log_{10}$) spores recovered per milliliter following heat[b] at			
	pH 7.5	pH 7.0	pH 6.5	pH 6.0
0	6.65	6.72	6.75	6.42
50	—	—	4.42	<2.00
100	—	5.85	<2.00	—
200	6.38	4.79	—	—
400	5.56	2.00	—	—
800	3.40	—	—	—
1200	<2.00	—	—	—

[a] Data from Roberts and Ingram (1966). Copyright by Blackwell Scientific Publications Ltd.

[b] Heat treatment was at 80°C for 20 min.

TABLE 13.3

Combinations of Factors Preserving Some Traditional Intermediate Moisture Foods[a]

Products	Inherent factors
Jams	a_w, pH, E_h, preservatives,[b] heat process
Meats	a_w, pH, (E_h), preservatives,[b] competitive flora,[c], storage temperature
Cakes and pastry	a_w, preservatives, heat process, storage temperature
Dried fruits	a_w, pH, preservatives, heat process
Frozen foods	a_w, storage temperature

[a] From Leistner and Rödel (1976).
[b] Preservatives such as sorbic acid in jams and nitrites in meats.
[c] Competitive flora including Lactobacillaceae, Streptococcaceae, molds.

The combined effects of salt (reduction in a_w), pH value, and concentration of nitrite, together with the mild heat process are critical to the stability and safety of cured meat products like pasteurized canned hams. The combined effects of two of these factor. on the growth ability of heated spores, pH value, and concentration of sodium nitrite, is illustrated in Table 13.2. Effectiveness of the nitrite increases about 10-fold with reduction in pH from 7 to 6, presumably as a result of increased concentration of undissociated nitrous acid. Similar interactive effects occur with pH and organic acid preservatives (Ingram and Mackey, 1976). The factors which together limit microbial growth in a range of Intermediate Moisture Foods are listed in Table 13.3.

Knowledge of the effectiveness of a wide range of such combinations would be particularly valuable in allowing predictions of the likely stability and safety of any foodstuff.

III. IMPLICIT PARAMETERS (MUTUAL INTERACTIONS IN MIXED POPULATIONS)

Within a microbial flora (determined primarily by factors such as pH, a_w, temperature, and processes applied) there are important biological attributes of individual organisms that influence the species that predominates. These include the individual growth rates of the microbial strains and mutual influence among species in mixed populations (Mossel and Ingram, 1955; Mossel, 1971, 1975).

A. Growth Characteristics

In a particular environment, an organism grows in a characteristic manner and at a characteristic rate. Length of lag phase, generation time, and total cell yield are ultimately determined by genetic factors. Even different strains of the same species may differ; hence caution is necessary in inferring from observations on one or two strains, what may happen with other strains of the same species. Since these characteristics are genetically determined, change is a rare event and only as a result of mutation.

Accumulation of metabolic products may limit growth of particular species. If the limiting metabolic product can be used as substrate by other species, these may take over partly or wholly, creating an "association" or "succession." As long as metabolically active organisms remain, they continue to interact, so that "dominance" in the flora is a dynamic process. These interactions are synergistic or antagonistic, according to their growth enhancing or inhibiting nature.

B. Mutual Influences

1. Competition

Antagonistic processes include direct predation and parasitism among organisms, and competition for space and nutrients. In food microbiology, competition is more important. For example, even though *S. aureus* is often found in small numbers in raw minced meat, there are never reports of outbreaks due to *Staphylococcus* enterotoxin in raw minced meat. The reason is that the *Pseudomonas–Acinetobacter–Moraxella* association always present in that substrate grows at a higher rate, even when temperatures appear favorable for staphylococci.

a. Depletion. Organisms of high metabolic activity may consume required nutrients, thus selectively impoverishing the environment leading to inhibition of growth of other organisms. Depletion in O_2 or accumulation of CO_2 favors facultative and obligate anaerobes; this occurs in vacuum-packed fresh meat held under refrigeration.

Staphylococci are particularly sensitive to nutrient depletion. Coliforms and *Pseudomonas* spp. may utilize amino acids essential for staphylococcal growth, thereby making them unavailable; even other species of Micrococcaceae can utilize nutrients more rapidly than staphylococci. Streptococci inhibit *Staphylococcus,* especially by exhausting nicotinamide (Iandolo *et al.*, 1965) or niacin and biotin (Haines and Harmon, 1973).

b. Change in Intrinsic Conditions. A change of the conditions of the substrate may either stimulate or decrease competitive effects. Salt concentrations above 3.5% increasingly inhibit the saprophytes, so as to decrease competition and allow staphylococci to grow (Peterson *et al.*, 1962a,b, 1964a). Sucrose inhibits psychrotrophic organisms and decreases their competitive effect, whereas dextrose (0.5%) stimulates saphrophytic growth, which in turn inhibits staphylococcal growth (Peterson *et al.*, 1964b). At temperatures above 20°C, the combined presence of sucrose and whole egg favor growth of staphylococci more than does either constituent alone, while repressing saprophytes (Peterson *et al.*, 1964c).

c. Antimicrobial Products. Substances of microbial origin able to affect the growth of other microbes include H^+ ions, CO_2, organic acids, peroxides, and antibiotics. These are discussed in other chapters of this Volume and by Hurst (1973).

Microbial growth tends to produce acid or alkaline conditions in foods. Acids are produced mainly by dissimilation of carbohydrates, whereas alkaline conditions are usually the consequence of production of ammonia or amines and breakdown of protein. Acid-producing bacteria common on foods include lactic acid bacteria, *Acetobacter, Gluconobacter,* and *Pediococcus.* Bacteria raising the pH include *Pseudomonas, Alcaligenes,* and *Acinetobacter–Moraxella.* As pH is one of the decisive factors for the development of microbial communities on food, even slight shifts can greatly influence the development of a microbial flora (Chapter 5). Tolerance of acid or alkaline conditions is usually, but not always, related to acidogenesis or alkalinogenesis. Thus *Acetobacter* and lactobacilli tolerate pH values that are inhibitive to other food bacteria.

Carbon dioxide is antimicrobial because it displaces oxygen and tends to create acid conditions. This is important in associations of microorganisms on foods packaged in gas-tight pouches. Bacterial growth depletes oxygen and generates CO_2, creating conditions which depress the aerobic acid intolerant *Acinetobacter–Moraxella–Pseudomonas* group and permit growth of more CO_2 tolerant Enterobacteriaceae and *Microbacterium thermosphactum* (*incertae sedis*). The capacity of acid formation varies among different species, even when the conditions are equal, e.g. homofermentative lactobacilli produce more acid than heterofermentatives do. Under comparative conditions streptobacteria formed 1% lactic acid, "atypical streptobacteria" 0.54%, and "beta-bacteria" 0.64%. The respective values for *Leuconostoc mesenteroides* are 0.36% and for *Pediococcus cerevisiae* 0.8%. Because of its strong acid production, *P. cerevisiae* is clearly antagonistic to spoilage organisms in cured meats. This effect has been used in the manufacture of various raw meat products (Volume II, Chapter 15).

Peroxide is formed during oxidative breakdown of lipids and also as a metabolite of microbial carbohydrate degradation. Hydrogen peroxide is such a minimal metabolic poison of bacteria that virtually all species (except some anaerobes) have mechanisms (e.g., catalase, superoxide dismutase) to deal with the substance. Antimicrobial effects occur when the concentration of peroxide is too high to be dealt with or when the catalase and peroxidase systems have been damaged.

Sensitivity to peroxide varies, and some of the more insensitive groups of bacteria, particularly the lactic acid bacteria (Wheater *et al.*, 1952) which produce significant amounts of H_2O_2, seem to gain a competitive advantage thereby. Certain strains accumulate H_2O_2 particularly at low temperatures (+5°C) (Dahiya and Speck, 1968a,b). This has been shown to suppress *Staphylococcus, Pseudomonas,* and other organisms.

Antibiotics usually interfere specifically with the cell metabolism (e.g., by repressing synthesis of proteins). Other substances, such as megacin and polypeptide antibiotics, attack the cytoplasmic membrane to render it permeable. Although many microbial antagonists have been given scientific names, few, if any, are chemically defined. Substances described hitherto are summarized in Table 13.4.

Few of these substances have been isolated and purified and their effect is sometimes difficult to distinguish from that of low molecular weight metabolites, acids, or peroxides. Few are of practical significance in food technology, e.g., nisin and natamycin (cf. Chapter 9).

The conditions of the milieu always determine whether or not inhibitory substances are synthesized. Furthermore, if these agents have been synthesized, it depends on the respective environment whether they interact with other organisms (Hurst, 1973).

2. *Growth Stimulation*

Growth-stimulating activities have been reported among the following food organisms: yeasts, micrococci, streptococci, lactobacilli, bacilli, and Enterobacteriaceae (Marx, 1972). Synergistic processes have primary significance in the buildup of a typical flora. There are many ways in which those symbioses operate.

1. Metabolic products of one microbe can be absorbed and utilized by others. Important examples include the synthesis by yeasts of B vitamins which stimu-

TABLE 13.4

Microbial Antagonists Produced by Various Microbial Families[a]

Family	Antibiotics
Enterobacteriaceae	Bacteriocin, colicin
Micrococcaceae	Staphyline, micrococcin, staphylococcin, aureocin
Streptococcaceae	Nisin, enterocin, diplococcin, streptostasin
Lactobacillaceae	Lactobacillin, lactocidin, lactolin
Pseudomonadaceae	Pyocin, fluocin
Bacillaceae	Subtilin, megacin, clostocin, boticin

[a] From Marx (1972).

late the growth of lactic acid bacteria (Devoyod and Desmazeaud, 1971); polysaccharides such as starch and cellulose are hydrolyzed by molds providing mono- and disaccharides for yeast growth (e.g., *koji*-type fermentation).

2. The changes in pH discussed above in terms of antimicrobial effects may promote growth of certain microorganisms; e.g., in natural fermentations, acid production establishes the dominance of acid-tolerant organisms such as the lactic acid bacteria; surface growth of mold in acid fermentation can raise the pH locally facilitating the growth of acid-sensitive spoilage organisms. Growth of molds on Cheddar cheese was found to raise the pH, thus permitting growth of staphylococci (Duitschaever and Irvine, 1971).

3. Similar growth-promoting effects can result from changes of E_h or a_w values. At warm temperatures, *Clostridium perfringens* can lower the redox potential in the tissues of freshly slaughtered animals so that even more obligately anaerobic deep-spoilage organisms can grow (Ingram and Dainty, 1971). In low moisture foods, lipolysis caused by xerophilic organisms such as molds can liberate water which then supports the growth of less xerotolerant bacteria.

4. The hydrolysis of polymers in plant and animal tissues is a common mechanism facilitating invasion of tissues by spoilage bacteria.

5. Substances inhibitory to some organisms are sometimes eliminated by breakdown by other microorganisms. Examples are degradation of preservatives (e.g., sorbic acid), hydrogen peroxide, phenol and other smoke constituents, destruction of antibiotics, and decomposition of organic disinfectants such as quarternary ammonium compounds (Mossel, 1975).

6. There are symbiotic associations, where maximum growth and normal metabolic activity are not developed, unless both partners are present. A typical example is the close association of *Lactobacillus bulgaricus* and *Streptococcus thermophilus* in yogurt cultures. It has been demonstrated that these organisms in mixed culture display greater metabolic activity than the sum of the two acting individually. In the presence of *L. bulgaricus, S. thermophilus* produced more acid than it did in pure culture; it also grew faster and reached higher numbers (Moon and Reinbold, 1976). This effect was presumed to be due to unidentified growth factors produced by *L. bulgaricus*.

There are similar effects among other lactic acid bacteria. *Streptococcus lactis* subsp. *diacetylactis* has been reported to stimulate growth of a great number of *Lactobacillus* spp. (Branen and Keenan, 1969).

Many of those interactions have been found among the bacteria (enterococci, *Leuconostoc,* and lactic streptococci) used for the manufacture of dairy products. Proteolytic action of *S. faecalis* on casein releases peptides that stimulate growth and acid production by lactic streptococci and lactobacilli, and CO_2 production by *Leuconostoc* (Devoyod and Desmazeaud, 1970, and 1971). Propionibacteria, too, are stimulated by lactobacilli (Nieuwenhof *et al.*, 1969).

Growth stimulation may also be undesirable. For example, aerobic bacteria such as *Streptococcus faecalis, S. durans,* coliforms, and *Bacillus subtilis* have been reported to stimulate growth of clostridia even under conditions not strictly anaerobic (Swartling and Lindgren, 1960).

3. Successions

Because of the complex of continuing interactions between environmental factors and microorganisms, a food at any one point in time carries a characteristic flora—its "association." The microbial profile changes continuously, and one association succeeds another in what is called a "succession."

Spoilage frequently involves a succession of organisms though this has sometimes been designated an "association." One of the best known is the microfloral sequence in unprocessed meat, poultry, and fish products held at temperatures below 5°-10°C, where a highly mixed microflora is superseded by a complex of gram-negative rodlike bacteria commonly referred to as *Pseudomonas-Acinetobacter-Moraxella*. In many cases *Pseudomonas* becomes the dominant genus. While it is known that individual rates of growth and efficiency of substrate utilization are factors in this succession, there are other mutual interactions that many contribute to the development of the succession.

A frequently observed general succession in foods stored in the absence of air is gram-negative rods, to streptococci, to lactobacilli. This happens for example in vacuum-packaged meat and cheese and in silage. It is directed by consumption of O_2 creating anaerobic conditions with rising CO_2 concentrations, leading to the production of acid. Another succession is that observed during the spoilage (heating) of grains (Chapter 23, Section III,B, 3).

Resident populations may also alter their surroundings so that they are replaced by species better suited to the modified habitat. In spoiling raw milk the initial flora depends on the production hygiene and consists mainly of the genera *Micrococcus, Lactobacillus, Streptococcus, Microbacterium, Bacillus, Corynebacterium,* and some gram-negative bacteria including Enterobacteriaceae and a few yeasts. The succession in milk depends largely on the storage temperature. At intermediate temperatures after a transient bactericidal phase where there is no microbial proliferation, proteolytic species with a very brief lag phase start to grow and a short period of alkalization occurs. *Streptococcus lactis, S. lactis* subsp. *diacetylactis,* and *S. cremoris* then begin to multiply, inducing a phase of primary acidification. Subsequently another period of alkalization occurs initiated by sporeformers. A second phase of acidification is caused by streptococci and lactobacilli. When the milk curdles spontaneously, a pH range is reached where the gram-negative flora are virtually eliminated. Because of the drop of the water activity, blastomycetes (*Endomyces lactis*) may supplant the lactobacilli in the outer layer of the curd. In the deeper layers the development of propionic and butyribacteria is favored by microaerophilic or anaerobic conditions. Finally, a heavy mold growth arises and organic acids are metabolized. Hence, the pH is raised and secondary contaminants will continue the decomposition, resulting in proteolysis or even putrefaction.

At refrigeration temperatures, the miscellaneous initial flora of raw milk, predominantly gram-positive, is gradually overgrown by psychrotrophic (gram-negative) organisms (e.g., *Pseudomonas, Alcaligenes*). Consequently, spoilage is proteolytic rather than glycolytic or due to lactic acid formation.

Generally speaking, high water activities, high protein content, and a lack of carbohydrates result in a gram-negative proteolytic flora which normally will result in putrefaction, particularly in foods of neutral pH. In foods with low a_w and/or pH, a gram-positive flora will be supported only when fermentable carbohydrates are present. Declining water activities favor the growth of yeasts and select in certain cases a pure mold flora. There are many transitions and some species occur almost in any biotope, such as *Micrococcus* and *Bacillus* species. These, however, seldom constitute a typical spoilage association.

Microbial interactions are clearly important in fermentations. There are two general types of vegetable fermentation: those common in the Orient involving molds in the initial stages (typified by *koji*) and bacterial lactic fermentation of vegetables common in Europe and North America.

In soy sauce fermentation, a mold (e.g., *Aspergillus oryzae*, or *A. soyae*) grows on soybeans or a mixture of soybeans and wheat saccharifying starch, making nutrients available for the later growth of lactic acid bacteria and yeasts. The growth of the mold is terminated when 18% w/v sodium chloride solution is added. The mixture (soy mash or *moromi*) undergoes a lactic acid and yeast fermentation for at least 1 year, during which the lactic bacteria lower the pH from 6.5 to 4.5 when yeasts start to grow and the lactobacilli gradually decline. Ultimately, the number of yeasts seems to decrease in very old mashes (Yong and Wood, 1976).

The lactic acid bacterial sequence in vegetable fermentations is well-defined. Initial mixed bacterial growth eliminates O_2, favoring growth of lactic acid bacteria dominated in the first stages by *Leuconostoc* which produces lactic acid, thereby reducing the pH and producing a substrate for the growth of lactobacilli. The succession is determined by a progressive increase in acidity.

4. Starter Cultures

Many fermentations depend on the use of "starter cultures," which may be pure or mixed, to establish the succession of microbial growth. The objective is to replace a wild flora with organisms that efficiently guide the fermentation to desired products (such as lactic acid, acetic acid, ethanol).

The substrate is commonly sterilized or pasteurized prior to the addition of the starter cultures in many milk products. In other foods, starters are used without prior heating as in pickles, raw sausage, and cured meats, but chemical treatments (SO_2) may also be used (e.g., grapes).

The metabolic products of starter cultures exert a selective pressure that controls undesired contaminants either from the initial flora, or from secondary contaminants. Thus the proteolytic spoilage flora and even some pathogens are suppressed (Hurst, 1972).

Starters include gram-positive bacteria, yeasts, and molds. The dairy industries utilize mainly lactobacilli, lactic streptococci, micrococci, pediococci, *Leuconostoc,* and propionibacteria. Examples of starter cultures are listed in Table 13.5.

Lactic cultures have been suggested to suppress the growth of spoilage and pathogenic organisms in fresh meat. For example, a mixture of *Streptococcus lactis* and *Leuconostoc cremoris* (*citrovorum*) reduced the gram-negative proteolytic (*Pseudomonas-Achromobacter*) and lipolytic flora in minced meat, which improved the keeping quality (Reddy *et al.,* 1975).

Similarly, the acid formation of irradiated *Pediococcus cerevisiae* cells was claimed to be effective in suppressing *Staphylococcus aureus, Clostridium botulinum* types A and B, *Salmonella* spp., and *C. perfringens* in minced meat, ham, and turkey meat (Al-Mashat, 1973).

The purity of starter cultures used commercially for the production of foods must be controlled. Contamination with bacteriophage may cause loss of starter efficiency either by causing cell lysis or undesirable genetic change. Impurities, toxic metabolites, and microbial contamination can present health hazards. Recommendations for the production and the use of starter cultures are summarized (Anonymous, 1974b):

1. General Requirements

Unwanted spread of the organisms either in the plant or in the environment must be prevented when microbes are cultivated on a large scale. This requires appropriate ventilation, sewage systems, and safe removal of the residues of the cultures. Antibiotics and related compounds used in therapy may not be detectable either in enzymes or in metabolites.

Careful training of the personnel involved is indispensable. Special care must be given to protective clothes, sanitary facilities, disinfection of equipment, and health control of the personnel.

2. Requirements for the Production of Starter Cultures

Commercial starters should contain nothing but those cultures, metabolites, and culture ingredients which are not hazardous to human health.

The cultures should not contain organisms other than the labeled species. The essential organisms must be labeled with full scientific names. Species which are

TABLE 13.5

Starter Cultures in Food Production[a]

Product	Microorganism	Remarks
Sour milk and cream	*Streptococcus lactis* *Streptococcus cremoris* *Streptococcus lactis,* subsp. *diacetylactis*	No pyogenic streptococci or enterococci should be present
Sour cream butter	*Leuconostoc cremoris* *Leuconostoc dextranicum*	
Quarg	*Lactobacillus casei*	
Yogurt	*Streptococcus thermophilus* *Lactobacillus bulgaricus*	
Kefir	*Lactobacillus lactis* *Lactobacillus brevis* *Streptococcus lactis* *Streptococcus cremoris* and alcohol forming yeasts not clearly identified	*Geotrichum candidum* should not occur in the culture
Cheese (in general)	*Lactobacillus casei* and other lactose-fermenting species	
Soft cheese with surface smear	*Arthrobacter globiformis* (*Brevibacterium linens*)	Smearing with starter cultures prevents mold growth
Hard cheese	*Propionibacterium jensenii* (*petersonii*) *Propionibacterium* (*freudenreichii*) subsp. *shermanii*	Pure cultures are used if the milk has been pasteurized
German blue cheese Gorgonzola Roquefort Stilton Wensleydale Dorset Danablue Mycella ''Marmor'' cheese ''Niva'' cheese ''Tyrol'' grey cheese Gammelost	*Penicillium roquefortii* and varieties thereof	Toxin formation in some strains Formation of antibiotics in some strains
Camembert Brie	*Penicillium caseicolum* Commercial names: *Penicillium*	*Penicillium camemberti* THOM is no longer used

(continued)

TABLE 13.5 (*continued*)

Product	Microorganism	Remarks
Coulommier Carree de l'Est Olivet Neufchatel and other acid curd cheese with mold formation	*album, Penicillium albidum, Penicillium candidum, Pencillium camemberti*	
Raw sausage	*Pediococcus cerevisiae* *Micrococcus aurantiacus* M53 Lactobacilli, such as *Lactobacillus plantarum* or *Lactobacillus brevis*	Improvement of color Microbial stabilization Lactic acid formation Drop of pH Improvement of consistency Prevent growth of undesired organisms
	Mixture of micrococci and lactobacilli *Debaryomyces kloeckeri* *Penicillium nalgiovensis*	Formation of salami aroma Improves appearance
Alcoholic fermentation products	*Saccharomyces cerevisiae* and related species	Particular precautions when fungal amylases are used
Organic acids	*Acetobacter* spp. *Lactobacillus* spp. *Aspergillus niger* *Corynebacterium glutamicum* *Penicillium luteum* *Penicillium chrysogenum* *Aspergillus fumaricus*	Some penicillia and aspergilli are suspected to form toxins or antibiotics
Fermented foods		
Sauerkraut	*Lactobacillus* species	
Juices from fruit and fruit pulps	*Aspergillus alleaceus* *Penicillium roquefortii* *Penicillium stoloniferum* *Rhizopus javanicus*	Cultures added should not contain toxin forming organisms

[a] From Anonymous (1974b). Copyright by Deutsche Forschungsgemeinschaft.

newly introduced should be registered and deposited in a recommended culture collection. It must be proven that the cultures are harmless if properly used. Any strain of a species suspected to be potentially pathogenic or toxinogenic as used must be tested individually. The tests must establish whether the objections are confirmed. Long-term experiments for chronic toxic effects must be carried out in addition to the routine tests usually applied for the well known toxicants.

3. Control

Starter cultures must be tested periodically to check whether the organisms labeled are present and contaminants are absent. Recognized institutions may issue valid certificates. Sampling and analysis should be carried out by the official food control agency.

14

Cleaning, Disinfection, and Hygiene

I. INTRODUCTION

Keeping food very hot or very cold is the most effective and efficient means of reducing microbiological contamination in the food industries (see Chapter 1). There are, however, situations where removal or destruction of microorganisms by chemical agents is preferred for technical or economic reasons. The labels, under which these products are sold, do not necessarily list all ingredients or the limitations for safe use. Therefore, some governmental organizations evaluate such formulations and approve those found acceptable for use under specified conditions (U.S. Department of Agriculture, 1977).

This chapter is based on information gathered from (a) various governmental publications, such as the Federal Register issued by the United States Government, (b) comprehensive reference works on germicides (Reddish, 1957; Benarde, 1970; Sykes, 1974; Block, 1977), (c) food hygiene and sanitation texts (Cook and Binsted, 1975; Guthrie, 1972; Hobbs and Gilbert, 1978; Litsky, 1973), and (d) several technical manuals and monographs (Hopkins and Schulze, 1958; Maurer, 1978; National Canners Association, 1975; Elliott, 1980).

The presentation is focused on those practices and sanitizing agents that are widely recognized to be effective and safe for use in the food processing and serving industries. Related subjects, such as methods of testing or bases for choosing particular chemical agents and procedures for their use, are discussed only briefly. More detailed information can be obtained from the literature cited.

The terminology used in this field seems prone to misunderstanding. Confusion arises because some of the key words have more than one meaning, and they carry different implications in various regions of the world. For example, "food sanitation" is commonly used in North America to describe measures taken to promote the safety and quality of foods; whereas, the United Kingdom prefers to describe these measures as "food hygiene," because "sanitation" is associated mainly with treatment and disposal of sewage or other wastes. Several associated technical terms have been defined in the Glossary at the end of the chapter because their precise meaning may differ in other contexts.

II. WATER

Food hygiene depends on water in several important ways. It is the solvent for application of cleaning agents and disinfectants and the carrier for the removal of soils. The level of available moisture, as measured by water activity (a_w), determines whether food spoilage will occur due to the growth of microorganisms (Chapter 4), and the presence of water enhances the lethal effect of heat on microorganisms during food processing (Chapter 1). Except when distilled under controlled conditions, water may contain a variety of impurities from the environment including dissolved gases, inorganic salts, soluble organic substances, and biota. The specific kinds and concentrations of these impurities are important with respect to potability of the water and its suitability for use in conjunction with food. For example, contaminated water is a common vehicle for the transmission of pathogenic microorganisms, especially the enteric bacteria, to man, either directly or through food. The microflora of water frequently includes food-spoilage organisms, such as the psychrotrophic bacteria, that shorten the shelf-life of refrigerated foods. Treatment of water supplies to kill these harmful microorganisms is a prerequisite for food hygiene.

''Hard'' water contains calcium, magnesium, and less frequently other ions (such as iron, manganese, aluminum, strontium, or zinc) at a total level of 60 ppm or more. Hard water reacts with soap, other anionic detergents, bicarbonate cleaners, and a variety of disinfectants to reduce their sanitizing actions. Hard water can be ''softened'' by addition of a sequestering agent, such as sodium tripolyphosphate, or by the lime-soda process, which precipitates and removes the hardness, or preferably by treatment of the water by the so called ''zeolite'' process which uses ion exchange resins to replace the alkaline metal ions with sodium (Elliott, 1980).

III. CLEANING AGENTS

Cleaning is a process that suspends or dissolves dirt, usually in water, from the object being cleaned. Its effectiveness can be enhanced by (a) application of energy in some form such as scrubbing, spraying, or turbulent flow, and (b) use of chemical adjuvants, known collectively as cleaners or cleaning agents, which may decrease surface tension and emulsify, peptize, suspend, or solubilize various types of soils (Bueltman, 1972; Zottola, 1973).

Frequent cleaning at 2- to 4-hr intervals of food-contact surfaces and associated work areas is necessary to remove food residues, grease films, mineral deposits, dust, and other dirt that may protect and nourish microorganisms. Cleaning is a useful method of decontamination, because more than 90% of the microorganisms are removed with the soil (Gilbert, 1970); however, the process cannot be relied on to kill the organisms. Disinfectants are sometimes combined with detergents, although this is not the best way of using a disinfectant, nor is there convincing evidence that one is needed in most cleaning operations. The presence of certain chlorinated compounds, for example, may improve the peptizing action of the cleaning mixture, but its alkalinity inhibits their disinfecting action. Treatment with heat or a chemical disinfectant *after* cleaning is more effective. The terms "detergent sterilizer" and "detergent sterilant" are misleading because they are not sterilizing agents.

Table 14.1 summarizes the kinds and amounts of agents that are commonly added to water to clean up food processing or serving operations. Certain cleaners can be used in combination, but others are antagonistic. For example, strong alkalis and strong acids or anionic and cationic wetting agents are not compatible as mixtures; but they can be used sequentially. Many wetting agents can also be mixed with either alkalis or acids to improve the detergent action.

In some food industries, which use or produce dry products (such as chocolate, flour, cake mixes, gelatin, dry milk, and peanut butter) water remaining after cleaning operations is intolerable because it permits caking, sticking to surfaces, chemical changes, and growth of microorganisms. After cleaning, special care is required to remove residual moisture by thorough draining and circulation of clean warm air. Where these measures are not feasible, alternative dry cleaning procedures may be useful. They include brushing, wiping, scraping, burning, vacuuming and treatment with nonaqueous solvents of low toxicity, such as ethyl alcohol, hexane, white mineral oil, 1,1,1-trichloroethane or Freon TF.

Because selection of the proper combination of cleaning agents and procedures can be rather complicated (Thomas, 1969), many users prefer to buy commercially formulated products which are sold under trade names at prices substantially above the cost of the ingredients. However, the suppliers usually provide their customers with services and advice about water hardness, pH adjustment, soil characteristics, equipment corrosion, and other factors that affect the cleaning process. Before buying a new cleanser, the purchaser would be well advised to compare competitive products in terms of their effectiveness under the expected conditions of use, potential hazards to employees or consumers, corrosiveness for equipment, and unit cost of the concentration specified for use (Seabaugh *et al.*, 1977). Incidentally, treatment of hard water with ion-exchange resins, which remove objectionable cations, will permit the use of cheaper, more effective cleaners that also contribute less phosphorus to the environment (Watrous, 1975).

Types, Functions, and Limitations of Cleaning Agents Used in the Food Industries[a]

Categories of aqueous cleaners	Approximate concentrations for use (%)[b]	Examples of chemicals used[c]	Functions	Limitations
Clean water	100	Usually contains dissolved air and soluble minerals in small amounts	Solvent and carrier for soils, as well as chemical cleaners	Hard water leaves deposit on surfaces Residual moisture may allow microbial growth on washed surfaces Promotes rusting of iron
Strong alkalis	1–5	Sodium hydroxide Sodium orthosilicate Sodium sesquisilicate	Detergents for fat and protein Precipitate water hardness Produce alkaline pH	Highly corrosive Difficult to remove by rinsing Irritating to skin and mucous membranes
Mild alkalis	1–10	Sodium carbonate Sodium sesquicarbonate	Detergents Buffers at pH 8.4 or above	Mildly corrosive High concentrations are irritating to skin
		Trisodium phosphate Sodium tetraborate	Water softeners	
Inorganic acids	0.5	Hydrochloric Sulfuric Nitric Phosphoric Sulfamic	Produce pH 2.5 or below Remove precipitates from surfaces	Very corrosive to metals, but can be partially inhibited by amines Irritating to skin and mucous membranes
Organic acids	0.1–2	Acetic Hydroxyacetic Lactic	Remove inorganic precipitates and other acid-soluble substances from surfaces	Moderately corrosive, but can be inhibited by various organic nitrogen compounds

(continued)

TABLE 14.1 (*continued*)

Categories of aqueous cleaners	Approximate concentrations for use (%)[b]	Examples of chemicals used[c]	Functions	Limitations
Organic acids (*continued*)	0.1–2 (*continued*)	Gluconic Citric Tartaric Levulinic Saccharic	Remove inorganic precipitates and other acid-soluble substances from surfaces	Moderately corrosive, but can be inhibited by various organic nitrogen compounds
Anionic wetting agents	0.15 or less	Soaps Sulfated alcohols Sulfated hydrocarbons Aryl-alkyl polyether sulfates Sulfonated amides Alkyl-aryl sulfonates	Wet surfaces Penetrate crevices and woven fabrics Effective detergents Emulsifiers for oils, fats, waxes, and pigments Compatible with acid or alkaline cleaners and may be synergistic	Some foam excessively Not compatible with cationic wetting agents
Nonionic wetting agents	0.15 or less	Polyethenoxyethers Ethylene oxide–fatty acid condensates Amine–fatty acid condensates	Excellent detergents for oil Used in mixtures of wetting agents to control foam	May be sensitive to acids
Cationic wetting agents	0.15 or less	Quaternary ammonium compounds	Some wetting effect Antibacterial action	Not compatible with anionic wetting agents Inactivated by many minerals and some other soils
Sequestering agents	Variable (depending on hardness of water)	Tetrasodium pyrophosphate Sodium tripolyphosphate Sodium hexametaphosphate Sodium tetraphosphate Sodium acid pyrophosphate	Form soluble complexes with metal ions such as calcium, magnesium, and iron to prevent film formation on equipment and utensils	Phosphates are inactivated by protracted exposure to heat Phosphates are unstable in acid solutions

		Ethylenediaminetetraacetic acid (sodium salt) Sodium gluconate with or without 3% sodium hydroxide	See also strong and mild alkalis above	
Abrasives	Variable	Volcanic ash Seismotite Pumice Feldspar Silica flour Steel wool Metal or plastic ''chore balls'' Scrub brushes	Removal of dirt from surfaces with scrubbing Can be used with detergents for difficult cleaning jobs	Scratch surfaces Particles may become imbedded in equipment and later appear in food Damage skin of workers
Chlorinated compounds	1	Dichlorocyanuric acid Trichlorocyanuric acid Dichlorohydantoin	Used with alkaline cleaners to increase peptizing of proteins and minimize milk stone deposits	Not germicidal because of high pH Concentrations vary depending on the alkaline cleaner and conditions of use
Amphoterics	1–2	Mixtures of a cationic amine salt or a quaternary ammonium compound with an anionic carboxy compound, a sulfate ester, or a sulfonic acid	Loosen and soften charred food residues on ovens or other metal and ceramic surfaces	Not suitable for use on food contact surfaces
Enzymes	0.3–1	Proteolytic enzymes produced in cultures of aerobic, spore-forming bacteria	Digest proteins, and other complex organic soils	Inactivated by heat Some people become hyper-sensitive to the commerical preparations Some have contained *Salmonella*

[a] Based largely on information assembled by Elliott (1980).
[b] Concentration of cleaning agent in solution as applied to equipment.
[c] Some regulatory agencies require prior approval (e.g., see U.S. Department of Agriculture, 1977).

IV. DISINFECTANTS

Chemicals that reduce pathogens and other microorganisms to an insignificant level are known as disinfectants. They are known also as "sanitizers" when applied to water supplies (Johnson, 1975), or to utensils and equipment in food-processing factories, retail markets, and food-service establishments (U.S. Department of Health, Education and Welfare, 1976a). Disinfectants are most useful in food industries where application of heat or cold is inadequate to cope with microbial contamination, as in poultry eviscerating and milk-processing operations. Rigorously cleaned surfaces harbor few microorganisms; hence, for many processing operations, particularly of raw foods, disinfection is unnecessary. Equipment used for ready-to-eat foods does require disinfection to destroy pathogens and other bacteria of public health significance. In addition, the destruction of spoilage microorganisms is an important factor in improving the shelf-life and quality of processed foods.

These chemicals cannot be depended on to sterilize the items to which they are applied, because bacterial spores and a few resistant vegetative cells are not readily killed (Davis and Keen, 1974). The quaternary ammonium compounds tend to be less effective on gram-negative bacteria than on gram-positive bacteria. In fact, the maximum concentrations that can be used without a final water rinse, may allow survival of some coliform organisms and pseudomonads. In addition, the germicidal activity depends on the conditions of use including concentration, time, temperature, pH, water hardness, amount and kind of organic matter present, type of surface, and kinds and concentration of microorganisms to be killed. These conditions influence not only the effectiveness of disinfection but the rate at which these solutions decline in potency. They may also determine how frequently sanitization procedures should be repeated. To be safe, food contact surfaces should be cleaned and sanitized with freshly made solutions before each use, and at intervals of 2 to 4 hr during use, where continuous operations are in progress. Experience, based on laboratory data, will permit more precise scheduling of the reagents and procedures to prevent build up of microorganisms on equipment and utensils. A number of authors (including Pelczar and Reid, 1965; Litsky, 1973; Lewis, 1978) have identified a dozen desirable characteristics of an "ideal" disinfectant that relate to potency, corrosiveness, persistence, toxicity, responses to temperature and pH, organoleptic properties, and cost.

As noted above, the adverse effect of food residues and other organic matter on most sanitizers makes thorough cleaning highly desirable before disinfection of equipment and utensils. However, proprietary mixtures of cleaning agents and disinfectants are frequently recommended as "detergent-sanitizers" for simultaneous cleaning and disinfection, especially in closed "clean-in-place" systems (see Section VI, B below). They often require more chemicals and longer exposure times but may be attractive to management because less labor is involved and mistakes are less likely.

Because of toxicity, corrosiveness, or production of off-flavors, many excellent hospital disinfectants cannot be used safely in ordinary food sanitation procedures. Formaldehyde, phenolic compounds, and salts of heavy metals are rarely used, and several others, including ethylene oxide and chlorine gas, are usually limited to factory processes such as decontamination of dry spices or chlorination of water, respectively.

Table 14.2 identifies the disinfectants and adjuvants in 21 types of sanitizing solutions that the United States Government has approved for use without a water rinse on food-processing equipment and utensils (U.S. Department of Health, Education and Welfare, 1977). The active ingredients include chlorine compounds, iodine, iodine compounds, and a variety of organic surfactants such as the quaternary ammonium compounds. The regulations specify the maximum permissible concentrations of the active ingredients and the conditions of use. They neither require nor prohibit rinsing with water after the equipment and utensils have been treated with these sanitizing solutions, provided adequate drainage occurs before the next use. Nevertheless, some jurisdictions may require rinsing with potable water as the final step of sanitizing food equipment and utensils. If the hazard of contaminating food with chemical residues from the sanitizing solution is perceived to exceed the hazard of contaminating it with microorganisms from the rinse water, then a final rinse is appropriate; otherwise, none is needed

One of the newer disinfectants, not included in Table 14.2, is glutaraldehyde, which has strong sporicidal activity and is growing in popularity for hospital disinfection (Thomas and Russell, 1974; Saitanu and Lund, 1975; Collins and Montalbine, 1976). There are insufficient data to determine whether this disinfectant can be used safely on food contact surfaces.

V. METHODS OF TESTING

An assortment of physical, chemical, and microbiological tests have been used to estimate (a) the concentration of active agents, (b) their relative potency under specified conditions as compared to a standard, (c) measurement of environmental factors (such as pH, water hardness, or soil residue) that affect detergency or antibacterial performance, (d) the extent of decontamination under either simulated or actual conditions of use, and (e) the residual levels of particular chemical or microbiological contaminants on treated surfaces and in the food that has been in contact with them.

Among the well-known bacteriological methods are (a) the phenol coefficient method in which the relative bactericidal potency of disinfectants is compared

TABLE 14.2

Some Sanitizing Solutions Listed in the United States Code of Federal Regulations[a] for Use on Food-Processing Equipment, Utensils, and Certain Other Food-Contact Surfaces

Category	Aqueous solution of:	Maximum concentration of active ingredient(s)	Conditions of use
1.	Potassium, sodium, or calcium hypochlorite with or without the bromides of potassium, sodium, or calcium	200 p.p.m.[b] available halogen calculated as chlorine	Adequate drainage after treatment. No final rinse required[c]
2.	Dichloroisocyanuric acid, trichloroisocyanuric acid, or the sodium or potassium salts of of these acids with or without the bromides of potassium, sodium, or calcium	100 p.p.m. of available halogen calculated as available chlorine	As in 1, above
3.	Potassium iodide, sodium *p*-toluenesulfonchloramide, and sodium lauryl sulfate	25 p.p.m. of titratable iodine; all constituents will be present at a level not in excess of the minimum required to produce their intended functional effect	As in 1, above
4.	Iodine, butoxy monoether of mixed ethylene propylene polyalkylene glycol, having a cloud point of 90°–100°C in 0.5% solution and an average molecular weight of 3300, and ethylene glycol monobutyl ether. Diethylene glycol monoethyl ether may be added as an optional ingredient	25 p.p.m. of available iodine. The adjuvants used with the iodine will not be in excess of the minimum amounts required to accomplish the intended functional effect	As in 1, above
5.	Iodine, hydroiodic acid α-(*p*-nonylphenyl)-ω-hydroxypoly(oxyethylene) (average mol wt 748) and/or polyoxyethylene–polyoxypropylene block polymers (minimum average mol wt 1900). Isopropyl alcohol may be added as an optional ingredient	As in 4, above	As in 1, above

6.	Iodine, sodium iodide, sodium dioctylsulfosuccinate, and polyoxyethylene–polyoxypropylene block polymers (having a minimum average mol wt of 1900)	As in 4, above,	
7.	Dodecylbenzenesulfonic acid, polyoxyethylene–polyoxypropylene block polymers (having a minimum average mol wt of 2800)	400 p.p.m. of dodecylbenzene–sulfonic acid and 80 p.p.m. of polyoxyethylene–polyoxypropylene block polymers (having a minimum average mol wt of 2800)	In addition to use on food-processing equipment and utensils, this solutions may be used on glass bottles and containers intended for holding milk; no final rinse required
8.	Iodine, butoxy monoether of mixed (ethylene–propylene) polyalkylene glycol, having a minimum average mol wt of 2400, and α-lauroyl-ω-hydroxypoly (oxyethylene) with an average 8–9 moles of ethylene oxide and an average molecular weight of 400	As in 4, above	In addition to use on food-processing equipment and utensils, this solution may be used on beverage containers, including milk containers or equipment. Rinse water treated with this solution can be recirculated as a preliminary rinse. It is not to be used as a final rinse
9.	*N*-Alkyl (C_{12}-C_{18}) benzyldimethylammonium chloride compounds having average mol wts of 351–380 and consisting principally of alkyl groups with 12–16 carbon atoms with or without not over 1% each of groups with 8 and 10 carbon atoms. Isopropyl alcohol may be added as an optional ingredient	200 p.p.m. of the active quaternary ammonium compound	As in 1, above
10.	Trichloromelamine and either sodium lauryl sulfate or dodecylbenzene sulfonic acid	Not more than sufficient trichloromelamine to produce 200 p.p.m. of available chlorine and either sodium lauryl sulfate at a level not in excess of the minimum required to produce its intended functional effect or not more than 400 p.p.m. of dodecylbenzene sulfonic acid	In addition to use on food-processing equipment and utensils and other food-contact articles, this solution may be used on beverage containers except milk containers or equipment. No final rinse required
11.	Equal amounts of *N*-alkyl (C_{12}-C_{18}) benzyldimethylammonium chloride and	200 p.p.m. of active quaternary ammonium compound	In addition to use on food-processing equipment and utensils, this solution may be used on

(*continued*)

TABLE 14.2 (*continued*)

Category	Aqueous solution of:	Maximum concentration of active ingredient(s)	Conditions of use
	N-alkyl (C_{12}-C_{18}) dimethylethylbenzyl-ammonium chloride having an average mol wt of 384		food-contact surfaces in public eating places. No final rinse required
12.	Sodium salt of sulfonated oleic acid, polyoxyethylene–polyoxypropylene block polymers having an average mol wt of 2000 and 27 to 31 moles of polyoxypropylene	200 p.p.m. of sulfonated oleic acid, soldium salt	In addition to use on food-processing equipment and utensils, this solution may be used on glass bottles and other glass containers intended for holding milk. All items treated with this sanitizing solution shall have a drain period of 15 min prior to use in contact with food. No final rinse required
13.	Iodine and alkyl (C_{12}-C_{15}) monoether of mixed (ethylene–propylene) polyalkylene glycol, having a cloud point of 70° C–77° C in 1% aqueous solution and an average mol wt of 807	As in 4, above	As in 1, above
14.	Iodine, butoxymonoether of mixed (ethylene–propylene) polyalkylene glycol, having a cloud point of 90°–100°C in 0.5% aqueous solution and an average mol wt of 3300 and polyoxyethylene–polyoxypropylene block polymers having a minimum average mol wt of 2000	As in 4, above	As in 1, above
15.	Lithium hypochlorite	290 p.p.m. of available chlorine and 30 p.p.m. of lithium	As in 1, above
16.	Equal amounts of *N*-alkyl (C_{12}-C_{18}) benzyldimethylammonium chloride and *N*-alkyl (C_{12}-C_{14}) dimethylethylbenzyl-	200 p.p.m. of active quaternary ammonium compound	As in 11, above

	ammonium chloride (having average weights of 377–384). with the optional adjuvant substances tetrasodium ethylenediaminetetra-acetate and/or α-(p nonylphenol)-ω-hydroxy-poly(oxyethylene) having an average poly(oxyethylene) content of 11 moles		
17.	Di-*N*-alkyl (C_8-C_{10}) dimethylammonium chlorides and isopropyl alcohol having average mol wt's of 332–361	150 p.p.m. of the active quaternary ammonium compound	As in 11, above
18.	*N*-alkyl (C_{12}-C_{18}) benzyldimethylammonium chloride, sodium metaborate, α-terpineol and α-[*p*-(1,1,3,3-tetramethylbutyl) phenyl]-ω-hydroxypoly(oxyethylene) produced with 1 mole of the phenol and 4–14 moles of ethylene oxide	200 p.p.m. of active quaternary ammonium compound and 66 p.p.m. of α-[*p*-(1,1,3,3,-tetramethylbutyl) phenyl]-ω-hydroxypoly-(oxyethylene)	As in 1, above
19.	Sodium dichloroisocyanurate and tetrasodium ethylenediaminetetraacetate	100 p.p.m. of available chlorine	As in 11, above
20.	*o*-Phenylphenol, *o*-benzyl-*p*-chlorophenol, *p*-tertiary amylphenol, sodium α-alkyl (C_{12}–C_{15})-ω-hydroxypoly (oxyethylene) sulfate with the poly(oxyethylene) content averaging one mole, potassium salts of coconut oil fatty acids, and isopropyl alcohol or hexylene glycol	800 p.p.m. of total active phenols consisting of 400 p.p.m. *o*-phenylphenol, 320 p.p.m. *o*-benzyl-*p*-chlorophenol and 80 p.p.m. *p*-tertiary amylphenol	For single use applications only. No final rinse required
21.	Sodium dodecylbenzenesulfonate	Not more than 430 p.p.m. and not less than 25 p.p.m. of sodium dodecylbenzenesulfonate	As in 1, above

[a] U.S. Department of Health, Education and Welfare (1977).
[b] p.p.m. = parts per million.
[c] In some countries, other than the United States, a final water rinse is compulsory after use of a sanitizing solution.

with that of phenol (Horwitz, 1975), (b) the qualitative suspension test of the German Society for Hygiene and Microbiology (Deutsche Gessellschaft für Hygiene und Mikrobiologie, 1972), (c) the quantitative suspension test of the Dutch Committee of Phytopharmacy (Commissie voor Fytofarmacie, 1975), (d) the Kelsey and Sykes (1969) test for evaluating hospital disinfectants, (e) the use-dilution test and its several modifications for determining the germicidal effectiveness of a sanitizing agent when applied under particular environmental conditions (Block, 1977), and (f) several methods of swabbing, rinsing, or direct agar contact used to demonstrate the extent of microbial destruction on treated surfaces of utensils and equipment (Gabis *et al.*, 1976; Litsky, 1973; Spooner and Sykes, 1972).

The need for methods specifically applicable to various food sanitation situations has generated many modified methods based on the use-dilution test. Examples inlcude the modified Weber–Black test (Chambers, 1956) to evaluate the effect of water hardness on disinfecting action, a method based on exposure of inoculated stainless steel strips to disinfectants (Mosley *et al.*, 1976), and a method for evaluating virucidal efficacy of disinfectants on surfaces (Gaustad *et al.*, 1974).

Where the same solution of disinfectant is used repeatedly, the deterioration of antibacterial potency can be checked either by a suitable chemical test such as the *N,N*-diethyl- *p*-phenylenediamine (DPD) (Rand *et al.*, 1976) method or the starch-iodine test for residual chlorine (National Canners Association, 1975), or by a bacteriological test such as the "in-use test" (Kelsey and Maurer, 1966, 1974; Maurer, 1972, 1978). The latter has been adapted from the clinical laboratory practice of checking discard jars in which used instruments (forceps, scissors) are immersed in disinfectant until they can be cleaned. One ml of disinfectant solution from the used jar is removed and diluted with neutralizing agent. (A 0.5% solution of sodium thiosulfate in 9 ml of nutrient broth is used for hypochlorite neutralization.) Ten drops of the diluted material are placed on two nutrient agar plates. One plate is incubated at 35°–37°C and the other at room temperature for 72 hr. Appearance of bacterial colonies indicates that the disinfectant is ineffective and its concentration should be increased.

VI. SANITATION OF FOOD-PROCESSING PLANTS

A. General Cleaning Requirements

In food-processing plants the main reason for cleaning both processing equipment and processing areas is to remove food debris and other soils which may contain food-poisoning or spoilage microorganisms and thus contaminate the food. A further reason is to maintain an environment of cleanliness that encourages a high standard of hygiene among personnel.

Cleaning programs should be designed to suit the needs of a particular product and process (Thomas, 1969). The type of detergent, its pH, and its use temperature will be governed by the sort of soil to be removed and by the material and construction of the equipment to be cleaned. The frequency of cleaning process equipment should take into account the initial levels and types of microbial contamination found on raw materials; whether these materials are able to support microbial growth and are likely to be held during processing at times and temperatures at which significant growth can occur; and the presence of "dead" areas within the processing equipment where microbial growth could occur and cause contamination of the product during processing (Benarde, 1970; Nickerson and Sinskey, 1972; Parker and Litchfield, 1962).

The basic stages and sequence of wet cleaning and disinfection follow.

1. *Remove gross soil* by the use of cold or hot water. The proper temperature of the water will depend on the type of soil to be removed and the nature of the equipment.

2. *Apply a chemical agent* which could be a detergent, acid, or alkali that has been found to emulsify or dissolve the debris and dirt adhering to the equipment.

3. *Scrub* the soiled surfaces, if necessary.

4. *Rinse off the suspended soil* using, as appropriate, hot or cold water.

5. *Apply hot (77°C) water or chemical disinfectant* to kill any remaining microorganisms. The preferred means to destroy residual microorganisms on surfaces is to raise the temperature of the surface of equipment with steam or hot water, and maintain it at or slightly above 77°C for 30 sec or more (Longrée, 1967; U.S. Department of Health, Education and Welfare, 1976a). In Great Britain, the Department of Health and Social Security (1978) recommends 80°C for 2 minutes.

A second-choice alternative, to be used if the heating step is not practicable, is to apply a chemical disinfectant after the equipment is visually clean. If any soil remains after cleaning, the cleaning step should be repeated, because virtually all disinfectants suitable for use in the food industry are rapidly inactivated by organic matter. Cleaning agents and disinfectants are most effective when applied in hot water, provided they are not volatilized or destroyed by the heat.

6. *Rinse* the disinfectant (if used) from the equipment using potable water; however some regulatory agencies do not require rinsing with water after equipment has been disinfected with the sanitizing solutions shown in Table 14.2. Before rinsing, all drainage points should be opened or parts stacked so that rapid drying of the equipment occurs in order to prevent growth of microorganisms on moist soil not removed during cleaning.

There are five different methods of cleaning. Based on convenience and efficiency, most persons responsible for cleaning would grade the effectiveness of the methods in the order given below. Grading the methods on cost alone is difficult.

B. Clean-in-Place (CIP) Systems

A CIP system includes a fully controlled consistent cleaning schedule that can be partially or fully automated. It also minimizes the shutdown time for cleaning and eliminates damage to equipment during dismantling for manual cleaning. However, a CIP system requires modification of some equipment and entails heavy capital investment for automation. It must be properly designed and used, since a makeshift system, or a poorly programmed cleaning operation, may leave pockets of soil where massive microbial growth can occur, or it may leave pools of cleaner or disinfectant where they will enter the flow of product. CIP systems are almost certainly most expensive to install but are most cost-effective when savings in labor and downtime are taken into account. Manual cleaning, on the other hand, requires little capital investment but substantially greater labor costs. Its effectiveness depends on much hard work and a high degree of supervision.

Equipment should be designed so that it can be cleaned in place by one of three methods.

1. *Circulation of cleaning fluids* through the equipment at a flow velocity high enough to remove all soil from internal surfaces. In practice, the minimum flow velocity required to clean pipelines is 1.5 m/sec; higher velocities, up to 3 m/sec, are required if dead ends, such as T-pieces, are present. This method is particularly applicable to pipelines and is used widely in breweries and dairies. It can also be applied to circuits containing pieces of equipment, such as plate heat exchangers, small mixing vessels, and some types of circulating pumps, provided that rotors and paddles, which would impede the main flow velocity during cleaning, are removed and cleaned manually.
2. *Cleaning by fixed pressure jets.* This method is used to clean pieces of equipment *in situ* by installing pressure jets which will ensure that all surfaces are subjected to an effective pressure spray or rinse. The pressures used can vary from 100 to 1400 lb/in.^{2} (approximately 7.0 to 98.4 kg/cm^2 or 7 to 95 atmospheres), and the distance between the jet and the surface to be cleaned depends upon the pressure and type of jet used. A common application of this system is the cleaning of moving conveyor belts during production or at fixed cleaning times.
3. *Revolving pressure jets.* The interior surfaces of tanks, vats, or other large containers are often inaccessible and therefore difficult to clean. Revolving jets

are sited so that all inner surfaces of the container are subjected to a pressure spray. These can either be permanently installed or inserted through an opening during the cleaning cycle.

C. Washing Machines

Many types of containers used for food processing can be cleaned in washing machines. Basically the machines carry out mechanically the six stages given above. Jets and sometimes brushes are used to remove the soil with the help of a detergent. Although disinfectants are often included in the final rinse to kill residual organisms, it is more usual to rely on hot water (>82°C) (U.S. Department of Health, Education and Welfare, 1976a).

In some mechanical washing machines high frequency waves (ultrasonic cleaning) are used for loosening adhesive soil from containers and particularly deposits from inaccessible areas in equipment, e.g., residues in filler valves. Minor (1959) has described the application of ultrasonics to the removal of beerstone films and residues from tubes, filler valves, and closing machines in a brewery. Normally the system consists of an ultrasonic generator which produces waves having frequencies of 3 to 4 kHz and transducers which convert the ultrasonic energy into mechanical vibrations. The parts are immersed in a detergent solution and the ultrasonic energy applied. Soil should be removed in a few seconds, and the parts are then rinsed in the normal way.

Washing machines give excellent results provided the jets and brushes are properly maintained, a suitable detergent is used to remove the soil from utensils, and the temperature and residence time are adequate at each stage.

D. Manual Cleaning Using Pressure Sprays

The equipment normally available is mobile and will supply cleaning fluids at high pressures by means of a variety of jets. The type, size, and working pressure of the jet is important and must be chosen to match the cleaning problem. As mentioned above, jets are available to operate at pressures from 100 to 1400 $lb/in.^2$ (7.0 to 98.4 kg/cm^2 or 7 to 95 atmospheres).

In large plants, it is possible to install a ring circuit with junctions conveniently located for connecting lances (flexible hoses with nozzles for directing jets of liquid). For large free-standing pieces of equipment and for floors and walls, this is a very effective and efficient means of cleaning. This system uses relatively small amounts of water, permits control of both pressure and working distance of the jet, and provides a convenient means of cleaning crevices and large surfaces. However, the spray may contaminate adjacent equipment unless used with care.

E. Manual Cleaning Using Brushes, Cloths, and Similar Equipment

In the absence of pressure sprays it may be necessary to clean free-standing equipment manually. Small equipment, containers, or parts can be cleaned by hand with brushes, by pressure jets, or in washing machines. For manual cleaning, it is convenient to use the three-tank system, as follows: (1) Rinse parts in running water to remove bulk soil. (2) Transfer parts to tank 1 which should contain a suitable detergent solution. Soak for a period to loosen the soil. (3) Transfer parts to tank 2 containing *clean* detergent solution and scrub parts clean. (4) Rinse parts in water, then transfer to tank 3, containing hot (77°C) water, for at least one-half minute (U.S. Department of Health, Education and Welfare, 1976a). In Great Britain, the Department of Health and Social Security [1978] recommends 80°C for two minutes.) Immerse parts in cages to avoid burning the hands. As a second-choice alternative, immerse parts in a chemical disinfectant solution, and allow contact for several minutes. (5) If a chemical disinfectant is used, rinse in clean water if a rinse is required. Stack the parts to drain and dry quickly. All cloths, brushes, sponges, or other cleaning aids used for manual cleaning must be washed frequently and disinfected either by heat or by soaking for at least 10 min in a solution containing 200 ppm of available chlorine or 400 ppm of quaternary ammonium compound.

Manual cleaning is adaptable to all types and sizes of equipment, but its effectiveness depends heavily on the integrity of the worker. Thorough scrubbing, frequent changes of solutions, and special care to avoid cross contamination are essential.

F. Dry Cleaning

Food factories producing dry products such as chocolate, flour, cake mixes, and dried milk, have difficulty in maintaining clean dry equipment. Generally they scrape or brush surfaces and use a vacuum cleaner. Ethyl alcohol (95%) or other solvents of low toxicity may be helpful in dissolving some residues.

Providing that surfaces are maintained dry, the risk of microbial growth is small. Where a limited amount of water is permissible for cleaning, the surfaces can be covered with a detergent foam, which is removed by vacuum. The surfaces can then be dried with warm air.

VII. SANITATION IN FOOD SERVICE ESTABLISHMENTS AND MARKETS

The purposes and basic steps of sanitation in restaurants and retail markets are essentially the same as those for food-processing plants (Kupchik and Katz,

1977: U.S. Department of Health, Education and Welfare, 1976a; see also above). Except for the use of dishwashing machines in food service establishments, the cleaning and disinfection procedures are generally performed manually with simple tools and a limited variety of materials. Table 14.3 suggests some cleaning and disinfectant solutions that are compatible with various food-contact surfaces and other parts of the work area. When properly applied, they facilitate the removal of soil without harm to the user or damage to the surface. The effective use of these materials requires both hot and cold running water of potable quality.

A general purpose cleaner, such as an anionic liquid detergents, is suitable for cleaning most floors, walls, windows, and other portions of the work area. It may be supplied commercially in a container that can be joined to a water tap so that a predetermined concentration is dispensed with the water. Attachment of such a device to the outlet of a public water supply is illegal in some jurisdictions

TABLE 14.3

Selection of Cleaners and Chemical Disinfectants for Use on Various Kinds of Food-Contact Surfaces[a]

Surface	Cleaning solution[b]	Disinfectant solution[b]	Selected references
Hands	Wash with soap and warm water (40°–50°C) *n*-propanol, or isopropanol	Iodophors or organic chlorine compounds; 70% ethanol	Sykes (1974); Wewalka *et al.* (1977); Litsky (1973)
Glass and ceramic-ware	Alkaline, nonionic detergents	Hypochlorites, organic chlorine compounds, iodophors, quaternary ammonium compounds, or ampholyte solutions	Sykes (1974); Goldenberg and Relf (1967)
Stainless steel	Alkaline, nonionic detergents; detergent disinfectants; or acid-detergent mixtures for removal of milkstone	As for glass and ceramic-ware above; iodophores are commonly used in the milk industry	Sykes (1974); Parker and Litchfield (1962)
Plastic, rubber, wood, paint	Alkaline, nonionic detergents, detergent disinfectants	As for glass and ceramic-ware above	Gilbert and Watson (1971)
Fabrics (cotton or synthetic fibers)	Hot (77°–80°C) water and alkaline, nonionic detergents	Hypochlorites or organic chlorine compounds (boiling or autoclaving may also be used to kill microorganisms)	Walter and Shillinger (1975)

[a] Refer also to Tables 14.1 and 14.2.

[b] Water is the diluent. Water alone, at 77°C or above is preferred over chemical disinfectants.

because of the potential for backsiphoning. However, vacuum breakers are available to solve such a problem.

Alkaline nonionic detergents or detergent sanitizers are commonly used to clean food-contact surfaces of equipment and utensils. In dishwashing machines, a powdered nonfoaming detergent should be used at the concentration and temperature recommended by the manufacturer as compatible with the operating characteristics of the machines (Guthrie, 1972). Washing is typically done at 60°–75° C and is followed by a rinse with hot water (75°–80° C) or with warm water 25°–50° C) plus a disinfectant such as one of those listed in Table 14.3. The clean equipment should be dried quickly and stored in a manner that prevents recontamination prior to reuse.

Special materials such as scouring powder, washing soda, or a grease solvent will be needed where heavy contamination with food or other organic residues occurs as in sinks and on some cooking utensils. Cleaners or cleaning tools that are excessively corrosive or abrasive are to be avoided.

The following additional materials, equipment, and procedures are suggested for use in manual cleaning operations.

1. Disposable paper should be used for scrubbing and wiping if available.
2. Nylon brushes, which can be dried quickly, should be rinsed well in water after use, shaken to remove excess water, and stored dry. They should be disinfected with hot water or autoclaved frequently.
3. Nondisposable cloths and sponges are difficult to disinfect and should be avoided. If used, they should be boiled, washed, and then dried quickly.
4. Mop heads should be detachable so that they can be enclosed in a cotton bag and washed in a machine at 70°–80° C for 10 min, then boiled or autoclaved and dried quickly.
5. In order to carry out the foregoing procedures, special facilities should be provided for (a) washing the equipment, (b) autoclaving or disinfection by heat, and (c) drying and storage where the equipment will remain clean.
6. Potable water is essential.

VIII. PHYSICAL FACILITIES AND SERVICES THAT ENHANCE FOOD HYGIENE

Food hygiene in factories, markets, and restaurants is determined not only by sanitary practices and technical procedures, but by the construction and layout of physical facilities, design of equipment, and the types of supporting services (Guthrie, 1972; Hobbs and Gilbert, 1978). Features that facilitate cleaning are just as important as those that contribute to the efficiency and convenience of operation. Equipment should be made of impervious materials that resist corrosion, and it should be designed to avoid dead ends, cracks, or crevices where microorganisms can accumulate and grow.

Wood, fabrics, corroded metals, or unglazed ceramics are far more difficult to clean and disinfect than stainless steel, nonporous plastics, glass, or glazed ceramics; therefore, the latter materials of construction are preferable, especially for food-contact surfaces (Gilbert and Watson, 1971). Aluminum may also be satisfactory, provided corrosive alkaline cleaners are avoided. Filled concrete is widely used for floors that receive heavy use. When smooth and evenly sloped to drains, such floors can be washed and dried with ease.

It is important to establish a pattern of operations that minimizes cross-contamination of the final product from any other part of the process. Potential sources of contamination, in addition to the workers' hands, are incoming raw material, inadequately cleaned equipment (such as meat slicers and cutting boards), dust and spray, dirty floors, and improperly maintained cleaning equipment such as mops and towels (Dempster, 1973; Dempster *et al.*, 1973; Gilbert, 1969; Gilbert and Maurer, 1968; Hess, 1973; Lehmann, 1965).

The greatest threat is transfer of contamination from raw to processed food, which may occur because of (a) lack of physical separation between foods at the early stages of ingredient preparation and those at the final steps of food processing, (b) assignment of duties in both of the above areas to an employee without applying the hygienic safeguards noted in Section IX below, (c) poor ventilation, including recirculation and inadequate filtration of air, as well as failure to maintain positive air pressure at the more vulnerable stages of processing, or (d) carelessness on the part of cleaning crews in using common tools and solutions to sanitize all parts of the establishment.

Equipment can become a major source of contamination unless (a) it is properly designed either to be cleaned in place or to be taken apart easily for cleaning, and (b) it is installed to prevent harborage of vermin. Nonportable equipment should be sealed at all points of contact with floor, wall, ceiling, or other equipment, or should be installed with 20–30 cm of clearance for ease of cleaning under, above, and around it.

The exclusion of pets (especially cats and dogs) and pests (such as insects, rodents, and birds) from food establishments is necessary because these creatures harbor pathogenic microorganisms that are transmissible through food to man. They are also potential sources of food-spoilage organisms (Smith, 1973). Pest control extends beyond the food-processing and serving areas, and it is especially important where there are storage or waste disposal facilities. The most effective steps are sealing (vermin proofing) of buildings and denial of access to food or shelter anywhere on the premises. Pesticides can be effective when used indoors or outdoors, if employees and the public are protected from these toxic chemicals.

Several supporting services are required to operate a clean food establishment. The principal items are

1. A continuous power source to provide heat, light, refrigeration and to operate equipment. Effective time–temperature control is the most important single contributor to food protection.
2. An abundant supply of hot and cold, potable, running water. Washing in clean water containing a detergent is the basic step in sanitizing procedures, without which the food industries could not operate.
3. A plumbing system that transports water to points of use such as toilets, sinks, and processing lines, and also removes sewage. Essential features are installation of traps on drains, construction that prevents cross-connections and backflow from sewerage to water lines, and provision of mixing valves for control of temperature and flow rate.
4. A ventilation system that distributes clean tempered air and removes aerosols created within the building. Filtration of the air can reduce the need for cleaning by removing dust. Proper humidity and temperature control, not only provides comfortable working conditions, but discourages growth of molds in films of moisture.
5. Facilities for sanitary collection and disposal of solid wastes. Impermeable tightly covered containers are essential.
6. Physical separation of raw materials from prepared food. Separate work areas, utensils, cleaning tools, and operating personnel are necessary to prevent contamination of cooked food by microorganisms from raw materials.

IX. EMPLOYEE HEALTH AND HYGIENIC PRACTICES

The benefits of cleaning and disinfection can be nullified by insanitary food handling during preparation for marketing or serving (Rajagopalan and Shiffman, 1974; National Academy of Sciences/National Research Council, 1975). In the United States, for example, poor personal hygiene was thought to be a contributory factor in 228 of 859 outbreaks of foodborne diseases for which contributory factors were reported during the 4-year period 1972–1975 (U.S. Department of Health, Education and Welfare, 1975, 1976b). Effective control requires an understanding on the part of employees and management that people, as well as their surroundings, are sources of contamination. Personal hygiene and good health of all workers are, therefore, important factors in reducing the transfer of microorganisms to the food. A management commitment is needed to train employees and to supervise them so that a high level of protection is maintained (Greene, 1974; Zottola, 1973). The following aspects are especially important (Hobbs and Gilbert, 1978; U.S. Department of Health, Education and Welfare, 1976a).

1. *Health:* Persons who are afflicted with a communicable disease that can be transmitted through food, or are carriers of microorganisms that cause such a disease, should not work where food is being processed or served. Each employee has an obligation to report to his supervisor evidence of illness such as boils, infected wounds, diarrhea, sore throat, respiratory congestion, or jaundice. Management, with medical advice, also has an obligation to exclude known carriers or ill persons from areas where they may contaminate food intended for public consumption. Identification of ambulatory cases or carriers is technically difficult, and their exclusion by sick leave or reassignment of duties is so expensive that a share of the cost may have to be paid from public funds. Routine medical examination of employees is unreliable because of the rapidity with which infection and recovery can occur. However, examination for a specific pathogen may be a necessary control measure after an employee has been ill or after an outbreak has occurred among consumers.

2. *Cleanliness:* The foremost requirement of personal hygiene is that the workers' hands be washed thoroughly with soap and warm water before starting work and after handling potentially contaminated materials, such as raw meat and poultry. Thus repeated washing will be necessary during the work period whenever raw food, soiled utensils, cleaning tools or other contaminated items have been handled. After engaging in any nonwork activity such as eating, drinking, smoking, blowing the nose, or using the toilet, the employee must wash his hands just before returning the work (Pether and Gilbert, 1971). Cross-contamination has occurred in some hospitals from the common use of soap and towels in washrooms frequented by many people, and the same can be true for the food industry (Symposium on Hygiene of the Hand, 1974; Rowland and Alder, 1972). The use in common of plastic brushes for cleaning fingernails may also be a questionable practice.

Washed hands should be rinsed in running potable water and then thoroughly dried with a clean single-service towel or a stream of warm air. Disinfectants, such as chlorhexidine or a quaternary ammonium compound, are sometimes added to hand soap, rinse water, or hand creams, but their benefits appear marginal with respect to reducing the numbers of bacteria on skin. They may act as preservatives for liquid soap or cream, which could otherwise become a source of contamination. However, bacterial decomposition of some quaternary ammonium compounds has been reported (Dean-Raymond and Alexander, 1977). Frequent cleaning of empty dispensers and filling with fresh liquid is probably a more reliable means of controlling growth than is the use of disinfectants.

Even the most rigorous washing is no guarantee that the hands will be free of objectionable microorganisms; therefore, gloves of rubber or plastic are sometimes worn when food must be handled. They are beneficial only when changed frequently and washed inside and out before being worn again. A feeling of awkwardness, the nuisance of changing, and the accumulation of perspiration on the hands are disadvantages that restrict their general use.

Since some managers and employees insist that direct use of the hands speeds up food preparation, the practice is likely to continue. Therefore, people who handle food should perform no other duties that may contaminate their hands, such as taking money, handling raw foods, picking up soiled dishes, or cleaning work surfaces and serving areas. While working, employees should avoid unhygienic habits such as fingering the nose or mouth, combing the hair, and smoking. As stated above, whenever contamination of the hands occurs for any reason they should be washed, rinsed, and dried before returning to work.

Changing from street clothes to a clean uniform when entering the work area tends to minimize the number and variety of microorganisms that may be shed from the employee's outer garments while he is working. The complete uniform includes a cover for the hair such as a net or hat, and footwear, usually rubber boots or safety shoes. Individually assigned lockers and freshly laundered uniforms in clean change rooms will encourage employees to put on clean clothing as often as necessary to maintain a high level of personal cleanliness.

One of the most effective ways to minimize contamination of food is to use clean utensils, whenever possible instead of the hands. Forks, spoons, tongs, spatulas, mechanical slicers,and disposable paper, metal foil, or plastic films are among the most useful devices for preventing food from coming into direct contact with the worker's hands. In order to use these utensils skillfully, most people need training by a person who has already learned the techniques. Many untrained individuals are so accustomed to manipulating food with their hands that they consider most utensils to be inefficient and unacceptable. Under the stress of a heavy workload, close supervision is necessary to prevent people from reverting to hand operations.

X. SAFETY PRECAUTIONS

All cleaning agents and disinfectants are, in some measure, toxic. Misuse may cause skin irritation, eye damage, gastrointestinal lesions, or even death from accidental ingestion of large quantities. Management has an obligation to ensure the safe storage and use of these materials. The first step toward control is a continuous evaluation of the cleaning and disinfection required in each area of the establishment. Specifications can then be formulated for supplies of reagents and equipment.

Development of standard operating procedures (SOPs) for routine cleaning and disinfection gives cleaning crews and their supervisors at all levels a common understanding of prescribed methods, thus improving safety and effectiveness. The SOPs may vary in technical details, depending on the knowledge and experience of the personnel who use them, but they should cover the following subjects.

1. Directions for preparation of "use solutions" from commercial products, including safety precautions, such as wearing goggles, gloves, respirators, and protective clothing while handling concentrated materials.
2. Identification of the uses for which each solution is appropriate.
3. Description of techniques and tools for application, including the proper temperature and time of contact, as well as the operation of washing machines or other equipment.
4. Criteria for renovation or discard and replacement of used solutions.
5. Methods of disposal of used solutions by dilution, chemical inactivation, biodegradation or other means.
6. Requirements for rinsing and/or drying, before the next use of the item being treated.
7. Directions for plain labeling and color coding, and isolated storage of cleaning agents and disinfectants to prevent their accidental substitution for food ingredients.
8. Designation of the kinds of tools such as mops, brushes, and sprayers that may be used, and directions for their maintenance.
9. Specifications for training of employees who perform cleaning and disinfection operations.
10. Requirements for evaluation of the effectiveness of cleaning and disinfection procedures including criteria, methods, and responsibilities of workers and management. Useful food sampling plans and microbiological methods for examination of samples have been recommended by the ICMSF (1974, 1978).

Conversion of the SOPs to safe and effective practice depends on the skilled performance of the sanitation crew, whose motivation will, in turn, reflect the support of management. It would be difficult to overstate the importance of such teamwork for food safety and quality.

XI. GLOSSARY

The definitions have been adapted from several sources including the British Standard Glossary of Terms Relating to Disinfectants (BS 5283, 1976), Webster's New Collegiate Dictionary (G & C Merriam Co., Springfield, Illinois, 1976) and various publications of the Codex Alimentarius Commission.

Definitions inconsistent with those offered by Codex have been avoided wherever practical. Although there is no official document containing a glossary of Codex definitions, useful guidance has been provided by the unpublished list of *Codes of Hygienic Practice Definitions* prepared by Mr. L. J. Erwin, Australia (no date).

1. *Adjuvant:* A compound that aids, facilitates, or enhances the functioning of another substance.
2. *Adulterate:* To make impure by the addition of a foreign, inferior, or harmful substance.
3. *Antisepsis:* The destruction or inhibition of microorganisms on living tissues having the effect of limiting or preventing the harmful results of infection. It is not a synonym for disinfection.
4. *Antiseptic:* A substance that destroys, checks the growth, or inhibits the action of microorganisms on living tissues having the effect of limiting or preventing the harmful results of infection. It is not a synonym for disinfection.
5. *Bactericide:* A chemical agent which under defined conditions is capable of killing bacteria, but not necessarily bacterial spores (see sporicide).
6. *Bacteriostasis:* Inhibition of growth, but not killing, of bacteria by chemicals such as some dyes and sulfonamides.
7. *Bacteriostat:* A chemical agent which under defined conditions induces bacteriostasis.
8. *Chemical sterilizing agent:* A misnomer used as a synonym for disinfectant.
9. *Clean (or cleanse):* To free from dirt or pollution; to free from contamination or disease.
10. *Cleaner (cleaning agent or cleanser):* A preparation for cleaning, or an implement or machine for cleaning.
11. *Cleaning:* The removal of food residues, dirt, grease or other objectionable matter
12. *Contamination:* The direct or indirect transmission of objectionable matter to a food product.
13. *Detergent:* A cleansing agent, as soap; any of numerous synthetic water-soluble or liquid organic preparations that are chemically different from soaps, but are able to emulsify oils, hold dirt in suspension, and act as wetting agents.
14. *Detergent–Sanitizer:* A mixture of chemicals formulated to clean and disinfect food-contact surfaces simultaneously.
15. *Detergent–Sterilizer:* A misnomer used as a synonym for sanitizer.
16. *Dirt:* A filthy or soiling substance (as mud, dust, or grime); excrement.
17. *Disinfect:* To kill infectious microorganisms outside the body by chemical or physical means. Some bacterial spores and a few highly resistant vegeta-

tive organisms may survive the process, but all other microorganisms are reduced to insignificant levels or eliminated completely.

18. *Disinfection:* The destruction of microorganisms by the use, on cleaned surfaces, of hygienically satisfactory chemical or physical agents or processes, so as to reduce the number of microorganisms to a level that will not cause harmful contamination of food that may come in contact with the disinfected surfaces.

19. *Disinfectant:* A chemical or physical agent that kills infectious microorganisms (see ''disinfect'' above).

20. *Emulsifier:* A surface-active agent that promotes the dispersion of small fat globules in water.

21. *Food hygiene:* A component of sanitary science that is concerned with the preparation, handling, packing, transport, storage, wrapping, or exposure for sale of food so as to ensure that the food is kept clean and fit for human consumption. It involves a knowledge of the habits of microorganisms in food so that their growth may be checked by heat, cold, pH, a_w, and preservatives. It also includes the design, maintenance, and cleanliness of establishments and their surroundings, of vehicles that transport food, and utensils and equipment with which food comes into contact. In its broadest sense, food hygiene may extend to the rearing, feeding, marketing, and slaughter of animals as well as the sanitary procedures designed to prevent bacteria of human origin from reaching foodstuffs.

22. *Fungicide:* A chemical agent which under defined conditions is capable of killing fungi including their spores.

23. *Fungistasis:* Inhibition of the growth or multiplication of fungi.

24. *Fungistat:* A chemical agent which under defined conditions induces fungistasis.

25. *Germicide:* A vague term which should be avoided. A substance capable of killing certain microorganisms.

26. *Hard water:* Water that contains alkaline metal ions, mainly calcium and magnesium, at levels above 60 ppm.

27. *Hot water:* Potable water at a minimum temperature of 77°C.

28. *Hygiene:* A science of the establishment and maintenance of health; conditions or practices (as of cleanliness) conducive to health.

29. *Infection:* The establishment of a pathogen in its host after invasion; the state produced by the establishment of an infective agent in or on a suitable host.

30. *Manager:* The person who manages the plant and is responsible for ensuring that the hygiene requirements are observed.

31. *Pathogen:* A specific causative agent (as a bacterium or virus) of disease.

32. *Peptize:* To cause to disperse in a medium; specifically, to bring proteins into colloidal solution.

33. *Pollute:* To make physically impure or unclean; to contaminate (an environment) especially with man-made waste.

34. *Potable water:* Water that is wholesome at the point of usage and is in accordance with the WHO requirements contained in the "International Standards for Drinking Water" (WHO, 1971). Pathogens are typically absent, but food-spoilage microorganisms are apt to be present.

35. *Residue:* The remainder of cleaning agent, disinfectant, or their derivatives on food equipment and utensils after sanitization.

36. *Sanitarian:* A specialist in sanitary science and public health, e.g., a milk sanitarian.

37. *Sanitary:* Adequately hygienic to ensure a safe, sound, wholesome product fit for human consumption.

38. *Sanitation:* The act or process of making sanitary; the promotion of hygiene and prevention of disease by maintenance of sanitary conditions.

39. *Sanitization:* A process for adequate treatment of surfaces that is effective in destroying vegetative cells of pathogenic bacteria and in substantially reducing the numbers of other microorganisms. Such treatment shall not adversely affect the product and shall be safe for the consumer.

40. *Sanitizer:* A chemical agent used for sanitization.

41. *Sequestering agent:* A substance that holds a metallic ion in solution, usually by inclusion in an appropriate coordination complex.

42. *Soft water:* Water that contains less than 60 ppm of calcium, magnesium, and other metallic ions that react with soap or other sanitizing compounds.

43. *Soil:* Something that spoils or pollutes (as food residues, refuse, dirt, sewage, or dung).

44. *Sporicide:* A chemical agent which under defined conditions is capable of killing bacterial spores.

45. *Sterile:* Free from all living organisms.

46. *Sterilization:* A process intended to destroy or remove all living organisms.

47. *Sterilize:* To free from living microorganisms.

48. *Sterilizing agent (Sterilant):* A chemical agent that destroys all living microorganisms.

Bibliography

Achtzehn, M. K. (1964). Zur Chemie und Mikrobiologie des Packpapiers. *Mitteilungsbl. GDCh (Ges. Dtsch. Chem.)-Fachgruppe Lebensmittelchem. Gerichtl. Chem.* **18,** 251-252.

Acott, K. M., and Labuza, T. P. (1975a). Inhibition of *Aspergillus niger* in an intermediate moisture food system. *J. Food Sci.* **40,** 137-139.

Acott, K. M., and Labuza, T. P. (1975b). Microbial growth response to water sorption preparation. *J. Food Technol.* **10,** 603-611.

Acton, J. C., and Dick, R. L. (1977). Cured pigment and color development in fermented sausage containing glucono-delta-lactone. *J. Food Prot.* **40,** 398-401.

Adler, N. (1965). Residual ethylene oxide and ethylene glycol in ethylene oxide sterilized pharmaceuticals. *J. Pharm. Sci.* **54,** 735-742.

Aharoni, Y., and Stadelbacher, G. J. (1973). The toxicity of acetaldehyde vapors to postharvest pathogens of fruits and vegetables. *Phytopathology* **63,** 544-545.

Albalas, B., and Roberts, T. A. (1977). The effect of sodium nitrite and sodium chloride on growth of food poisoning bacteria at intermediate temperatures. *Acta Microbiol. Hell.* **22,** 286-293.

Alderton, G., and Snell, N. S. (1963). Base exchange and heat resistance in bacterial spores. *Biochem. Biophys. Res. Commun.* **10,** 139-143.

Alford, J. A., Smith, J. L., and Lilly, H. D. (1971). Relationship of microbial activity to changes in lipids of foods. *J. Appl. Bacteriol* **34,** 133-146.

Al-Mashat, A. J. H. (1973). Control of food-poisoning bacteria by irradiated *Pediococcus cerevisiae* cells. *Diss. Abstr. Int. B* **33,** 4850.

Amano, K., Shibasaki, I., Yokoseki, M., and Kawabata, T. (1968). Preservation of fish sausage with tylosin, furylfuramide and sorbic acid. *Food Technol.* **22,** 881-885.

Amerine, M. A., and Joslyn, M. A. (1940). Commercial production of table wines. *Calif. Agric. Exp. Stn., Bull.* No. 639, 143 pp.

Anderson, E. S., Humphreys, G. O., and Willshaw, G. A. (1975). The molecular relatedness of R factors in Enterobacteria of human and animal origin. *J. Gen. Microbiol.* **91,** 376-382.

Anderson, G. E., and Whitehead, J. A. (1974). The validity of the methylene blue reduction test in the grading of ice-cream. *J. Appl. Bacteriol.* **37,** 487-492.

Anderson, J. D., Gillespie, W. A., and Richmond, M. H. (1973). Chemotherapy and antibiotic-resistance transfer between enterobacteria in the human gastro-intestinal tract. *J. Med. Microbiol.* **6,** 461-473.

Anderson, J.M., and Baird-Parker, A.C. (1975). A rapid and direct plate method for enumerating *Escherichia coli* biotype I in food. *J. Appl. Bacteriol.* **39,** 111-117.

Anellis, A., Lubas, J., and Rayman, M. M. (1954). Heat resistance in liquid eggs of some strains of the genus *Salmonella*. *Food Res.* **19,** 377–395.

Anellis, A., Berkowitz, D., and Kemper, D. (1973). Comparative resistance of nonsporogenic bacteria to low-temperature gamma irradiation. *Appl. Microbiol.* **25,** 517–523.

Angelotti, R., Foter, M. J., and Lewis, K. H. (1961). Time-temperature effects on salmonellae and staphylococci in foods. 1. Behavior in refrigerated foods. 2. Behavior at warm holding temperatures. *Am. J. Public Health* **51,** 76–88.

Anonymous (1969). "Report of the Joint Committee on the Use of Antibiotics in Animal Husbandry and Veterinary Medicine." HM Stationery Off., London.

Anonymous (1971). Aseptic cartoning will give four-month life to fruit drinks. *Packag. News* Nos. 1 and 4.

Anonymous (1973). Conclusions and recommendations of the microbiological session. *Proc. Int. Symp. Nitrite Meat Prod., Zeist, Neth.* (B. Krol and B. J. Tinbergen, eds.), pp. 125–126. Pudoc, Wageningen, Netherlands.

Anonymous (1974a). Bestimmung der Oberflächenkeimzahl (Bakterien, Schimmelpilze, Hefen und coliforme Keime) auf night saugfähigen Packstoffen. *Milchwissenschaft* **29,** 670–673.

Anonymous (1974b). "Mikroorganismen für die Lebensmitteltechnik." Komm. Ernährungsforsch., Komm. Prüf. Fremder Stoff Lebensm., Gemeinsame Mitt., Dtsch.Forschungsgem., Bonn-Bad Godesberg.

Ashworth, J., Jarvis, B., Rawlinson, A. P., and Wood, J. M. (1974). "The Replacement of Sulphur Dioxode as the Microbial Preservative in British Fresh Sausages," Res. Rep. No. 207. Food Res. Assoc., Leatherhead, England.

Ayres, J. C., and Denisen, E. L. (1958). Maintaining freshness of berries using selected packaging materials and antifungal agents. *Food Technol.* **12,** 562–567.

Babcock, G. F., Berryhill, D. L., and Marsh, D. H. (1973). R-factors of *Escherichia coli* from dressed beef and humans. *Appl. Microbiol.* **25,** 21–23.

Bacq, Z. M., and Alexander, P. (1961). "Fundamentals of Radiobiology." (2nd Ed.) Pergamon, Oxford.

Bacus, J. N., and Deibel, R. H. (1972). Nitrite burn in fermented sausage. *Appl. Microbiol.* **24,** 405–408.

Baer, E. F., Gilden, M. M., Wienke, C. L., and Mellitz, M. B. (1971). Comparative efficiency of two enrichment and four plating media for isolation of *Staphylococcus aureus*. *J. Assoc. Off. Anal. Chem.* **54,** 736–738.

Bailey, C., James, S. J., Kitchell, A. G., and Hudson, W. R. (1974). Air-, water-, and vacuum-thawing of frozen pork legs. *J. Sci. Food Agric.* **25,** 81–97.

Baird-Parker, A. C. (1971). Factors affecting the production of bacterial food poisoning toxins. *J. Appl. Bacteriol.* **34,** 181–197.

Baird-Parker, A. C., and Baillie, M. A. H. (1973). The inhibition of *Clostridium botulinum* by nitrite and sodium chloride. *Proc. Int. Symp. Nitrite Meat Prod.,* Zeist. (B. Krol and B. J. Tinbergen, eds.), pp. 77–90. Pudoc, Wageningen, Netherlands.

Baird-Parker, A. C., and Davenport, E. (1965). The effect of recovery medium on the isolation of *Staphylococcus aureus* after heat treatment and after storage of frozen or dried cells. *J. Appl. Bacteriol.* **28,** 390–402.

Baird-Parker, A. C., and Freame, B. (1967). Combined effect of water activity, pH and temperature on the growth of *Clostridium botulinum* from spore and vegetative cell inocula. *J. Appl. Bacteriol.* **30,** 420–429.

Baird-Parker, A. C., and Holbrook, R. (1971). The inhibition and destruction of cocci. *In* Inhibition and Destruction of the Microbial Cell" (W. B. Hugo, ed.), Academic Press, New York.

Baird-Parker, A. C., Boothroyd, M., and Jones, E. (1970). The effect of water activity on the heat resistance of heat sensitive and heat resistant strains of salmonellae. *J. Appl. Bacteriol.* **33,** 515-522.

Banwart, G. J., and Ayres, J. C. (1957). The effect of pH on the growth of *Salmonella* and functional properties of liquid egg white. *Food Technol.* **11,** 244-246.

Barach, J. T., Adams, D. M., and Speck, M. L. (1974). Recovery of heated *Clostridium perfringens* type A spores on selective media. *Appl. Microbiol.* **28,** 793-797.

Baran, W. L., and Stevenson, K. E. (1975). Survival of selected pathogens during processing of a fermented turkey sausage. *J. Food Sci.* **40,** 618-620.

Bard, J., and Townsend, W. E. (1971). Meat curing. *In* "The Science of Meat and Meat Products" (J. F. Price and B. S. Schweigert, eds.), pp. 452-483. Freeman, San Francisco, California.

Barkai-Golan, R., and Aharoni, Y. (1976). The sensitivity of food spoilage yeasts to acetaldehyde vapors. *J. Food Sci.* **41,** 717-718.

Barnes, E. M., and Impey, C. S. (1968). Symposium on growth of micro-organisms at extremes of temperature. Psychrophilic spoilage bacteria of poultry. *J. Appl. Bacteriol.* **31,** 97-107.

Barnes, E. M., and Ingram, M. (1955). Changes in the oxidation-reduction potential of the sterno-cephalicus muscle of the horse after death in relation to the development of bacteria. *J. Sci. Food Agric.* **6,** 448-455.

Barnes, E. M., and Ingram, M. (1956). The effect of redox potential on the growth of *Clostridium welchii* strains isolated from horse muscle. *J. Appl. Bacteriol.* **19,** 117-128.

Barnes, E. M., Despaul, J. E., and Ingram, J. (1963). The behaviour of a food-poisoning strain of *Clostridium welchii* in beef. *J. Appl. Bacteriol.* **26,** 415-427.

Barnett, H. W., Nordin, H. R., Bird, H. D., and Rubin, L. J. (1965). A study of factors affecting the flavour of cured ham. *Proc. Annu. Meet. Eru. Meat Res. Workers, 11th, Belgrade.*

Barrile, J. C., and Cone, J. F. (1970). Effect of added moisture on the heat resistance of *Salmonella anatum* in milk chocolate. *Appl. Microbiol.* **19,** 177-178.

Bartl, V. (1973). Semi-preserved foods: General microbiology and food poisoning. *In* "The Microbiological Safety of Food" (B. C. Hobbs and J. H. B. Christian, eds.), pp. 89-106. Academic Press, New York.

Bauernfeind, J. C., and Pinkert, D. M. (1970). Food processing with added ascorbic acid. *Adv. Food Res.* **18,** 219-315.

Bean, E. L. (1959). Ozone production and costs. *In* "Ozone Chemistry and Technology," (Advances in Chemistry, Ser. 21 (H. A. Leedy, ed.) pp. 430-442. Am. Chem. Soc., Washington, D.C.

Bean, P. G., and Roberts, T. A. (1974). The effect of pH, sodium chloride and sodium nitrite on heat resistance of *Staphylococcus aureus* and growth of damaged cells in laboratory media. *Proc. Int. Congr. Food Sci. Technol., 4th* **3,** 93-102.

Bee, G. R., and Denny, C. B. (1958). "Construction and Use of a Vacuum Micro-Leak Detector for Metal Containers." Natl. Canners Assoc., Washington, D.C.

Bell, R. N. (1971). The nomenclature and manufacture of phosphates. *Symp. Phosphates Food Process.* (J. M. Deman and P. Melnychyn, eds.), pp. 24-37. AVI, Westport, Connecticut.

Bell, T. A., and Etchells, J. L. (1952). Sugar and acid tolerance of spoilage yeasts from sweet cucumber pickles. *Food Technol.* **6,** 468-472.

Bem, Z., Hechelmann, H., and Leistner, L. (1976). The bacteriology of DFD-meat. *Fleischwirtschaft* **56,** 985-987.

Benarde, M. A. (1970). "Disinfection." Dekker, New York.

Ben-Yehoshua, S., and Krinsky, P. (1968). Gas chromatography of ethylene oxide and its toxic residues. *J. Gas Chromatogr.* **6,** 350-351.

Bergner, K. G. (1962). Lebensmittelchemische Fragen bei der Anwendung von Kunststoffen. *Dtsch. Lebensm.-Rundsch.* **58,** 101.

Berry, J. L. (1975). A French method of aseptically filling milk in plastic bottles in a new way. *Nordeur. Mejeri-Tidsskr.* **41,** No. 5, 166-171.

Beuchat, L. R. (1973). Interacting effects of pH, temperature and salt concentration on growth and survival of *Vibrio parahaemolyticus*. *Appl. Microbiol.* **25,** 844-846.

Beuchat, L. R., and Lechowich, R. V. (1968). Effect of salt concentration in the recovery medium on heat-injured *Streptococcus faecalis*. *Appl. Microbiol.* **16,** 772-776.

Bier, J. W. (1976). Experimental anisakiasis: cultivation and temperature tolerance determinations. *J. Milk Food Technol.* **39,** 132-137.

Binkerd, E. F., and Kolari, O. E. (1975). The history and use of nitrate and nitrite in the curing of meat. *Food Cosmet. Toxicol.* **13,** 655-661.

Block, S. S., ed. (1977). "Disinfection, Sterilization, and Preservation," 2nd Ed. Lea & Febiger, Philadelphia, Pennsylvania.

Blood, R. M. (1975). Lactic acid bacteria in marinated herring. *Food Sci. Technol. Abstr.* **7,** No. 11 R 656.

Board, R. G. (1969). The microbiology of the hen's egg. *Adv. Appl. Microbiol.* **11,** 245-281.

Böhme, C.-F. (1969). Folien für die Verpackung von Fleisch und Fleischerzeugnissen. *Fette, Seifen, Anstrichm.* **71,** 499-506.

Böhme, C.-F. (1971). Die optimale Verpackung von Frischfleischportionen. *Verpack.-Rundsch.* **21,** 594-606.

Booth, G. H., Cooper, A. W., and Robb, J. A. (1968). Bacterial degradation of plasticized PVC. *J. Appl. Bacteriol.* **31,** 305-310.

Borrie, P., and Barrett, J. (1961). Dermatitis caused by penicillin in bulked milk supplies. *Br. Med. J.* **ii,** 1267.

Bradshaw, J. G., Peeler, J. T., and Twedt, R. M. (1975). Heat resistance of ileal loop reactive *Bacillus cereus* strains isolated from commercially canned food. *Appl. Microbiol.* **30,** 943-945.

Braithwaite, P. J., and Perigo, J. A. (1971). The influence of pH, water activity and recovery temperature on the heat resistance and outgrowth of *Bacillus* spores. *In* "Spore Research, 1971" (A. N. Barker, G. W. Gould, and J. Wolf, eds.), pp. 289-302. Academic Press, New York.

Branen, A. L., and Keenan, T. W. (1969). Growth stimulation of *Lactobacillus* species by lactic streptococci. *Appl. Microbiol.* **17,** 280-285.

Bridges, A. E., Olivo, J. P., and Chandler, V. L. (1956). Relative resistance of microorganisms to cathode rays II. Yeasts and molds. *Appl. Microbiol.* **4,** 147-149.

Briston, J. (1971). Plastics for food packaging—1. *Food Manuf.* **46**(¾), 71-75.

Briston, J. (1976). New plastic materials for packaging films. *Packag. Rev.* **96(3),** 71, 73-74.

Broadwater, W. T., Hoehn, R. C., and King, P. H. (1973). Sensitivity of three selected bacterial species to ozone. *Appl. Microbiol.* **26,** 391-393.

Brooks, C., Miller, E. V., Bratley, C. O., Cooley, J. S., Mook, D. V., and Johnson, H. B. (1932). Effect of solid and gaseous carbon dioxide upon transit diseases of certain fruits and vegetables. *U.S. Dep. Agric., Tech. Bull.* No. 318.

Brown, A. D. (1974). Microbial water relations: features of the intracellular composition of sugar-tolerant yeasts. *J. Bacteriol.* **118,** 769-777.

Brown, M. H. (1977). Microbiology of the British fresh sausage. Ph.D. Thesis, Univ. of Bath, Bath, England.

Brown, M. R. W., and Melling, J. (1971). Inhibition and destruction of microorganisms by heat. *In* "Inhibition and Destruction of the Microbial Cell." (W. B. Hugo, ed.), Academic Press, New York.

Brown, W. (1922). On the germination and growth of fungi at various temperatures and in various concentrations of oxygen and carbon dioxide. *Ann. Bot.* **36,** 257-283.

Bruch, C. W. (1961). Gaseous sterilization. *Annu. Rev. Microbiol.* **15,** 245-262.

Bruch, C. W. (1972). Sterilization of plastics: toxicity of ethylene oxide residues. *In* "Industrial

Sterilization'' (G. B. Phillips and W. S. Miller, eds.) pp. 49-77. Duke Univ. Press, Durham, North Carolina.

Bruch, C. W., and Koesterer, M. G. (1961). The microbicidal activity of gaseous propylene oxide and its application to powdered or flaked foods. *J. Food Sci.* **26,** 428-435.

Buchanan, R. E., and Gibbons, N. E., eds. (1974). ''Bergey's Manual of Determinative Bacteriology,'' 8th Ed. Williams & Wilkins, Baltimore, Maryland.

Bueltman, C. G. (1972). Detergents: Phosphates, surfactants, and enzymes. *Assoc. Food Drug Off. U.S., Q. Bull.* **36,** 31-34.

Busta, F. F. (1976). Practical implications of injured microorganisms in food. *J. Milk Food Technol.* **39,** 138-145.

Buttiaux, R. (1973). Microbiologie et décongélation. *Rev. Gén. Froid* **64,** 591-592.

Byrne, A. F., Rayman, M. M., and Schneider, M. D. (1955). Methods for the detection and estimation of numbers of *Salmonella* in dried eggs and other food products. *Appl. Microbiol.* **3,** 368-372.

Callow, E. H. (1934). Gas storage of bacon. *G.B. Dep. Sci. Ind. Res., Food Invest. Board, Rep. 1933* p. 71.

Canada, J. C., Strong, D. H., and Scott, L. G. (1964). Response of *Clostridium perfringens* spores and vegetative cells to temperature variation. *Appl. Microbiol.* **12,** 273-276.

Carlson, H. J., Olynyk, P., Duncan, H., and Smolin, A. (1951). The chemical, physical and antibiotic properties of methyl gallate isolated from *Koelrenteria paniculata. Antibiot. Chemother.* (*Washington, D.C.*) **1,** 431-442.

Cassens, R. G., Woolford, G., Lee, S. H., and Goutefongea, R. (1976). Fate of nitrite in meat. *Proc. Int. Symp. Nitrite Meat Prod., 2nd, Zeist, Neth.* (B. J. Tinbergen and B. Krol, eds.), pp. 95-100. Pudoc, Wageningen, Netherlands.

Castell, C. H., and MacCallum, W. A. (1950). The value of temperatures close to freezing on the storage of fish. *J. Fish Res. Board Can.* **8,** 111-116.

Castellani, A. G., and Niven, C. F., Jr. (1955). Factors affecting the bacteriostatic action of sodium nitrite. *Appl. Microbiol.* **3,** 154-159.

Cavett, J. J. (1962). The microbiology of vacuum-packed sliced bacon. *J. Appl. Bacteriol.* **25,** 282-289.

Cavett, J. J. (1968). The effects of newer forms of packaging on the microbiology and storage life of meats, poultry and fish. *Prog. Ind. Microbiol.* **7,** 77-123.

Chambers, C. W. (1956). A procedure for evaluating the efficiency of bactericidal agents. *J. Milk Food Technol.* **19,** 183-187.

Chichester, D. F., and Tanner, F. W. (1972). Antimicrobial food additives. *In* ''Handbook of Food Additives'' 2nd Ed. (T. E. Furia, ed.) pp. 115-184. Chem. Rubber Publ. Co., Cleveland, Ohio.

Cho, I. C., and Bratzler, L. J. (1970). Effect of sodium nitrite on flavor of cured pork. *J. Food Sci.* **35,** 668-670.

Christian, J. H. B., and Hall, J. M. (1972). Water relations of *Salmonella oranienburg:* accumulation of potassium and amino acids during respiration. *J. Gen. Microbiol.* **70,** 497-506.

Christian, J. H. B., and Scott, W. J. (1953). Water relations of salmonellae at 30°C. *Aust. J. Biol. Sci.* **6,** 565-573.

Christian, J. H. B., and Stewart, B. J. (1973). Survival of *Staphylococcus aureus* and *Salmonella oranienburg* in dried foods, as influenced by water activity and oxygen. *In* ''The Microbiological Safety of Food'' (B. C. Hobbs and J. H. B. Christian, eds.), pp. 107-119. Academic Press, New York.

Christian, J. H. B., and Waltho, J. A. (1962). Solute concentrations within cells of halophilic and nonhalophilic bacteria. *Biochim. Biophys. Acta* **65,** 506-508.

Christiansen, L. N., and Foster, E. M. (1965). Effect of vacuum packaging on growth of *Clostridium botulinum* and *Staphylococcus aureus* in cured meats. *Appl. Microbiol.* **13,** 1023-1025.

Christiansen, L. N., Deffner, J., Foster, E. M., and Sugiyama, H. (1968). Survival and outgrowth of *Clostridium botulinum* type E spores in smoked fish. *Appl. Microbiol.* **16**, 133–137.

Christiansen, L. N., Tompkin, R. B., Shaparis, A. B., Johnston, R. W., and Kautter, D. A. (1975). Effect of sodium nitrite and nitrate on *Clostridium botulinum* growth and toxin production in a summer style sausage. *J. Food Sci.* **40**, 488–490.

Christiansen, L. N., Tompkin, R. B., and Shaparis, A. B. (1978). Fate of *Clostridium botulinum* in perishable canned cured meat at abuse temperature. *J. Food Prot.* **41**, 354–355.

Christophersen, J. (1968). Effect of freezing and thawing on the microbial population of foodstuffs. *In* "Low Temperature Biology of Foodstuffs" (J. Hawthorn and E. J. Rolfe, eds.). Pergamon, Oxford.

Chung, B. H., Cannon, R. Y., and Smith, R. C. (1976). Influence of growth temperature on glucose metabolism of a psychrotrophic strain of *Bacillis cereus*. *Appl. Environ. Microbiol.* **31**, 39–45.

Chung, K. C., and Goepfert, J. M. (1970). Growth of *Salmonella* at low pH. *J. Food Sci.* **35**, 326–328.

Cieplinska, T., and Zychowicz, C. (1974). Effect of *Lactobacillus acidophilus* on salmonella and shigella organisms under laboratory conditions. *Pediat. Pol.* **49**, 991–996. (In Polish).

Clark, C. W., and Ordal, Z. J. (1969). Thermal injury and recovery of *Salmonella typhimurium* and its effect on enumeration procedures. *Appl. Microbiol.* **18**, 332–336.

Clark, D. S., and Lentz, C. P. (1969). The effect of carbon dioxide on the growth of slime-producing bacteria on fresh beef. *Can. Inst. Food Technol. J.* **2**, 72–75.

Clark, D. S., and Lentz, C. P. (1972). Use of carbon dioxide for extending shelf-life of prepackaged beef. *Can. Inst. Food Sci. Technol. J.* **5**, 175–178.

Clark, D. S., and Lentz, C. P. (1973). Use of mixtures of carbon dioxide and oxygen for extending shelf-life of prepackaged fresh beef. *Can. Inst. Food Sci. Technol. J.* **6**, 194–196.

Clark, D. S., Lentz, C. P., and Roth, L. A. (1976). Use of carbon monoxide for extending shelf-life of prepackaged fresh beef. *Can. Inst. Food Sci. Technol. J.* **9**, 114–117.

Claydon, T. J. (1975). A membrane-filter technique to test for the significance of sublethally injured bacteria in pasteurized milk. *J. Milk Food Technol.* **38**, 87–88.

Cleland, J. E., and Fetzer, W. R. (1944). Moisture absorptive power of starch hydrolyzates. *Ind. Eng. Chem.* **36**, 552–555.

Collins, F. M., and Montalbine, V. (1976). Mycobactericidal activity of gluteraldehyde solutions. *J. Clin. Microbiol.* **4**, 408–412.

Collins-Thompson, D. L., Hurst, A., and Kruse, H. (1973). Synthesis of enterotoxin B by *Staphylococcus aureus* strain S6 after recovery from heat injury. *Can. J. Microbiol.* **19**, 1463–1468.

Collins-Thompson, D. L., Chang, P. C., Davidson, C. M., Larmond, E., and Pivnick, H. (1974). Effect of nitrite and storage temperature on the organoleptic quality and toxinogenesis by *Clostridium botulinum* in vacuum-packaged side bacon. *J. Food Sci.* **39**, 607–609.

Commissie voor Fytofarmacie (1975). Waardebepaling van desinfectiemiddelen voor materiaalonts metting in Ziekenhuizen Sept. 1974. *Zentralbl. Bakteriol., Parasitenkd., Infektionskr. Hyg. Abt. 1: Orig., Reihe B* **160**, 541–550.

Cook, D. J., and Binsted, R. (1975). "Food Processing Hygiene." Food Trade Press, London.

Cooke, E. M., Hettiaratchy, I. G. T., and Buck, A. C. (1972). Fate of ingested *Escherichia coli* in normal persons. *J. Med. Microbiol.* **5**, 361–369.

Corry, J. E. L. (1973). The water relations and heat resistance of microorganisms. *Prog. Ind. Microbiol.* **12**, 73–108.

Corry, J. E. L., Kitchell, A. G., and Roberts, T. A. (1969). Interactions in the recovery of *Salmonella typhimurium* damaged by heat or gamma radiation. *J. Appl. Bacteriol.* **32**, 415–428.

Cousins, C. M. (1976). The inactivation of vegetative microorganisms by chemicals in the dairying industry. *In* "Inhibition and Inactivation of Vegetative Microbes" (F. A. Skinner and W. B.

Hugo, eds.), Society of Applied Bacteriology Symposia Series, No. 5, pp. 13–30. Academic Press, New York.
Coyne, F. P. (1932). The effect of carbon dioxide on bacterial growth with special reference to the preservation of fish. Part I. *J. Soc. Chem. Ind.* **51,** 119T–121T.
Coyne, F. P. (1933). The effect of carbon dioxide on bacterial growth. *Proc. R. Soc. London, Ser. B* **113,** 196–217.
Cramer, J. A., and Prestegard, J. H. (1977). NMR studies of pH-induced transport of carboxylic acids across phospholipid vesicle membranes. *Biochem. Biophys. Res. Commun.* **75,** 295–301.
Critchfield, F. E., and Johnson, J. B. (1957). Colorimetric determination of ethylene oxide by conversion to formaldehyde. *Anal. Chem.* **29,** 797–800.
Cronan, J. E., Jr., and Gelmann, E. P. (1975). Physical properties of membrane lipids: biological relevance and regulation. *Bacteriol. Rev.* **39,** 232–256.
Cronan, J. E., Jr., and Vagelos, P. R. (1972). Metabolism and function of the membrane phospholipids of *Escherichia coli. Biochim. Biophys. Acta* **265,** 25–60.
Cross, C. K., and Ziegler, P. (1965). A comparison of the volatile fractions from cured and uncured meat. *J. Food Sci.* **30,** 610–614.
Crowther, J. S., Holbrook, R., Baird-Parker, A. C., and Austin, B. L. (1976). Role of nitrite and ascorbate in the microbiological safety of vacuum-packed sliced bacon. *Proc. Int. Symp. Nitrite Meat Prod., 2nd, Zeist, Neth.* (B. J. Tinbergen and B. Kroll, eds.), pp. 13–20. Pudoc, Wageningen, Netherlands.
Curran, H. R. (1935). The influence of some environmental factors upon the thermal resistance of bacterial spores. *J. Infect. Dis.* **56,** 196–202.
Curran, H. R., and Evans, F. R. (1937). The importance of enrichments in the cultivation of bacterial spores previously exposed to lethal agencies. *J. Bacteriol.* **34,** 179–189.
Dahiya, R. S., and Speck, M. L. (1968a). Inhibition of *Staphylococcus aureus* by lactic acid bacteria. *Bacteriol. Proc.* p. 46. Abstr. G156.
Dahiya, R. S., and Speck, M. L. (1968b). Hydrogen peroxide formation by lactobacilli and its effect on *Staphylococcus aureus. J. Dairy Sci.* **51,** 1568–1572.
Dainty, R. H., Shaw, B. G., De Boer, K. A., and Scheps, E. S. J. (1975). Protein changes caused by bacterial growth on beef. *J. Appl. Bacteriol.* **39,** 73–81.
Dakin, J. C. (1957). "Microbiological Keeping Quality of Pickles and Sauces," Res. Rep. No. 79. Food Res. Assoc., Leatherhead, England.
Dakin, J. C., and Radwell, J. Y. (1971). Lactobacilli causing spoilage of acetic acid preserves. *J. Appl. Bacteriol.* **34,** 541–545.
D'Aoust, J.-Y., and Evans, A. (1978). Suspect case of human salmonellosis from a dry cured meat product—Ontario. *Can. Dis. Weekly Rep.* **4**(7), 27–28.
Datta, N. (1969). Drug resistance and R factors in the bowel bacteria of London patients before and after admission to hospital. *Br. Med. J.* **ii,** 407–411, (May 17).
Davidson, C. M., and Webb, G. (1972). The behaviour of salmonellae in vacuum-packed cooked cured meat products. *Proc. Annu. Meet. Eur. Meat Res. Workers, 18th, Guelph, Can.* pp. 27–41.
Davies, W. L., Davis, J. G., Dearden, D. V., and Mattick, A. T. R. (1934). Studies in Cheddar cheese II. The effect of controlled oxidation-reduction potential on ripening. III. The rôle of rennin, pepsin, and lactobacilli. *J. Dairy. Res.* **5,** 144–152.
Davis, E. M., and Keen, S. R. (1974). Municipal wastewater bacteria capable of surviving chlorination. *Health Lab. Sci.* **11,** 268–274.
Davis, J. G. (1932). Studies in Cheddar cheese I. The oxidation-reduction potentials of ripening cheddar cheese. *J. Dairy Res.* **3,** 241–253.
Davis, J. G. (1975). The microbiology of yoghurt. *In* "Lactic Acid Bacteria in Beverages and Food" (J. G. Carr, C. V. Cutting, and G. C. Whiting, eds.), pp. 245–263. Academic Press, New York.

Dean-Raymond, D., and Alexander, M. (1977). Bacterial metabolism of quaternary ammonium compounds. *Appl. Environ. Microbiol.* **33,** 1037-1041

Deibel, R. H., and Evans, J. B. (1957). "Nitrite burn" in cured meat products—particularly in fermented sausages. *Am. Meat Inst. Found. Bull.* **32,** 5-13.

Deibel, R. H., Niven, C. F., Jr., and Wilson, G. D. (1961). Microbiology of meat curing. III. Some microbiological and related technological aspects in the manufacture of fermented sausages. *Appl. Microbiol.* **9,** 156-161.

Dempster, J. D., Reid, S. N., and Cody, O. (1973). Sources of contamination of cooked, ready to eat, cured and uncured meats. *J. Hyg.* (Cambr.) **71,** 815.

Dempster, J. F. (1973). A note on hygiene of meat mincing machines. *J. Hyg.* (Cambr.) **71,** 739.

Department of Health and Social Security (undated). "Processing of Pasteurized Large Canned Hams." Advis. Memo. Dep. Health Soc. Security, London, U.K.

Department of Health and Social Security (1978). Letter of 13 Dec., 1978 signed by J. D. Henry, addressed to Codex Committee on Food Hygiene, Cleaning and Disinfection Appendix to Codex General Principles of Food Hygiene, Reference A/C 252/117.

Desrosier, N. W., and Rosenstock, H. M. (1960). "Radiation Technology in Food Agriculture and Biology." AVI, Westport, Connecticut.

Deutsche Gessellschaft für Hygiene and Mikrobiologie (1972). "Richtlinien für die Prüfung Chemischer Desinfektionsmittell," 3rd Ed. Fischer, Stuttgart.

Devoyod, J. J., and Desmazeaud, M. (1970). Les associations microbiennes dans le fromage de Roquefort. I. Action des entérocoques vis-à-vis des streptocoques lactiques et des leuconostoc. *Lait* **50,** 374-390.

Devoyod, J. J., and Desmazeaud, M. (1971). Les associations microbiennes dans le fromage de Roquefort. III. Action des entérocoques et des levures fermentant le lactos vis-à-vis des lactobacilles. *Lait* **51,** 399-415.

Diener, U. L., and Davis, N. D. (1967). Limiting temperature and relative humidity for growth and production of aflatoxin and free fatty acids by *Aspergillus flavus* in sterile peanuts. *J. Am. Oil Chem. Soc.* **44,** 259-263.

Dobson, A., and Bullen, J. J. (1964). A method for the control of E_h and pH during bacterial growth. *J. Gen. Microbiol.* **35,** 169-174.

Dolman, C. E. (1957). The epidemiology of meat-borne diseases. *In* "Meat Hygiene," Monograph Series, No. 33, pp. 11-108. World Health Organ., Geneva.

Donald, J. R., Jones, C. L., and MacLean, A. R. M. (1924). The effect of carbonation on bacteria in beverages. *Am. J. Public Health* **14,** 122-128.

Dorner, W., and Thoni, M. (1936). Zur Biologie der Propionsaurebakterien. *Landwirtsch. Jahrb. Schweiz* **50,** 889-898.

Douglas, F., and Rigby, G. J. (1974). The effect of oxygen on the germination and outgrowth of *Clostridium butyricum* spores and changes in the oxidation-reduction potential of cultures. *J. Appl. Bacteriol.* **37,** 251-259.

Douglas, F., Hambleton, R., and Rigby, G. J. (1973). An investigation of the oxidation-reduction potential and of the effect of oxygen on the germination and outgrowth of *Clostridium butyricum* spores, using platinum electrodes. *J. Appl. Bacteriol.* **36,** 625-633.

Dowdell, M. J., and Board, R. G. (1971). The microbial associations in British fresh sausages. *J. Appl. Bacteriol.* **34,** 317-337.

Draudt, H. N. (1963). The meat smoking process: a review. *Food Technol.* **17,** 1557-1562.

Drechsel, H. (1965). Zum Einfluss von Knoblauchbestandteilen auf die Rohwurstflora. Vet. Med. Dissertation, Univ. Munich, Munich.

Dubey, J. P. (1974). Effect of freezing on the infectivity of *Toxoplasma* cysts to cats. *J. Am. Vet. Med. Assoc.* **165,** 534-536.

Duitschaever, C. L., and Irvine, D. M. (1971). A case study: Effect of mold on growth of coagulase positive staphylococci in Cheddar cheese. *J. Milk Food Technol.* **34,** 583.

Dukes, E. P. (ed.) (1969) Sorbic acid. (etc) *In* "Kirk-Othmer's Encyclopedia of Chemical Technology," 2nd Ed., Vol. 18, p. 589-599. Wiley, New York.

Duncan, C. L., and Foster, E. M. (1968). Effect of sodium nitrite, sodium chloride and sodium nitrate on germination and outgrowth of anaerobic spores. *Appl. Microbiol.* **16,** 406-411.

Eddy, B. P. (1960). The use and meaning of the term "psychrophilic". *J. Appl. Bacteriol.* **23,** 189-190.

Eddy, B. P., and Ingram, M. (1956). A salt-tolerant denitrifying *Bacillus* strain which "blows" canned bacon. *J. Appl. Bacteriol.* **19,** 62-70.

Eddy, B. P., and Ingram, M. (1962). The occurrence and growth of staphylococci on packed bacon, with special reference to *Staphylococcus aureus. J. Appl. Bacteriol.* **25,** 237-247.

Effenberger, G., and Schotte, K. (1971). "Verpackung von Fleisch und Fleischwaren. Fleischforschung und Praxis," Publ. No. H. 3a,b. Verlag Rheinhess, Alzey, West Germany.

Einerson, M. A., and Reineccius, G. A. (1977). Inhibition of warmed-over flavor in retorted turkey by antioxidants formed during processing. *J. Food Process. Preserv.* **1,** 279-291.

Eklund, M. W., Wieler, D. I., and Poysky, F. T. (1967). Outgrowth and toxin production of nonproteolytic type B *Clostridium botulinum* at 3.3 to 5.6°C. *J. Bacteriol.* **93,** 1461-1462.

El-Badawi, A. A., Cain, R. F., Samuels, C. E., and Anglemeier, A. F. (1964). Color and pigment stability of packaged refrigerated beef. *Food Technol.* **18,** 753-757.

El-Bisi, H. M., and Ordal, Z. J. (1956). The effect of sporulation temperature on the thermal resistance of *Bacillus coagulans* var. *thermoacidurans. J. Bacteriol.* **71,** 10-16.

Eller, C., Edwards, F. F., and Wynne, E. S. (1968). Sporicidal action of auto-oxidized ascorbic acid for *Clostridium. Appl. Microbiol.* **16,** 349-354.

Elliker, P. R., and Frazier, W. C. (1938). Influence of time and temperature of incubation on heat resistance of *Escherichia coli. J. Bacteriol.* **36,** 83-98.

Elliott, R. P. (1947). Evaluation of surface pH as a freshness index for fish fillets. *Food Res.* **12,** 87-98.

Elliott, R. P. (1980). Chapter VI—The microbiology of sanitation; Chapter VII—Cleaning and sanitizing. *In* "Principles of Food Processing Sanitation." (B. Denny, ed.) Food Processors Inst., Washington, D.C.

Elliott, R. P., and Michener, H. D. (1965). Factors affecting the growth of psychrophilic microorganisms in foods. A review. *U.S. Dep. Agric., Tech. Bull.* No. 1320.

Elliott, R. P., and Straka, R. P. (1964). Rate of microbial deterioration of chicken meat at 2°C after freezing and thawing. *Poult. Sci.* **43,** 81-86.

Elliott, R. P., Straka, R. P., and Garibaldi, J. A. (1964). Polyphosphate inhibition of growth of pseudomonads from poultry meat. *Appl. Microbiol.* **12,** 517-522.

Emard, L. D., and Vaughn, R. H. (1952). Selectivity of sorbic acid media for the catalase negative lactic acid bacteria and clostridia. *J. Bacteriol.* **63,** 487-494.

Empey, W. A., and Scott, W. J. (1939). Investigations on chilled beef. Part 1. Microbial contamination acquired in the meat works. *Bull. Counc. Sci. Ind. Res. Aust.* No. 126.

Empey, W. A., and Vickery, J. R. (1933). The use of carbon dioxide in the storage of chilled beef. *J. Counc. Sci. Ind. Res.* **6,** 233-243.

Erdman, I. E., Thatcher, F. S., and MacQueen, K. F. (1961a). Studies on the irradiation of microorganisms in relation to food preservation. I. The comparative sensitivities of specific bacteria of public health significance. *Can. J. Microbiol.* **7,** 199-205.

Erdman, I. E., Thatcher, F. S., and MacQueen, K. F. (1961b). Studies on the irradiation of microorganisms in relation to food preservation. II. Irradiation resistant mutants. *Can. J. Microbiol.* **7,** 207-215.

Ernst, R. R. (1972). Ethylene oxide gaseous sterilization for industrial applications. *In* "Industrial Sterilization" (G. B. Phillips and W. S. Miller, eds.), pp. 181-208. Duke Univ. Press, Durham, North Carolina.

Ernst, R. R., and Shull, J. J. (1962). Ethylene oxide gaseous sterilization I. Concentration and temperature effects. *Appl. Microbiol.* **10,** 337–341.

Eskin, N. A. M., Henderson, H. M., and Townsend, R. J. (1971). "Biochemistry of Foods," pp. 160–163. Academic Press, New York.

Esty, J. R., and Meyer, K. F. (1922). The heat resistance of spores of *B. botulinus* and allied anaerobes. XI. *J. Infect. Dis.* **31,** 650–663.

Etchells, J. L., Fleming, H. P., and Bell, T. A. (1975). Rep., Dep. Food Sci., North Carolina State Univ., Raleigh. *Food Sci. Technol. Abstr.* **7,** No. 11, J 1729.

Evans, J. B., and Niven, C. F., Jr. (1951). Nutrition of the heterofermentative lactobacilli that cause greening of cured meat products. *J. Bacteriol.* **62,** 599–603.

Ewell, A. W. (1950). Ozone and its application in food preservation. *Refrig. Eng.* **58,** 4 unnumbered pages between 874 and 875.

Falkow, S. (1975). "Infectious Multiple Drug Resistance." Pion Advanced Biochemistry Series. Pion, London.

Fantasia, L. D., Mestrandrea, L., Schrade, J. P., and Yager, J. (1975). Detection and growth of enteropathogenic *Escherichia coli* in soft ripened cheese. *Appl. Microbiol.* **29,** 179–185.

Farhat, B. Y., and Koburger, J. A. (1975). Potential for aflatoxin production on smoked mullet. *J. Milk Food Technol.* **38,** 634.

Farrall, A. W. (1976). "Food Engineering Systems. Vol. 1: Operations." AVI Westport, Connecticut.

Farrell, J., and Rose, A. H. (1968). Cold Shock in a mesophilic and a psychrophilic pseudomonad. *J. Gen. Microbiol.* **50,** 429–439.

Fay, J. P., and Farías, R. N. (1976). Chilling cells enhances the bactericidal action of fatty acids on *Escherichia coli*. *Appl. Environ. Microbiol.* **31,** 153–157.

Fennema, O. R., Powrie, W. D., and Marth, E. H. (1973). "Low-Temperature Preservation of Foods and Living matter." Dekker, New York.

Fetner, R. H., and Ingols, R. S. (1959). Bactericidal activity of ozone and chlorine against *Escherichia coli* at 1°C. *In* "Ozone Chemistry and Technology," Advances in Chemistry Series, No. 21, pp. 370–374. Am. Chem. Soc., Washington, D.C.

Firstenberg-Eden, R., Rosen, B., and Mannheim, C. H. (1977). Death and injury of *Staphylococcus aureus* during thermal treatment of milk. *Can. J. Microbiol.* **23,** 1034–1037.

Flowers, R. S., and Adams, D. M. (1976). Spore membrane(s) as the site of damage within heated *Clostridium perfringens* spores. *J. Bacteriol.* **125,** 429–434.

Flowers, R. S., Martin, S. E., Brewer, D. G., and Ordal, Z. J. (1977). Catalase and enumeration of stressed *Staphylococcus aureus* cells. *Appl. Environ. Microbiol.* **33,** 1112–1117.

Food and Agriculture Organization/International Atomic Energy Agency/World Health Organization Expert Committee (1977). "Wholesomeness of Irradiated Food." Tech. Rep. Ser. 604. World Health Organ., Geneva.

Food and Agriculture Organization/World Health Organization (1971). A Review of the Technological Efficacy of Some Antimicrobial Agents. Joint FAO/WHO Expert Committee on Food Additives, 14th. Rep. FAO Nutrition Rep. Ser. 48C; WHO Tech. Rep. Ser. 1970, No. 462.

Food and Agriculture Organization/World Health Organization (FAO/WHO) (1973). "List of Additives Evaluated for the Safety-in-Use in Foods," Joint FAO/WHO Food Standards Programme, Codex Alimentarius Commission CAC/FAL-1-1973. World Health Organ., Geneva.

Food and Agriculture Organization/World Health Organization (FAO/WHO) (1977). "Microbiological Specifications for Foods," Rep. 2nd Joint FAO/WHO Consultation. Feb. 21–Mar. 2, World Health Organ., Geneva.

Foster, J. W., and Davis, J. B. (1949). Carbon dioxide inhibition of anaerobic fumarate formation in the mold *Rhizopus nigricans*. *Arch. Biochem.* **21,** 135–142.

Foster, R. D., and Mead, G. C. (1976). Effect of temperature and added polyphosphate on the survival of salmonellae in poultry meat during cold storage. *J. Appl. Bacteriol.* **41,** 505–510.

Fowler, G. C., and McCann, B. (1972). The use of nisin in the food industry. *Food Ind. S. Afr.* **25,** 49, 51, 55.

Fox, J. B., Jr., and Ackerman, S. A. (1968). Formation of nitric oxide myoglobin: mechanisms of the reaction with various reductants. *J. Food Sci.* **33,** 364–370.

Fox, J. B., Jr., and Nicholas, R. A. (1974). Nitrite in meat. Effect of various compounds on loss of nitrite. *J. Agric. Food Chem.* **22,** 302–306.

Fraenkel-Conrat, H. (1944). Action of 1,2 epoxides on proteins. *J. Biol. Chem.* **154,** 227–238.

Frank, H. A., Reid, A., Santo, L. M., Lum, N. A., and Sandler, S. T. (1972). Similarity in several properties of psychrophilic bacteria grown at low and moderate temperatures. *Appl. Microbiol.* **24,** 571-574.

Fränkel, C. (1889). Die Einwirkung der Kohlensäure auf die Lebenstätigkeit der Mikroorganismen. *Z. Hyg.* **5,** 332–362.

Frazier, W. C. (1958). "Food Microbiology," 2nd Ed. McGraw-Hill, New York.

Freeman, B. M., and Bridges, B. A. (1960). Suitability of various plating media for counting bacteria after exposure to gamma radiation. *Int. J. Appl. Radiat. Isot.* **8,** 136–138.

Freese, E., Sheu, C. W., and Galliers, E. (1973). Function of lipophilic acids as antimicrobial food additives. *Nature (London)* **241,** 321–325.

Fuchs, H. (1958). Wirken flüssige Gewürzauszüge, die aus natürlichen Gewürzen hergestellt worden sind, antibakteriell? Vet. Med. Dissertation, Univ. West Berlin.

Fujimaki, M., Emi, M., and Okitani, A. (1975). Fate of nitrite in meat-curing model systems composed of myoglobin, nitrite and ascorbate. *Agric. Biol. Chem.* **39,** 371–377.

Gabis, D. A., Vesley, D., and Favero, M. S. (1976). Sampling, equipment, supplies and environment. *In* "Compendium of Methods for the Microbiological Examination of Foods" (M. L. Speck, ed.), pp. 95–104. Am. Public Health Assoc., Washington, D.C.

Galbraith, H., Miller, T. B., Paton, A. M., and Thompson, J. K. (1971). Antibacterial activity of long chain fatty acids and the reversal with calcium-magnesium, ergocalciferol and cholesterol. *J. Appl. Bacteriol.* **34,** 803–813.

Gallagher, I. H. C. (1971). Occurrence of waxes in *Acinetobacter*. *J. Gen. Microbiol.* **68,** 245–247.

Gardner, G. A. (1971). Microbiological and chemical changes in lean wiltshire bacon during aerobic storage. *J. Appl. Bacteriol.* **34,** 645–654.

Gaustad, J. W., McDuff, C. R., and Hatcher, H. J. (1974). Test method for the evaluation of virucidal efficacy of three common liquid surface disinfectants on a simulated environmental surface. *Appl. Microbiol.* **28,** 748–752.

Gelda, C. S., Mathur, A. O., and Stersky, A. K. (1974). The retarding effect of the anti-fungal agent pimaricin on the growth of *Aspergillus* species and aflatoxin production in raw ground peanuts. *Int. Congr. Food Sci. and Technol., 4th Food Sci., Tech. Abstr.* No. 9a, p. 207.

Genigeorgis, C. (1974). Recent developments on staphylococcal food poisoning. *Proc. Annu. Food Hyg. Symp., 16th, Teachers Food Hyg. Coll. Vet. Med.* pp. 34–93.

Genigeorgis, C. (1976). Quality control for fermented meats. *J. Am. Vet. Med. Assoc.* **169,** 1220–1228.

Genigeorgis, C., and Prucha, J. (1971). Production of enterotoxin C in processed meats. *Bacteriol. Proc.* p. 18, Abstr. A102.

Genigeorgis, C., and Sadler, W. W. (1966). Effect of sodium and pH on enterotoxin B production. *J. Bacteriol.* **92,** 1383–1387.

Genigeorgis, C., Riemann, H., and Sadler, W. W. (1969). Production of enterotoxin B in cured meats. *J. Food Sci.* **34,** 62–68.

Genigeorgis, C., Martin, S., Franti, C. E., and Riemann, H. (1971a). Initiation of staphylococcal growth in laboratory media. *Appl. Microbiol.* **21,** 934–939.

Genigeorgis, C., Savoukidis, M., and Martin, S. (1971b). Initiation of staphylococcal growth in processed meat environments. *Appl. Microbiol.* **21,** 940–942.

Georgala, D. L., and Davidson, C. M. (1970). Food package. Br. Patent 1,199,998.

Georgala, D. L., and Hurst, A. (1963). The survival of food poisoning bacteria in frozen foods. *J. Appl. Bacteriol.* **26,** 346–358.

George, E., Jr., and Olson, J. C., Jr. (1960). Growth of staphylococci in condensed skim milk at low and moderate temperatures. *J. Dairy Sci.* **43,** 852.

Gibbons, N. E., Rose, D., and Hopkins, J. W. (1954). Bactericidal and drying effects of smoking on bacon. *Food Technol.* **8,** 155–157.

Giggard, E. D., and Gottschall, P. B. (1958). Heat preservation of pressure dispensed food products. *Food Technol.* **12,** 324–329.

Gilbert, J., and Knowles, M. E. (1975). The chemistry of smoked foods: a review. *J. Food Technol.* **10,** 245–261.

Gilbert, R. J. (1969). Cross-contamination by cooked meat slicing machines and cleaning cloths. *J. Hyg.* (Cambr.) **67,** 249–254.

Gilbert, R. J. (1970). Comparison of materials used for cleaning equipment in retail food premises, and of two methods for the enumeration of bacteria on cleaned equipment and work surfaces. *J. Hyg.* (Cambr.) **68,** 221–232.

Gilbert, R. J., and Maurer, I. M. (1968). The hygiene of slicing machines, carving knives and can-openers. *J. Hyg.* (Cambr.) **66,** 439–450.

Gilbert, R. J., and Watson, H. M. (1971). Some laboratory experiments on various meat preparation surfaces with regard to surface contamination and cleaning. *J. Food Technol.* **6,** 163–170.

Gill, C. O. (1975). Effect of growth temperature on the lipids of *Pseudomonas fluorescens*. *J. Gen. Microbiol.* **89,** 293–298.

Gill, C. O. (1976). Substrate limitation of bacterial growth at meat surfaces. *J. Appl. Bacteriol.* **41,** 401–410.

Giolitti, G., and Cantoni, C. (1966). A medium for the isolation of staphylococci from foodstuffs. *J. Appl. Bacteriol.* **29,** 395–398.

Gladstone, G. P., Fildes, P., and Richardson, G. M. (1935). Carbon dioxide as an essential factor in the growth of bacteria. *Br. J. Exp. Pathol.* **16,** 335–348.

Goepfert, J. M., and Biggie, R. A. (1968). Heat resistance of *Salmonella typhimurium* and *Salmonella seftenberg* 775W in milk chocolate. *Appl. Microbiol.* **16,** 1939–1940.

Goepfert, J. M., and Hicks, R. (1969). Effect of volatile fatty acids on *Salmonella typhimurium*. *J. Bacteriol.* **97,** 956–958.

Goepfert, J. M., and Kim, H. U. (1975). Behavior of selected food-borne pathogens in raw ground beef. *J. Milk Food Technol.* **38,** 449–452.

Goepfert, J. M., Spira, W. M., and Kim H. U. (1972). *Bacillus cereus:* Food poisoning organism. A review. *J. Milk Food Technol.* **35,** 213–227.

Goldenberg, N., and Relf, C. J. (1967). Use of disinfectants in the food industry. *J. Appl. Bacteriol.* **30,** 141–147.

Gomez, R. F., Sinskey, A. J., Davies, R., and Labuza, T. P. (1973). Minimal medium recovery of heated *Salmonella typhimurium* LT2. *J. Gen. Microbiol.* **74,** 267–274.

Gooding, C. M., Melnick, D., Lawrence, R. L., and Luckmann, F. H. (1955). Sorbic acid as a fungistatic agent for foods *IX*. Physicochemical considerations in using sorbic acid to protect foods. *Food Res.* **20,** 639–648.

Gough, B. J., and Alford, J. A. (1965). Effect of curing agents on the growth and survival of food-poisoning strains of *Clostridium perfringens*. *J. Food Sci.* **30,** 1025–1028.

Gould, G. W. (1964). Effect of food preservatives on the growth of bacteria from spores. *In* "Microbial Inhibitors in Food" (N. Molin and A. Erichsen, eds.), pp. 17–24. Almqvist & Wiksell, Stockholm.

Gounelle, H., and Szakvary, A. (1966). L'utilisation des antibiotiques hors la préscription médicale. *Bull. Acad. Natl. Med., Paris* **150,** 149–156.

Graham, E. (1950). Packaging whipping cream in pressurized containers. *Food Technol.* **4,** 225-229.

Graham, W. E. (1958). The history and background of pressurized foods. *Food Technol.* **12,** 317-319.

Graves, R. C., and MacLaury, D. W. (1962). The effects of temperature vapor pressure and absolute humidity on bacterial contamination of shell eggs. *Poult. Sci.* **41,** 1219-1225.

Gray, T. R. G., and Postgate, J. R., eds. (1976). The Survival of Vegetative Microbes. *Symp. Soc. Gen. Microbiol, 26th* Cambridge Univ. Press, London and New York.

Greenberg, R. A., Silliker, J. H., and Fatta, L. D. (1959). The influence of sodium chloride on toxin production and organoleptic breakdown in perishable cured meat inoculated with *Clostridium botulinum. Food Technol.* **13,** 509-511.

Greenberg, R. A., Bladel, B. O., and Zingelmann, W. J. (1965). Radiation injury of *Clostridium botulinum* spores in cured meat. *Appl. Microbiol.* **13,** 743-748.

Greene, F. J., ed. (1974). "Applied Foodservice Sanitation: A Certification Coursebook." Natl. Inst. Foodserv. Ind., Chicago, Illinois.

Griffith, C. L., and Hall, L. A. (1938). Sterilization process. U.S. Patent 2,107,697.

Griffith, C. L., and Hall, L. A. (1940). Sterilization process. U.S. Patent 2,189,947.

Griffiths, R. P., and Haight, R. D. (1973). Reversible heat injury in the marine psychrophilic bacterium *Vibrio marinus* MP-1. *Can. J. Microbiol.* **19,** 557-561.

Gripon, J. C., and Bergère, J. L. (1973). Penetration and stability of pimaricin in soft cheeses. *Lait* **52,** 428-443.

Grover, D. W. (1947). The keeping properties of confectionery as influenced by its water vapour pressure. *J. Soc. Chem. Ind.* **66,** 201-205.

Gudkov, A. V., Trofimova, T. I., Dolidze, G. G., Lyubimova, L. A., Sileva, M. N., and Blagushnia, R. F. (1973). Comparative characteristics of nisin preparation (in Russian). *Antibiotiki* **18,** 162-165.

Guinée, P. A. M. (1970). Resistance transfer to the resident intestinal *Escherichia coli* of rats. *J. Bacteriol.* **102,** 291-292.

Guinée, P., Ugueto, N., and van Leeuwen, N. (1970). *Escherichia coli* with resistance factors in vegetarians, babies and non-vegetarians. *Appl. Microbiol.* **20,** 531-535.

Guinvarch, P. (1959). Three years of ozone sterilization of water in Paris. *In* "Ozone Chemistry and Technology" (H. A. Leedy, ed.), Advances in Chemistry Series, No. 21, Am. Chem. Soc., Washington, D.C., pp. 416-429.

Gundermann, K. O., and Glück, S. (1971). Untersuchungen zur Überlebensdauer Von Bakterien auf Oberflächen und der Möglichkeiten ihrer Beeinflussung. *Arch. Hyg.* **154,** 480-487.

Guthrie, R. K. (1972). Control of microorganisms. *In* "Food Sanitation," (R. K. Guthrie, ed.) pp. 61-87. AVI, Westport, Connecticut.

Hadden, J. P., Ockerman, H. W., Cahill, V. R., Parrett, N. A., and Borton, R. J. (1975). Influence of sodium nitrite on the chemical and organoleptic properties of comminuted pork. *J. Food Sci.* **40,** 626-630.

Haenni, E. O., Affens, W. A., Lento, H. G., Yeomans, A. H., and Fulton, R. A. (1959). New nonflammable formulations for sterilizing sensitive materials. *Ind. Eng. Chem.* **51,** 685-688.

Hahn, F. E., ed. (1976). "Antibiotics and Chemotherapy, Vol. 20: Acquired Resistance of Microorganisms to Chemotherapeutic Drugs. Karger, Basel.

Haines, R. B. (1933). The influence of carbon dioxide on the rate of multiplication of certain bacteria as judged by viable counts. *J. Soc. Chem. Ind.* **52,** 13T-17T.

Haines, W. C., and Harmon, L. G. (1973). Effect of selected lactic acid bacteria on growth of *Staphylococcus aureus* and production of enterotoxin. *Appl. Microbiol.* **25,** 436-441.

Hall, L. P. (1969). "The Effects of Freezing, Storing and Mishandling on Microbial Populations in Frozen Vegetables," Vol. 1. Campden Food Preserv. Res. Assoc., Chipping Campden, England.

Hamm, R. (1971). Interaction between phosphates and meat proteins. *Symp. Phosphates Food Proc.* (J. M. Deman and P. Melnychyn, eds.), pp. 65–82. AVI. Westport, Connecticut.

Hamm, R. (1977). Analysis of smoke and smoked foods. *Pure Appl. Chem.* **49,** 1655–1666.

Hammond, S. M., and Carr, J. C. (1976). The antimicrobial activity of SO_2—with particular reference to fermented and non-fermented fruit juices. *In* "Inhibition and Inactivation of Vegetative Microbes" (F. A. Skinner and W. B. Hugo, eds.), Society of Applied Bacteriology Symposia Series, No. 5, pp. 89–110. Academic Press, New York.

Handford, P. M., and Gibbs, B. M. (1964). Antibacterial effects of smoke constituents on bacteria isolated from bacon. *In* "Microbial Inhibitors in Food" (N. Molin and A. Erichsen, eds.), pp. 333–346. Almqvist & Wiksell, Stockholm.

Hanke, M. E., and Bailey, J. H. (1945). Oxidation-reduction potential requirements of *Clostridium welchii* and other clostridia. *Proc. Soc. Exp. Biol. Med.* **59,** 163–166.

Hanna, M. O., Stewart, J. C., Zink, D. L., Carpenter, Z. L., and Vanderzant, C. (1977). Development of *Yersinia enterocolitica* on raw and cooked beef and pork at different temperatures. *J. Food Sci.* **42,** 1180–1184.

Hannan, R. S. (1955). Scientific and technological problems in using ionising radiations for the preservation of food. *G. B., Dep. Sci. Ind. Res., Food Invest. Board, Spec. Rep.* No. 61.

Hansen, N. H. (1960). Quality deterioration and bacterial growth in prepacked bacon. *Dan. Meat Res. Inst. Publ.* No. 28.

Hansen, N. H., and Riemann, H. (1963). Factors affecting the heat resistance of nonsporing organisms. *J. Appl. Bacteriol.* **26,** 314–333.

Hardy, P. W., Hoffman, H. T., Jr., and Wheaton, E. (1961). Packaging of food products in the new free-piston pressure container. I. Cheese spread. *Food Technol.* **15,** 499–502.

Hargreaves, L. L., Jarvis, B., Rawlinson, A. P., and Wood, J. M. (1975). "The Antimicrobial Effect of Spices, Herbs, and Extracts from these and other Food Plants." Sci. Tech. Surv. No. 88. Food Res. Assoc., Leatherhead, England.

Harmon, S. M., and Kautter, D. A. (1976). Estimating population levels of *Clostridium perfringens* in foods based on alphatoxin. *J. Milk Food Technol.* **39,** 107–110.

Harper, J. C. (1976). "Elements of Food Engineering." AVI, Westport, Connecticut.

Harrison, D. E. F. (1972). Physiological effects of dissolved oxygen tension and redox potential on growing populations of microorganisms. *J. Appl. Chem. Biotechnol.* **22,** 417–440.

Hartman, M. W., Powers, J. J., and Pratt, D. E. (1963). Bacterial permeability of selected food packaging films. *Food Technol.* **17,** 1172–1174.

Hauschild, A. H. W. (1979). Microbial problems in food safety with particular reference to *Clostridium botulinum. In* "The Safety of Foods" (H. Graham, ed.), 2nd ed., AVI Westport, Connecticut, pp. 68–107.

Hauschild, A. H. W., and Hilsheimer, R. (1974a). Evaluation and modifications of media for enumeration of *Clostridium perfringens, Appl. Microbiol.* **27,** 78–82.

Hauschild, A. H. W., and Hilsheimer, R. (1974b). Enumeration of food-borne *Clostridium perfringens* in egg yolk-free tryptose-sulfite-cycloserine agar. *Appl. Microbiol.* **27,** 521–526.

Hausler, W. J., ed. (1972). "Standard Methods for the Examination of Dairy Products," 13th Ed. Am. Public Health Assoc., Washington, D.C.

Hays, G. L., Burroughs, J. D., and Warner, R. C. (1959). Microbiological aspects of pressure packaged foods. II. The effect of various gases. *Food Technol.* **13,** 567–570.

Heinert, H., van der Wall, G., and Brehmer, H. (1976). Lysozym-Ursache von unspezifischen Reaktionen im mikrobiologischen Hemmstofftest. *Arch. Lebensmittelhyg.* **27,** 55–66.

Heiss, R., and Eichner, K. (1969). Stand der Forschung der Verpackung von Frischfleisch. I. Mitt.: Beiträge zur Erzielung einer optimalen Verpackung von frischem Rindfleisch. *Fleischwirtschaft* **49,** 757–764.

Heldman, D. R. (1975). "Food Process Engineering." AVI, Westport, Connecticut.

Henry, M., Joubert, L., and Goret, P. (1954). Mécanisme biochimique de l'action du nitrite dans la conservation des viandes. Conditions physiochimiques favorables à son action bactériostatique. *Ct. Rd. Soc. Biol.* **148,** 819–821.

Hentges, D. J., and Maier, B. R. (1970). Inhibition of *Shigella flexneri* by the normal intestinal flora III. Interactions with *Bacteroides fragilis* strains *in vitro*. *Infect. Immunol.* **2,** 364–370.

Herring, H. K. (1973). Effect of nitrite and other factors on the physico-chemical characteristics and nitrosamine formation in bacon. *Proc. Meat Ind. Res. Conf., Am. Meat Inst./Am. Meat Sci. Assoc., Chicago, Ill.* pp. 47–60.

Hersom, A. C., and Hulland, E. D. (1964). "Canned Foods." 5th Ed. Chem. Publ. Co., New York.

Hess, E. (1934). Effects of low temperatures on the growth of marine bacteria. *Contrib. Can. Biol. Fish.* **8,** 489–505.

Hess, E. (1973). Hygiene during meat production. *In* "The Microbiological Safety of Food" (B. C. Hobbs and J. H. B. Christian, eds.), pp. 3–6. Academic Press, New York.

Hewitt, L. F. (1950). "Oxidation-reduction Potentials in Bacteriology and Biochemistry," 6th Ed. Livingstone, Edinburgh.

Himmelfarb, P., El-Bisi, H. M., Read, R. B., Jr., and Litsky, W. (1962). Effect of relative humidity on the bactericidal activity of propylene oxide vapor. *Appl. Microbiol.* **10,** 431–435.

Hobbs, B. C. (1969). *Clostridium perfringens* and *Bacillus cereus* infections. *In* "Food-Borne Infections and Intoxications" (H. Riemann, ed.), pp. 131–173. Academic Press, New York.

Hobbs, B. C., and Gilbert, R. J. (1978). "Food Poisoning and Food Hygiene," 4th Ed. Arnold, London.

Høyem, T. (1976). Experiences with nitrite and nitrate ban in Norway. *Proc. Int. Symp. Nitrite Meat Prod., 2nd, Zeist, Neth.* (B. J. Tinbergen and B. Kroll, eds.), pp. 293–297. Pudoc, Wageningen, Netherlands.

Hollaender, A. (1942). "Aerobiology." Am. Assoc. Adv. Sci., Lancaster, Pennsylvania.

Hollaender, A., Stapleton, G. E., and Martin, F. L. (1951). X-ray sensitivity of *E. coli* as modified by oxygen tension. *Nature (London)* **167,** 103–104.

Hollenbeck, C. M. (1977). Novel concepts in technology and design of machinery for production and application of smoke in the food industry. *Pure Appl. Chem.* **49,** 1687–1702.

Hopkins, S. E., and Schulze, H. W. (1958). "The Practice of Sanitation." Williams & Wilkins, Baltimore, Maryland.

Horwitz, W., ed. (1975). "Official Methods of Analysis of the Association of Official Analytical Chemists," 12th Ed. Assoc. Offic. Anal. Chem., Washington, D.C.

Hou, C. T., Ciegler, A., and Hesseltine, C. W. (1971). Tremorgenic toxins from penicillia. III Tremortin production by *Penicillium* species on various agricultural commodities. *Appl. Microbiol.* **21,** 1101–1103.

Huber, W. G. (1971). The impact of antibiotic drugs and their residues. *Adv. Vet. Sci. Comp. Med.* **15,** 101–132.

Huffman, D. L. (1974). Effect of gas atmospheres on microbial quality of pork. *J. Food Sci.* **39,** 723–725.

Huffman, D. L., Davis, K. A., Marple, D. N., and McGuire, J. A. (1975). Effect of gas atmospheres on microbial growth, color and pH of beef. *J. Food Sci.* **40,** 1229–1231.

Huhtanen, C. N. (1975). Use of pH gradient plates for increasing the acid tolerance of salmonellae. *Appl. Microbiol.* **29,** 309–312.

Hunter, D. R., and Segel, I. H. (1973). Effect of weak acids on amino acid transport by *Penicillium chrysogenum:* Evidence for a proton or charge gradient as the driving force. *J. Bacteriol.* **113,** 1184–1192.

Hurst, A. (1981). NISIN. *Adv. Appl. Microbial.* **27,** 85–123.

Hurst, A. (1973). Microbial antagonism in foods. *Can. Inst. Food Sci. Technol. J.* **6,** 80–90.

Hurst, A. (1977). Bacterial injury: A review. *Can. J. Microbiol.* **23,** 935–944.

Hurst, A., Hughes, A., Beare-Rogers, J. L., and Collins-Thompson, D. L. (1973). Physiological studies on the recovery of salt-tolerance by *Staphylococcus aureus* after sublethal heating. *J. Bacteriol.* **116,** 901-907.

Hurst, A., Hughes, A., Collins-Thompson, D. L., and Shah, B. G. (1974a). Relationship between loss of magnesium and loss of salt tolerance after sublethal heating of *Staphylococcus aureus*. *Can. J. Microbiol.* **20,** 1153-1158.

Hurst, A., Hughes, A., and Collins-Thompson, D. L. (1974b). The effect of sublethal heating on *Staphylococcus aureus* at different physiological ages. *Can. J. Microbiol.* **20,** 765-768.

Hurst, A., Hendry, G. S., Hughes, A., and Paley, B. (1976). Enumeration of sublethally heated staphylococci in some dried foods. *Can. J. Microbiol.* **22,** 677-683.

Iandolo, J. J., and Ordal, Z. J. (1966). Repair of thermal injury of *Staphylococcus aureus*. *J. Bacteriol.* **91,** 134-142.

Iandolo, J. J., Ordal, Z. J., and Witter, L. D. (1964). The effect of incubation temperature and controlled pH on the growth of *Staphylococcus aureus* MF 31 at various concentrations of NaCl. *Can. J. Microbiol.* **10,** 808-811.

Iandolo, J. J., Clark, C. W., Bluhm, L., and Ordal, Z. J. (1965). Repression of *Staphylococcus aureus* in associative culture. *Appl. Microbiol.* **13,** 646-649.

Idyll, C. P. (1956). "How to Use Sodium Bisulfite to Control 'Black Spot' on Shrimp," Special Services Bull. No. 12. Fl. State Board Conserv. Mar. Lab. Univ. Miami, Coral Gables.

Ikegami, Y., Okaya, C., Sawayama, Z., Shimoda, Y., Mori, D., and Oku, M. (1970). Gaseous spoilage by butyric anaerobes in canned fruits. I. In canned mandarin orange. *Kanzume Jiho* **49,** 993-996. (*Food Sci. Technol. Abstr.* **5,** 5J 656.)

Ingraham, J. L. (1958). Growth of psychrophilic bacteria. *J. Bacteriol.* **76,** 75-80.

Ingraham, J. L., and Stokes, J. L. (1959). Psychrophilic bacteria. *Bacteriol. Rev.* **23,** 97-108.

Ingram, M. (1951). The effect of cold on microorganisms in relation to food. *Proc. Soc. Appl. Bacteriol.* **14,** 243-260.

Ingram, M. (1958). General factors predisposing towards spoilage by yeasts. *In* "The Chemistry and Biology of Yeasts" (A. H. Cook, ed.), pp. 603-633. Academic Press, New York.

Ingram, M. (1960). An influence of carbon source on the resistance of a yeast to benzoic acid. *Ann. Inst. Pasteur Lille* **11,** 167-178.

Ingram, M. (1962a). Microbiological principles in prepacking meats. *J. Appl. Bacteriol.* **25,** 259-281.

Ingram, M. (1962b). I. Post mortem changes in redox potential and pH of meat. II. The importance of redox potential in the microbiology of meat. III. The importance of pH in the microbiology of meat. Medlemsbl. Nor. Veterinaerforen. **8,** Suppl. 1-7, 8-18, and 19-32.

Ingram, M. (1969). Sporeformers as food spoilage organisms. *In* "The Bacterial Spore" (G. W. Gould and A. Hurst, eds.), pp. 549-610. Academic Press, New York.

Ingram, M. (1971). The microbiology of food pasteurization. *In* "Pastorisering av Livesmedel," S.I.K. Rep. No. 292. Sven. Inst. Konserv. Forskning, Göteborg.

Ingram, M. (1972). Meat chilling—the first reason why. *In* "Meat Chilling—Why and How," pp. 1.1-1.13. *ARC Meat Res. Inst. Symp. (2nd.)* Langford, Bristol, U.K.

Ingram, M. (1973). The microbiological effects of nitrite. *Proc. Int. Symp. Nitrite Meat Prod., Zeist, Neth.* (B. Krol and B. J. Tinbergen, eds.), pp. 63-75. Pudoc, Wageningen, Netherlands.

Ingram, M., and Barnes, E. M. (1954). Sterilization by means of ozone. *J. Appl. Bacteriol.* **17,** 246-271.

Ingram, M., and Dainty, R. H. (1971). Changes caused by microbes in spoilage of meats. *J. Appl. Bacteriol.* **34,** 21-39.

Ingram, M., and Farkas, J. (1977). Microbiology of foods pasteurized by ionising radiation. *Acta Aliment. Acad. Sci. Hung.* **6**(2), 123-185.

Ingram, M., and Haines, R. B. (1949). Inhibition of bacterial growth by pure ozone in the presence of nutrients. *J. Hyg.* (Cambr.) **47,** 146-158.

Ingram, M., and Mackey, B. M. (1976). Inactivation by cold. *In* "Inhibition and Inactivation of Vegetative Microbes" (F. A. Skinner and W. B. Hugo, eds.), Society of Applied Bacteriology Symposia Series, No. 5, pp. 111-151. Academic Press, New York.

Ingram, M., and Roberts, T. A. (1971). Application of the "D-concept" to heat treatments involving curing salts. *J. Food Technol.* **6,** 21-28.

Ingram, M., Ottoway, F. J. H., and Coppock, J. B. M. (1956). The preservative action of acid substances in food. *Chem. Ind. (London)* No. 42, 1154-1163.

Insalata, N. F. (1952). CO_2 versus beverage bacteria. *Food Eng.* **24,** 84-85, 190.

Institute of Food Technologists (1975). Sulfites as food additives. A scientific status summary. *Food Technol.* **29**(10), 117-120.

International Atomic Energy Agency (1963). *Radiat. Control Salmonellae Food Feed Prod., Panel Proc., 1962* STI/Doc/10/22. IAEA, Vienna.

International Atomic Energy Agency (1966). Food Irradiat., Proc. Int. Symp., Karlsruhe STI/PUB/127. IAEA, Vienna.

International Atomic Energy Agency (1967). *Microbiol. Probl. Food Preserv. Irradiat., Panel Proc., 1966* STI/PUB/168. IAEA, Vienna.

International Atomic Energy Agency (IAEA) (1968). *Elim. Harmful Organisms Food Feed Irradiat., Panel Proc. 1967* STI/PUB/200. IAEA, Vienna.

International Atomic Energy Agency (IAEA) (1970a). "Training Manual on Food Irradiation Technology and Techniques," Tech. Rep. Ser., No. 114, STI/Doc/10/114. IAEA, Vienna

International Atomic Energy Agency (IAEA) (1970b). "Microbiological Specifications and Testing Methods for Irradiated Foods," Tech. Rep. Ser., No. 104, STI/Doc/10/104. IAEA, Vienna.

International Atomic Energy Agency (1970c). *Preserv. Fish Irradiat., Panel Proc., 1968* STI/PUB/196. IAEA, Vienna

International Atomic Energy Agency (1973a). *Factors Influencing Econ. Appl. Food Irradiat., Panel Proc.* June, 1971 STI/PUB/331. IAEA, Vienna.

International Atomic Energy Agency (1973b). *Radiat. Preserv. Food, Proc. Symp., Bombay, 1972* STI/PUB/317. IAEA, Vienna.

International Commission on Microbiological Specifications for Foods (ICMSF) (1974). "Microorganisms in Foods. 2: Sampling for Microbiological Analysis: Principles and Specific Applications." Univ. of Toronto Press, Toronto.

International Commission on Microbiological Specifications for Foods (ICMFS) (1978). "Microorganisms in Foods. 1. Their Significance and Methods of Enumeration," 2nd Ed. Univ. of Toronto Press, Toronto.

International Dairy Federation (1970). "Detection of Penicillin in Milk by a Disk Assay Technique," Publ. No. 57. Gen. Secretariat, Brussels.

International Institute of Refrigeration (1971). "Recommendations for the Processing and Handling of Frozen Foods," 2nd Ed. Annexe Bull. Inst. Int. Froid, Paris.

International Standards Organization (ISO) (1977). Int. Stand. 4831. Assoc. Fr. Normal., Paris.

Ito, K. A., Chen, J. K., Lerke, P. A., Seeger, M. L., and Unverferth, J. A. (1976). Effect of acid and salt concentration in fresh-pack pickles on the growth of *Clostridium botulinum* spores. *Appl. Environ. Microbiol.* **32,** 121-124.

Ivey, F. J., and Robach, M. C. (1978). Effect of sorbic acid and sodium nitrite on *Clostridium botulinum* outgrowth and toxin production in canned comminuted pork. *J. Food Sci.* **43,** 1782-1785.

Ivey, F. J., Shaver, K. J., Christiansen, L. N., and Tompkin, R. B. (1978). Effect of potassium sorbate on toxinogenesis by *Clostridium botulinum* in bacon. *J. Food Prot.* **41,** 621-625.

Jackson, H. (1974). Loss of viability and metabolic injury of *Staphylococcus aureus* resulting from storage at 5°C. *J. Appl. Bacteriol.* **37,** 59-64.

Jacob, H. E. (1970). Redox potential. *In* "Methods Microbiol," Vol. 2, pp. 91-123. Academic Press, London.

Jakobsen, M., and Jensen, H. C. (1975). Combined effect of water activity and pH on the growth of butyric anaerobes in canned pears. *Lebensm.-Wiss. Technol.* **8,** 158–160.

Jakobsen, M., Filtenborg, O., and Bramsnaes, F. (1972). Germination and outgrowth of the bacterial spore in the presence of different solutes. *Lebensm.-Wiss. Technol.* **5,** 159–162.

Janssen, D. W., and Busta, F. F. (1973). Injury and repair of several *Salmonella* serotypes after freezing and thawing in milk solids. *J. Milk Food Technol.* **36,** 520–522.

Jarolmen, H., and Kemp, G. (1969). R factor transmission *in vivo*. *J. Bacteriol.* **99,** 487–490.

Jarvis, B., and Berridge, N. J. (1969). The application of microbial enzymes which inactivate antibiotics: a review. *Chem. Ind. (London)* **48,** 1721–1724.

Jarvis, B., and Burke, C. S. (1977). Practical and legislative aspects of the chemical preservatives of food. *In* "Inhibition and Inactivation of Vegetative Microbes" (F. A. Skinner and W. B. Hugo, eds.), pp. 345–367. Academic Press, New York.

Jarvis, B., and Morisetti, M. D. (1969). The use of antibiotics in food preservation. *Int. Biodeterior. Bull.* **5,** 39–61.

Jarvis, B., Rhodes, A. C., King, S. E., and Patel. M. (1976). Sensitization of heat-damaged spores of *Clostridium botulinum* type B to sodium chloride and sodium nitrite. *J. Food Technol.* **11,** 41–50.

Jay, J. M. (1970). "Modern Food Microbiology," pp. 27, 31. Van Nostrand-Reinhold, New York.

Jaye, M., Kittaka, R. S., and Ordal, Z. J. (1962). The effect of temperature and packaging material on the storage life and bacterial flora of ground beef. *Food Technol.* **16**(4), 95–98.

Jensen, L. B. (1943). Action of hardwood smoke on bacteria in cured meats. *Food Res.* **8,** 377–387.

Jezeski, J. J., and Olsen, R. H. (1962). The activity of enzymes at low temperatures. *Proc. Low Temp. Microbiol. Symp., 1961* pp. 139–155. Campbell Soup Co., Camden, New Jersey.

Johnson, J. D. (ed.) (1975). "Disinfection-Water and Waste-Water." Ann Arbor Sci. Publ., Ann Arbor, Michigan.

Johnston, M. A., and Loynes, R. (1971). Inhibition of *Clostridium botulinum* by sodium nitrite as affected by bacteriological media and meat suspensions. *Can. Inst. Food Technol. J.* **4,** 179–184.

Johnston, M. A., Pivnick, H., and Samson, J. M. (1969). Inhibition of *Clostridium botulinum* by sodium nitrite in a bacteriological medium and in meat. *Can. Inst. Food Technol. J.* **2,** 52–55.

Jones, G. W., and Kennedy, R. E. (1930). Extinction of ethylene oxide fumes with carbon dioxide. *Ind. Eng. Chem.* **22,** 146–147.

Juffs, H. S. (1976). Effects of temperature and nutrients on proteinase production by *Pseudomonas fluorescens* and *Ps. aeruginosa* in broth and milk. *J. Appl. Bacteriol.* **40,** 23–31.

Jukes, T. H. (1973). Public health significance of feeding low levels of antibiotics to animals. *Adv. Appl. Microbiol.* **16,** 1–30.

Juven, B. J. (1976). Bacterial spoilage of citrus products at pH lower than 3.5. *J. Milk Food Technol.* **39,** 819–822.

Kabara, J. J., Conley, A. J., and Truant, J. P. (1972). Relationship of chemical structure and antimicrobial activity of alkyl amides and amines. *Antimicrob. Agents Chemother.* **2,** 492–498.

Kaess, G., and Weidemann, J. F. (1968). Ozone treatment of chilled beef. *J. Food Technol.* **3,** 325–334.

Kaess, G., and Weidemann, J. F. (1973). Effects of ultraviolet irradiation on the growth of microorganisms on chilled beef slices. *J. Food Technol.* **8,** 59–70.

Karmas, E. (1975). "Fresh Meat Technology." Noyes Data Corp., Park Ridge, New Jersey.

Kashket, E. R., and Wong, P. T. S. (1969). The intracellular pH of *Escherichia coli*. *Biochim. Biophys. Acta* **193,** 212–214.

Kates, M., and Baxter, R. M. (1962). Lipid composition of mesophilic and psychrophilic yeasts

(*Candida* species) as influenced by environmental temperature. *Can. J. Biochem. Physiol.* **40,** 1213–1227.

Kautter, D. A. (1964). *Clostridium botulinum* type E in smoked fish. *J. Food Sci.* **29,** 843–849.

Kavanagh, F. (1963). "Analytical Microbiology." Academic Press, New York.

Kefford, J. (1948). Effect of ozone on microbial growth on beef muscle, and on the flavour of beef fat. *J. Counc. Sci. Ind. Res.* **21,** 116–121.

Kelsey, J. C., and Maurer, I. M. (1966). An in-use test for hospital disinfectants. *Mon. Bull. Minist. Health Public Health Lab. Serv.* **25,** 180.

Kelsey, J. C., and Maurer, I. M. (1974). An improved (1974) Kelsey–Sykes test for disinfectants. *Pharm. J.* **213,** 528–530.

Kelsey, J. C., and Sykes, G. (1969). A new test for the assessment of disinfectants with particular reference to their use in hospitals. *Pharm. J.* **202,** 607–609.

Kemp, J. D. (1974). Nitrate and nitrite substitutes in meat curing. *Food Prod. Dev.* **8**(8), 64–70.

Kenis, P. R., and Morita, R. Y. (1968). Thermally induced leakage of cellular material and viability of *Vibrio marinus* a psychrophilic marine bacterium. *Can. J. Microbiol.* **14,** 1239–1244.

Killeffer, D. H. (1930). Carbon dioxide preservation of meat and fish. *Ind. Eng. Chem.* **22,** 140–143.

Kim, T. K., and Stephens, J. F. (1972). Drug resistance and transferable drug resistance of *Escherichia coli* isolated from "Ready-to-cook" broilers. *Poult. Sci.* **51,** 1165–1170.

King, A. D., Jr., and Nagel, C. W. (1967). Growth inhibition of a *Pseudomonas* by carbon dioxide. *J. Food Sci.* **32,** 575–579.

King, A. D., Jr., and Nagel, C. W. (1975). Influence of carbon dioxide upon the metabolism of *Pseudomonas aeruginosa*. *J. Food Sci.* **40,** 362–366.

Kiser, J. S., Gale, G. O., and Kemp, G. A. (1971). Antibiotics as feedstuff additives: the risk-benefit equation for man. *CRC Crit. Rev. Toxicol.* **1,** 55–92.

Kissinger, J. C., and Zaika, L. L. (1978). Effect of major spices in Lebanon bologna on acid production by starter culture organisms. *J. Food Prot.* **41,** 429–431.

Kleniewski, A. (1970). Les problèmes posés par la perméabilité des matériaux d'emballage aux agents physico-chimiques et microbiologiques. *Emballages* **40,** 378–381, 384–393.

Kligler, I. J., and Guggenheim, K. (1938). The influence of vitamin C on the growth of anaerobes in the presence of air, with special reference to the relative significance of E_h and O_2 in the growth of anaerobes. *J. Bacteriol.* **35,** 141–156.

Kodicek, E. (1956). The effect of unsaturated fatty acids, of vitamin D and other sterols on Gram-positive bacteria. *In* "Biochemical Problems of Lipids" (G. Popjók and E. le Breton, eds.), pp. 401– 406. Butterworth, London.

Koh, W. Y., Morehouse, C. T., and Chandler, V. L. (1956). Relative resistances of microorganisms to cathode rays. I. Non-sporeforming bacteria. *Appl. Microbiol.* **4,** 143–146.

Koser, S. A., and Skinner, W. W. (1922). Viability of the colon-typhoid group in carbonated water and carbonated beverages. *J. Bacteriol.* **7,** 111–121.

Kraft, A. A., and Ayres, J. C. (1952). Post-mortem changes in stored meats. IV. Effect of packaging materials on keeping quality of self-service meats. *Food Technol.* **6,** 8–12.

Kristensen, M. (1974). Isolation of radiation resistant organisms from the environments of cobalt-60 sources: preliminary reports. *In* "Experiences in Radiation Sterilisation of Medical Products," Proc. Working Group Meet., Riso, 1972. Tech. Rep. Ser., No. 159, pp. 43–50. IAEA, Vienna.

Kupchik, G. J., and Katz, J. (1977). Potential health hazards associated with reusable foodservice utensils. *J. Food Prot.* **40,** 132–134.

Labadie, J., Fournaud, J., and Dumont, B. L. (1975). Relations entre le pH et la microflore des viandes hachées de bovins. *Ann. Technol. Agric.* **24**(2), 193–203.

Labuza, T. P., Acott, K., Tatini, S. R., Lee, R. Y., Flink, J., and McCall, W. (1976). Water activity determination: a collaborative study of different methods. *J. Food Sci.* **41,** 910–917.

Lafaix, C., Camerlynck, P., and Rey, M. (1969). Étude de la sensibilité aux antibiotiques des entérobacteries endémiques a Dakar; mise en évidence de polyrésistances transférables. *Med. Afr. Noire* **16,** 125-130.

Lamprecht, E. C., and Elliott, M. C. (1974). Death rate of *Salmonella oranienburg* in fish meals as influenced by autoxidation treatment. *J. Sci. Food Agric.* **25,** 1329-1338.

Langner, H. J., Teufel, U., Seigert, M., and Frommhold, M. (1973). Der chemische und mikrobiologische Nachweis von Antibiotika. *Fleischwirtschaft* **53,** 243-246, 249.

Larkin, J. M., and Stokes, J. L. (1968). Growth of psychrophilic microorganisms at subzero temperatures. *Can. J. Microbiol.* **14,** 97-101.

Larsson, K., Norén, B., and Odham, G. (1975). Antimicrobial effect of simple lipids and the effect of pH and positive ions. *Antimicrob. Agents Chemother.* **8,** 733-736.

Leach, R. H. (1962). The osmotic requirements for growth of *Mycoplasma. J. Gen. Microbiol.* **27,** 345-354.

Lechowich, R. V., and Ordal, Z. J. (1962). The influence of sporulation temperature on the heat resistance and chemical composition of bacterial spores. *Can. J. Microbiol.* **8,** 287-295.

Ledward, D. A., Nicol, D. J., and Shaw, M. K. (1971). Microbiological and colour changes during ageing of beef. *Food Technol. Aust.* **23,** 30-32.

Lee, S. H., Cassens, R. G., and Sugiyama, H. (1978a). Factors affecting inhibition of *Clostridium botulinum* in cured meats. *J. Food Sci.* **43,** 1371-1374.

Lee, S. H., Cassens, R. G., Winder, W. C., and Fennema, O. R. (1978b). Factors affecting the formation of nitrate from added nitrite in model systems and cured meat products. *J. Food Sci.* **43,** 673-676.

Lehmann, W. (1965). Das Vorkommen von Salmonellen in Fleischverarbeiten den Betrieben und die sich daraus ergebenden hygienischen Schwerpunkte. *Fleischmeister* **11,** 255.

Leistner, L. (1956). Hüllendefekte bei Rohwürsten in Kunstdärmen auf Zellulosebasis. *Fleischwirtschaft* **8,** 182-184.

Leistner, L. (1974). Mikrobiologie der Verpackung von Fleisch und Fleischwaren. *Fleischwirtschaft* **54,** 1036-1039.

Leistner, L., and Rödel, W. (1975). The significance of water activity for microorganisms in meats. *In* "Water Relations of Foods" (R. B. Duckworth, ed.), pp. 309-323. Academic Press, New York.

Leistner, L., and Rödel, W. (1976). The stability of intermediate moisture foods with respect to microorganisms. *In* "Intermediate Moisture Foods" (R. Davies, G. G. Birch, and K. J. Parker, eds.), pp. 120-137. Appl. Sci. Publ., London.

Lerche, M., Goerttler, V., and Rievel, H. (1957). "Lehrbuch der tierärztlichen Lebensmittelüberwachung." Verlag Schaper, Hannover.

Levetzow, R., and Weise, E. (1974). Zur Rückstandsuntersuchung im Rahmen der amtlichen Fleischuntersuchung. *Schlacht-Viehof Zg.* **74,** 329-337.

Levy, S. B., FitzGerald, G. B., and Macone, A. B. (1976a). Spread of antibiotic-resistant plasmids from chicken to chicken and from chicken to man. *Nature (London)* **260,** 40-42.

Levy, S. B., FitzGerald, G. B., and Macone, A. B. (1976b). Changes in intestinal flora of farm personnel after introduction of a tetracycline-supplemented feed on a farm. *N. Engl. J. Med.* **295,** 583-588.

Lewis, K. H. (1978). Disinfectants. *In* "Encyclopedia of Food Science" (M. S. Peterson and A. H. Johnson, eds.), pp. 000-000. AVI, Westport, Connecticut.

Lewis, N. F., Alur, M. D., Nadkarni, A. R., Gaonkar, S. G., and Kumta, U. S. (1973). Selective control of spoilage organisms in seafoods by gamma irradiation. *Radiat. Preserv. Food, Proc. Symp., Bombay, 1972* STI/PUB/317, pp. 201-218. IAEA, Vienna.

Licciardello, J. J., Nickerson, J. T. R., and Goldblith, S. A. (1970). Recovery of salmonellae from irradiated and unirradiated foods. *J. Food Sci.* **35,** 620-624.

Lightbown, J. W., and de Rossi, P. (1965). The identification and assay of mixtures of antibiotics by electrophoresis in agar gel. *Analyst* **90,** 89-98.

Litsky, B. Y. (1973). "Food Service Sanitation." Mod. Hosp. Press, Chicago, Illinois; McGraw-Hill, New York.

Locker, R. H., Davey, C. L., Nottingham, P. M., Haughey, D. P., and Law, N. H. (1975). New concepts in meat processing. *Adv. Food Res.* **21,** 157-222.

Longrée, K. (1967). "Quantity Food Sanitation," p. 245. Wiley (Interscience), New York.

Losikoff, M. E. (1978). Establishment of a heat inactivation curve for *Clostridium botulinum* 62A toxin in beef broth. *Appl. Environ. Microbiol.* **36,** 386-388.

Lotter, L. P., and Leistner, L. (1978). Minimal water activity for enterotoxin A production and growth of *Staphylococcus aureus*. *Appl. Environ. Microbiol.* **36,** 377-380.

Lubieniecki-von Schelhorn, M. (1964). Investigations of the distribution of preservatives between fat and water in foods. *In* "Microbial Inhibitors in Food" (N. Molin and A. Erichsen, eds.), pp. 139-144. Almqvist & Wiksell, Stockholm.

Lubieniecki-von Schelhorn, M. (1972-1974). Vermehrung und Absterben von Mikroorganismen in Abhängigkeit vom Milieu unter besonderer Berücksichtigung technologischer Einflüsse. I bis IV. *Chem., Mikrobiol., Technol. Lebensm.* **1,** 89-95, 138-144; **2,** 26-32; **3,** 138-147.

McDougall, J. (1971). Packaging materials: The interrelationship between the micro-climate within the package, shelf-life and quality. *Food Technol. N.Z.* Apr., pp. 20-24.

MacLeod, R. A., and Calcott, P. H. (1976). Cold shock and freezing damage to microbes. *Symp. Soc. Gen. Microbiol.* **26,** 81-109.

MacLeod, R. A., Smith, L. D. H., and Gelinas, R. (1966). Metabolic injury to bacteria. I. Effect of freezing and storage on the requirements of *Aerobacter aerogenes* and *Escherichia coli* for growth. *Can. J. Microbiol.* **12,** 61-72.

MacNeil, J. H., and Mast, M. G. (1973). Frankfurters without nitrates or nitrites. *Food Prod. Dev.* 7(2), 36-40.

Macris, B. J. (1975). Mechanisms of benzoic acid uptake by *Saccharomyces cerevisiae*. *Appl. Microbiol.* **30,** 503-506.

Magoon, C. A. (1926). Studies on bacterial spores. I. Thermal resistance as affected by age and environment. *J. Bacteriol.* **11,** 253-283.

Makashev, A. P. (1959). Carbon dioxide as a means of prolonging the storage life of chilled fish products. *Prog. Refrig. Sci. Technol.* **3,** 251-255.

Marr, A. G., and Ingraham, J. L. (1962). Effect of temperature on the composition of fatty acids in *Escherichia coli*. *J. Bacteriol.* **84,** 1260-1267.

Marriott, N. G., Smith, G. C., Hoke, K. E., Carpenter, Z. L., and West, R. L. (1977). Long-distance transoceanic shipments of fresh beef. *J. Food Sci.* **42,** 316-320.

Marshall, B. J., Ohye, D. F., and Christian, J. H. B. (1971). Tolerance of bacteria to high concentrations of NaCl and glycerol in the growth medium. *Appl. Microbiol.* **21,** 363-364.

Marth, E. H. (1966). Antibiotics in foods—naturally occurring, developed and added. Residue Rev. **12,** 65-161.

Marx, M. (1972). Untersuchungen zur Mikrobiocoenose von Lebensmittelmikroorganismen (antagonistische und synergistische Wechselwirkungen zwischen Enterobacteriaceae, Pseudomonadaceae, Staphylokokken, Enterokokken, Bazillen und Clostridien). Vet. Med. Dissertation, Univ., West Berlin.

Matches, J. R., and Liston, J. (1968). Growth of salmonellae on irradiated and non-irradiated sea foods. *J. Food Sci.* **33,** 406-410.

Matsuyama, A. (1973). Present status of food irradiation research in Japan with special reference to microbiological and entomological aspects. *Radiat. Preserv. Food, Proc. Symp., Bombay, 1972* STI/PUB/317, pp. 261-278. IAEA, Vienna.

Maunder, D. T., Folinazzo, J. F., and Killoran, J. J. (1968). Bio-test method for determining integrity of flexible packages of shelf-stable foods. *Food Technol.* **22,** 615-618.

Maurer, I. M. (1972). Glassware and magic pots. *Lab. News* **41,** 16-24.

Maurer, I. M. (1978). "Hospital Hygiene," 2nd Ed. Arnold, London.

Maxcy, R. B., and Tiwari, N. P. (1973). Irradiation of meats for public health protection. *Radiat. Preserv. Food, Proc. Symp., Bombay, 1972* STI/PUB/317, pp. 491-503. IAEA, Vienna.

Mayr, G., and Kaemmerer, H. (1959). Fumigation with ethylene oxide. *Food Manuf.* **34,** 169-170.

Mayr, G. E., and Suhr, H. (1973). Preservation and sterilization of pure and mixed spices. *Proc. Conf. Spices, Trop. Prod. Inst., London Sch. Pharm.* pp. 201-207.

Mazur, P. (1966). Physical and chemical basis of injury in single-celled microorganisms subjected to freezing and thawing. *In* "Cryobiology" (H. T. Meryman, ed.), pp. 213-315. Academic Press, New York.

Measures, J. C. (1975). Role of amino acids in osmoregulation of non-halophilic bacteria. *Nature (London)* **257,** 398-400.

Meers, P. D., and Goode, D. (1965). *Salmonella typhi* in corned beef. *Lancet* **7382,** 426-428, Feb. 20.

Meester, J. (1973). Nitrate and nitrite allowances in meat products. *Proc. Int. Symp. Nitrite Meat Prod., Zeist, Neth.* (B. Krol and B. J. Tinbergen, eds.), pp. 265-268. Pudoc, Wageningen, Netherlands.

Mercer, H. D. (1975). Antimicrobial drugs in food-producing animals: Control mechanisms of governmental agencies. *Vet. Clin. North Am.* **5,** 3-34.

Meryman, H. T. (1974). Freezing injury and its prevention in living cells. *Annu. Rev. Biophys. Bioeng.* **3,** 341-363.

Michael, G. T., and Stumbo, C. R. (1970). Ethylene oxide sterilization of *Salmonella senftenberg* and *Escherichia coli:* Death kinetics and mode of action. *J. Food Sci.* **35,** 631-634.

Michener, H. D., and Elliott, R. P. (1964). Minimum growth temperatures for food-poisoning, fecal-indicator, and psychrophilic microorganisms. *Adv. Food Res.* **13,** 349-396.

Miller, G. W., and Evans, H. J. (1956). Inhibition of plant cytochrome oxidase by bicarbonate. *Nature (London)* **178,** 974-976.

Miller, I., and Kandler O. (1967). Eiweissabbau und Anreicherung freier Aminosäuren durch Milchsäurebakterien in Milch. II. Die Anreicherung freier Aminosäuren durch Thermobakterien. *Milchwissenschaft* **22,** 469-480.

Miller, S., Burkhardt, B., Ehrlich, R., and Peterson, R. J. (1959). Disinfection and sterilization of sewage by ozone. *In* "Ozone Chemistry and Technology," (H. A. Leedy, ed.) pp. 381-387. Advances in Chemistry Series **21.** Am. Chem. Soc., Washington, D. C.

Ministry of Agriculture, Fisheries and Food (1970). "The Disinfection and Disinfestation of Poultry Houses," Minist. Agric., Fish. Food, Advis. Leafl. No. 514 (Rev.). HM Stationery Off., London.

Ministry of Agriculture, Fisheries and Food (1971). "Hatching Egg Hygiene for More and Better Chicks," Minist. Agric., Fish. Food, Agric. Dev. Advis. Serv. HM Stationery Off., London.

Minks, D., and Stringer, W. C. (1972). The influence of aging beef in vacuum. *J. Food Sci.* **37,** 736-738.

Minor, E. K. (1959). Processor now cleans with ultrasonics. *Food Eng.* **31**(6), 72-73.

Minor, T. E., and Marth, E. H. (1970). Growth of *Staphylococcus aureus* in acidified pasteurized milk. *J. Milk Food Technol.* **33,** 516-520.

Minor, T. E., and Marth, E. H. (1972). Fate of *Staphylococcus aureus* in cultured buttermilk, sour cream and yogurt during storage. *J. Milk Food Technol.* **35,** 302-305.

Mocquot, G., and Ducluzeau, R. (1968). The influence of cold storage of milk on its microflora and its suitability for cheesemaking. *In* "Low Temperature Biology of Foodstuffs" (J. Hawthorne and E. J. Rolfe, eds.), **4,** 235–250. Pergamon, Oxford.

Möhler, K. (1973). Formation of curing pigments by chemical, biochemical or enzymatic reactions. *Proc. Int. Symp. Nitrite Meat Prod. Zeist, Neth.* (B. Krol and B. J. Tinbergen, eds.), pp. 13–19. Pudoc, Wageningen, Netherlands.

Möhler, K., and Baumann, M. (1971). Bildung von Pökelfarbstoff im Blut durch die Einwirkung von Natriumnitrit auf Hämoglobin. *Z. Lebensm.-Unters.-Forsch.* **147,** 258–266.

Moerman, P. C. (1972). Mould control in meat products using pimaricin. *Voedingsmiddelen Technol.* **3,** 261–264.

Mol, J. H. H., and Timmers, C. A. (1970). Assessment of the stability of pasteurized comminuted meat products. *J. Appl. Bacteriol.* **33,** 233–247.

Molin, N. (1964). The action of N-α-acyldipeptides on microorganisms. *In* "Microbial Inhibitors in Food" (N. Molin and A. Erichsen, eds.), pp. 47–58. Almqvist & Wiksell, Stockholm.

Montgomery, H. B. S. (1958). Effect of storage conditions on the incidence of *Gloesporium* rots of apple fruits. *Nature (London)* **182,** 737–738.

Moon, N. J., and Reinbold, G. W. (1976). Commensalism and competition in mixed cultures of *Lactobacillus bulgaricus* and *Streptococcus thermophilus. J. Milk Food Technol.* **39,** 337–341.

Moorhouse, E. C. (1969). Transferable resistance in enterobacteria isolated from urban infants. *Br. Med. J.* **ii,** 405–407.

Moran, T. (1938). Gas storage of meat and eggs. *Food Res.* **3,** 149–154.

Morita, R. Y. (1975). Psychrophilic bacteria. *Bacteriol. Rev.* **39,** 144–167.

Morris, H. A. L. (1952). Effects due to residual nitrite in the canning of meat. *Analyst* **77,** 98.

Morris, J. G. (1976). Oxygen and the obligate anaerobe. *J. Appl. Bacteriol.* **40,** 229–244.

Mosley, E. B., Elliker, P. R., and Hays, H. (1976). Destruction of food spoilage, indicator and pathogenic organisms by various germicides in solution and on a stainless steel surface. *J. Milk Food Technol.* **39,** 830–836.

Moss, C. W., and Speck, M. L. (1966). Release of biologically active peptides from *Escherichia coli* at subzero temperatures. *J. Bacteriol.* **91,** 1105–1111.

Mossel, D. A. A. (1970). The role of microbiology and hygiene in the manufacturing of margarine. *In* "Margarine Today." (J. W. E. Coenen, R. Feron, and D. A. A. Mossel, eds.). pp. 104–105. Brill Publ., Leiden, The Netherlands.

Mossel, D. A. A. (1971). Symposium on microbial changes in foods: Physiological and metabolic attributes of microbial groups associated with foods. *J. Appl. Bacteriol.* **34,** 95–119.

Mossel, D. A. A. (1975). Occurrence, prevention and monitoring of microbial quality loss of foods and dairy products. *Crit. Rev. Environ. Control* **5,** 1–139.

Mossel, D. A. A., and Corry, J. E. L. (1977). Detection and enumeration of sublethally injured pathogenic and index bacteria in foods and water processed for safety. *Alimenta-Sonderausgabe Mikrobiol.* pp. 19–34.

Mossel, D. A. A., and Ingram, M. (1955). The phsyiology of the microbial spoilage of foods. *J. Appl. Bacteriol.* **18,** 232–268.

Mossel, D. A. A., and Ratto, M. A. (1970). Rapid detection of sublethally impaired cells of Enterobacteriaceae in dried foods. *Appl. Microbiol.* **20,** 273–275.

Mossel, D. A. A., Jongerius, E., and Koopman, M. J. (1965). Sur la nécessité d'une revivication au préalable pour le dénombrement des Enterobacteriaceae dans les aliments déshydrates, irradiés ou non. *Ann. Inst. Pasteur, Lille* **16,** 119–125.

Mossel, D. A. A., Harrewijn, G. A., and Nesselrooy-van Zadelhoff, C. F. M. (1974). Standardization of the selective inhibitory effect of surface active compounds used in media for the detection of Enterobacteriaceae in foods and water. *Health Lab. Sci.* **11,** 260–267.

Mottram, D. A., and Rhodes, D. N. (1973). Nitrite and the flavour of cured meat. I. *Proc. Int. Symp. Nitrite Meat Prod., Zeist, Neth.* (B. Krol and B. J. Tinbergen, eds.), pp. 161-171. Pudoc, Wageningen, Netherlands.

Mourges, R., and Auclair, J. (1973). Durée de conservation à 4°C et 8°C du lait pasteurisé conditionné aseptiquement. *Lait* **53,** 481-490.

Moussa, R. S., Keller, N., Curiat, G., and De Man, C. (1973). Comparison of five media for the isolation of coliform organisms from dehydrated and deep frozen foods. *J. Appl. Bacteriol.* **36,** 619-624.

Mukai, F., Hawryluk, I., and Shapiro, R. (1970). Mutagenic specificity of sodium bisulfite. *Biochem. Biophys. Res. Commun.* **39,** 983-988.

Munro, A. L. S. (1970). Measurement and control of pH values. *Methods Microbiol.* **2,** 39-89.

Murrell, W. G., and Scott, W. J. (1966). The heat resistance of bacterial spores at various water activities. *J. Gen. Microbiol.* **43,** 411-425.

Nagy, R. (1959). Application of ozone from sterilamp in control of mold bacteria and odors. *In* "Ozone Chemistry and Technology," Advances in Chemistry Series, No. 21, (H. A. Leedy, ed.), pp. 57-65. Am. Chem. Soc., Washington, D. C.

Nakamura, M., and Dawson, D. A. (1962). Role of suspending and recovery media in the survival of frozen *Shigella sonnei*. *Appl. Microbiol.* **10,** 40-43.

National Academy of Sciences/National Research Council (1975). "Prevention of Microbial and Parasitic Hazards Associated with Processed Foods." Natl. Acad. Sci., Washington, D.C.

National Canners Association (1975). Container closure evaluation. *In* "Canned Foods, Principles of Thermal Process Control and Container Closure Evaluation," pp. 165-206. Natl. Canners Assoc., Berkeley, California.

Neal, A. L., Weinstock, J. O., and Lampen, J. O. (1965). Mechanisms of fatty acid toxicity for yeast. *J. Bacteriol.* **90,** 126-131.

Neale, E. K., and Chapman, G. B. (1970). Effect of low temperature on the growth and fine structure of *Bacillus subtilis*. *J. Bacteriol.* **104,** 518-528.

Neitzert, W. A. (1974). Kunststoffolien für die Verpackung von Fleisch und Fertiggerichten. *Neue Verpackung* **26,** 72-80.

Neitzert, W. A. (1979). *Plastverarbeiter* **30,** 773-774.

Newton, K. G., Harrison, J. C. L., and Smith, K. M. (1977). The effect of storage in various gaseous atmospheres on the microflora of lamb chops held at −1°C. *J. Appl. Bacteriol.* **43,** 53-59.

Ng, H. (1966). Heat sensitivity of 300 *Salmonella* isolates. *Rep. Collaborators Conf. Destr. Salmonellae, West. Reg. Res. Lab., U.S. Dep. Agric., Albany, Calif.* ARS 74-37, pp. 39-41.

Ng, H., Vaughn, R. H., Stewart, G. F., Nagel, C. W., and Simpson, K. (1957). Antibiotics in poultry meat preservation: Development of resistance among spoilage organisms. *Appl. Microbiol.* **5,** 331-333.

Nickerson, J. T., and Sinskey, A. J. (1972). "Microbiology of foods and food processing." Am. Elsevier, New York.

Nieuwenhof, F. F. J., Stadhouders, J., and Hup, G. (1969). Stimulating effect of lactobacilli on the growth of propionibacteria in cheese. *Neth. Milk Dairy J.* **23,** 287-289.

Niinivaara, F. P., and Pohja, M. S. (1956). Über die Reifung der Rohwurst. I. Die Veränderungen der Bakterienflora während der Reifung. *Z. Lebensm.-Unters. -Forsch.* **104,** 413-422.

Niskanen, A., and Nurmi, E. (1976). Effect of starter culture on staphylococcal enterotoxin and thermonuclease production in dry sausage. *Appl. Environ. Microbiol.* **31,** 11-20.

Niven, C. F., Jr. (1958). Microbiological aspects of radiation preservation of food. *Annu. Rev. Microbiol.* **12,** 507-524.

Nordin, H. R. (1969). The depletion of added sodium nitrite in ham. *Can. Inst. Food Technol. J.* **2,** 79-85.

Nordin, H. R., Burke, T., Webb, G., Rubin, L. J., and van Binnendyk, D. (1975). Effect of pH, salt and nitrite in heat processed meat on destruction and out-growth of P.A. 3679. *Can. Inst. Food Sci. Technol. J.* **8,** 58–66.

North, W. R., Jr. (1961). Lactose pre-enrichment method for isolation of *Salmonella* from dried egg albumen. Its use in a survey of commercially produced albumen. *Appl. Microbiol.* **9,** 188–195.

Northolt, M. D. (1972). Composition of ox bile and of commercial preparations. *Antonie van Leeuwenhoek; J. Microbiol. Serol.* **38,** 632.

Northolt, M. D., Verhulsdonk, C. A. H., Soentoro, P. S. S., and Paulsch, W. E. (1976). Effect of water activity and temperature on aflatoxin production by *Aspergillus parasiticus. J. Milk Food Technol.* **39,** 170–174.

Nurmi, E. (1966). Effect of bacterial inoculations on characteristics and microbial flora of dry sausage. Thesis, Univ. of Helsinki, Helsinki.

Oblinger, J. L., and Kraft, A. A. (1973). Oxidation-reduction potential and growth of *Salmonella* and *Pseudomonas fluorescens. J. Food Sci.* **38,** 1108–1112.

O'Brien, R. W., and Morris, J. G. (1971). Oxygen and the growth and metabolism of *Clostridium acetobutylicum. J. Gen. Microbiol.* **68,** 307–318.

O'Donovan, D. C. (1965). Treatment with ozone. *J. Am. Water Works Assoc.* **57,** 1167–1192.

Ogilvy, W. S., and Ayres, J. C. (1951a). Post-mortem changes in stored meats. II. The effect of atmospheres containing carbon dioxide in prolonging the storage life of frankfurters. *Food Technol.* **5,** 97–102.

Ogilvy, W. S., and Ayres, J. C. (1951b). Post-mortem changes in stored meats. III. The effects of atmospheres containing carbon dioxide in prolonging the storage lofe of frankfurters. *Food Technol.* **5,** 300–303.

Ogilvy, W. S., and Ayres, J. C. (1953). Post-mortem changes in stored meats. V. Effect of carbon dioxide on microbial growth on stored frankfurters and characteristics of some microorganisms isolated from them. *Food Res.* **18,** 121–130.

Ohye, D. F., and Christian, J. H. B. (1967). Combined effects of temperature, pH and water activity on growth and toxin production by *C. botulinum* types A, B and E. *In* "Botulism 1966" (M. Ingram and T. A. Roberts, eds.), pp. 217–223. Chapman & Hall, London.

Olczyk, C. (1968). The effect of chemicals modifying the oxidation-reduction potential on the growth of a thermophilic strain of *Bacillus subtilis. Acta Microbiol. Pol.* **17,** 9–16.

O'Leary, V., and Solberg, M. (1976). Effect of sodium nitrite inhibition on intracellular thiol groups and on the activity of certain glycolytic enzymes in *Clostridium perfringens. Appl. Environ. Microbiol.* **31,** 208–212.

Olley, J. (1976). "Temperature Indicators, Temperature Integrators, Temperature Function Integrators and the Food Spoilage Chain," pp. 125–132. Int. Inst. Refrig., Sydney.

Olsen, C. Z. (1976). Chemical composition and application of smoke flavor. *Proc. Annu. Meet. Eur. Meat Res. Workers, 22nd, Malmo* 2(F-7:1).

Olsen, R. H., and Jezeski, J. J. (1963). Some effects of carbon source, aeration, and temperature on growth of a psychrophilic strain of *Pseudomonas fluorescens. J. Bacteriol.* **86,** 429–433.

Olsman, W. J., and Krol, B. (1972). Depletion of nitrite in heated meat products during storage. *Proc. Annu. Meet. Eur. Meat Res. Workers, 18th, Guelph, Can.*

Olson, J. C., Jr., and Sanders, A. C. (1975). Penicillin in milk and milk products: Some regulatory and public health consideration. *J. Milk Food Technol.* **38,** 630–633.

Olson, J. C., Jr., Macy, H., and Halvorson, H. O. (1952). Thermal death-time studies of coliform bacteria in milk. *Minn. Agric. Exp. Stn., Tech. Bull.* No. 202.

Onishi, H. (1963). Osmophilic yeasts. *Adv. Food Res.* **12,** 53–94.

Orillo, C. A., Sison, E. C., Luis, M., and Pederson, C. S. (1969). Fermentation of Philippine vegetable blends. *Appl. Microbiol.* **17,** 10–13.

Pappas, H. J., and Hall, L. A. (1952). The control of thermophilic bacteria. *Food Technol.* **6,** 456-458.

Parekh, K. G., and Solberg, M. (1970). Comparative growth of *Clostridium perfringens* in carbon dioxide and nitrogen atmospheres. *J. Food Sci.* **35,** 156-159.

Park, C. E., Stankiewicz, Z. K., Johnston, M. A., and Todd, E. C. D. (1972). The toxic effect of 4-chloro-2-cyclopentylphenyl β-D-galactopyranoside in lactose broth on the growth of *Shigella* in mixed culture with *Escherichia coli. Can. J. Microbiol.* **18,** 1743-1748.

Park, C. E., Rayman, M. K., Szabo, R., and Stankiewicz, Z. K. (1976). Selective enrichment of *Shigella* in the presence of *Escherichia coli* by use of 4-chloro-2-cyclopentylphenyl β-D-galactopyranoside. *Can. J. Microbiol.* **22,** 654-657.

Park, H. S., and Marth, E. H. (1972). Inactivation of *Salmonella typhimurium* by sorbic acid. *J. Milk Food Technol.* **35,** 532-539.

Park, H. S., Marth, E. H., and Olson, N. F. (1973). Fate of enteropathogenic strains of *Escherichia coli* during the manufacture and ripening of Camembert cheese. *J. Milk Food Technol.* **36,** 543-546.

Parker, M. E., and Litchfield, J. H. (1962). "Food Plant Sanitation." Reinhold, New York.

Partmann, W. (1975). The effects of freezing and thawing on food quality. *In* "Water Relations of Foods" (R. B. Duckworth, ed.), pp. 505-537. Academic Press, New York.

Partmann, W., Bomar, M., Hajek, M., Bohling, H., and Schlaszus, H. (1975). Über die Eignung von Gasgemischen mit 20% CO_2 für die Lagerung von Rindfleisch. *Fleischwirtschaft* **55,** 1441-1451.

Partmann, W., Bomar, M., Hajek, M., Bohling, H., and Schlaszus, H. (1976). Lagerungsversuche mit vorverpacktem Rindfleisch in kontrollierten Gasatmospheren mit hohem Kohlendioxidgehalt unter weitgehendem Ausschluss von Sauerstoff. *Fleischwirtschaft* **56,** 550-554.

Patterson, J. T., and Gibbs, P. A. (1977). Incidence and spoilage potential of isolates from vacuum-packaged meat of high pH value. *J. Appl. Bacteriol.* **43,** 25-38.

Patterson, J. T., and Gibbs, P. A. (1978). Some microbiological considerations applying to the conditioning, ageing, and vacuum packaging of lamb. *J. Food Technol.* **13,** 1-13.

Pelczar, M. J., Jr., and Reid, R. D. (1965). "Microbiology," 2nd Ed. McGraw-Hill, New York.

Pepper, R. E., Buffa, N. T., and Chandler, V. L. (1956). Relative resistances of microorganisms to cathode rays. III. Bacterial spores. *Appl. Microbiol.* **4,** 149-152.

Perigo, J. A., and Roberts, T. A. (1968). Inhibition of clostridia by nitrite. *J. Food Technol.* **3,** 91-94.

Perigo, J. A., Whiting, E., and Bashford, T. E. (1967). Observations on the inhibition of vegetative cells of *Clostridium sporogenes* by nitrite which has been autoclaved in a laboratory medium discussed in the context of sublethally processed cured meats. *J. Food Technol.* **2,** 377-397.

Peterson, A. C., and Gunderson, M. F. (1960). Role of psychrophilic bacteria in frozen food spoilage. *Food Technol.* **14,** 413-417.

Peterson, A. C., Black, J. J., and Gunderson, M. F. (1962a). Staphylococci in competition. I. Growth of naturally occurring mixed populations in precooked frozen foods during defrost. *Appl. Microbiol.* **10,** 16-22.

Peterson, A. C., Black, J. J., and Gunderson, M. F. (1962b). Staphylococci in competition. II. Effect of total numbers and proportion of staphylococci in mixed cultures on growth in artificial culture medium. *Appl. Microbiol.* **10,** 23-30.

Peterson, A. C., Black, J. J., and Gunderson, M. F. (1964a). Staphylococci in competition. III. Influence of pH and salt on staphylococcal growth in mixed populations. *Appl. Microbiol.* **12,** 70-76.

Peterson, A. C., Black, J. J., and Gunderson, M. F. (1964b). Staphylococci in competition. IV. Effect of starch and kind and concentration of sugar on staphylococcal growth in mixed populations. *Appl. Microbiol.* **12,** 77-82.

Peterson, A. C., Black, J. J., and Gunderson, M. F. (1964c). Staphylococci in competition. V.

Effect of eggs, eggs plus carbohydrates, and lipids on staphylococcal growth. *Appl. Microbiol.* **12,** 83–86.

Pether, J. V. S., and Gilbert, R. J. (1971). The survival of salmonellas on finger-tips and transfer of the organisms to foods. *J. Hyg.* (Cambr.) **69,** 673.

Pfeiffer, E. H. (1972). Hygienic evaluation of packaging material for food (plastics compared with cardboard). *Zentralbl. Bakteriol., Parasitenkd., Infektionskr. Hyg., Abt. 1: Orig., Reihe B* **156,** 4–5, 446–455.

Pflug, I. J., and Schmidt, C. F. (1968). Thermal destruction of microorganisms. *In* "Disinfection, Sterilization and Preservation" (C. A. Lawrence and S. S. Block, eds.), pp. 63–105. Lea & Febiger, Philadelphia, Pennsylvania.

Phillips, C. R. (1968). Gaseous sterilization. *In* "Disinfection Sterilization and Preservation" (C. A. Lawrence and S. S. Black, eds.), pp. 669–685. Lea & Febiger, Philadelphia, Pennsylvania.

Phillips, C. R., and Kaye, S. (1949). The sterilizing action of gaseous ethylene oxide I. *Am. J. Hyg.* **50,** 270–279.

Pietzsch, O., Kretschmer, F. J., and Bulling, E. (1975). Vergleichsuntersuchungen über Salmonella-Anreich erungsverfahren. *Zentralbl. Bakteriol., Parasitenkd., Infektionskr. Hyg., Abt. 1: Orig., Reihe A* **232,** 232–246.

Pirt, S. J. (1975). "Principles of Microbe and Cell Cultivation." Blackwell, Oxford.

Pitt, J. I. (1974). Resistance of some food spoilage yeasts to preservatives. *Food Technol. Aust.* **26**(6), 238–239, 241.

Pitt, J. I. (1975). Xerophilic fungi and the spoilage of foods of plant origin. *In* "Water Relations of Foods' (R. B. Duckworth, ed.), pp. 273–307. Academic Press, New York.

Pivnick, H. (1970). The inhibition of heat-damaged spores of *Cl. botulinum* by sodium chloride and sodium nitrite. *Symp. Microbiol. Semi-Preserved Foods, Prague* 17 pp.

Pivnick, H., and Barnett, H. (1965). Effect of salt and temperature on toxinogenesis by *Clostridium botulinum* in perishable cooked meats vacuum-packed in air-impermeable plastic pouches. *Food Technol.* **19,** 1164–1167.

Pivnick, H., and Bird, H. (1965). Toxinogenesis by *Clostridium botulinum* types A and E in perishable cooked meats vacuum-packed in plastic pouches. *Food Technol.* **19,** 1156–1164.

Pivnick, H., and Chang, P. C. (1973). Perigo effect in pork. *Proc. Int. Symp. Nitrite Meat Prod., Zeist, Neth.* (B. Krol and B. J. Tinbergen, eds.), pp. 111–116. Pudoc, Wageningen, Netherlands.

Pivnick, H., and Petrasovits, A. (1973). A rationale for the safety of canned shelf-stable cured meat: protection = destruction + inhibition. *Proc. Annu. Meet. Eur. Meat Res. Workers, 19th*, pp. 1086–1095. Paris, France.

Pivnick, H., and Thacker, C. (1970). Effect of sodium chloride and pH on initiation of growth by heat-damaged spores of *Clostridium botulinum. Can. Inst. Food Technol. J.* **3,** 70–75.

Pivnick, H., Rubin, L. J., Barnett, H. W., Nordin, H. R., Ferguson, P. A., and Perrin, C. H. (1967). Effect of sodium nitrite and temperature on toxinogenesis by *Clostridium botulinum* in perishable cooked meats vacuum-packed in air-impermeable plastic pouches. *Food Technol.* **21,** 204–206.

Pivnick, H., Barnett, H. W., Nordin, H. R., and Rubin, L. J. (1969). Factors affecting the safety of canned, cured, shelf-stable luncheon meat inoculated with *Clostridium botulinum. Can. Inst. Food Technol. J.* **2,** 141–148.

Pivnick, H., Johnston, M. A., Thacker, C., and Loynes, R. (1970). Effect of nitrite on destruction and germination of *Clostridium botulinum* and putrefactive anaerobes 3679 and 3679h in meat and in buffer. *Can. Inst. Food Technol. J.* **3,** 103–109.

Pivovarov, Y. P., and Kukhto, A. I. (1970). Effect of canned food pH on growth of *Bacillus cereus. Vopr. Pitan.* **29**(4), 72–75. [*Food Sci. Technol. Abstr.* **2,** 12 B 101 (1970).]

Placzek, E., and Witter, K.-T. (1972). Bakteriologische Oberflächenuntersuchung von Einmalpackungen. *Verpack.-Rundsch.* No. 6, 742–748.

Preonas, D. L., Nelson, A. I., Ordal, Z. J., Steinberg, M. P., and Wei, L. S. (1969). Growth of *Staphylococcus aureus* MF31 on the top and cut surfaces of southern custard pies. *Appl. Microbiol.* **18,** 68–75.

Proctor, B. E., and Goldblith, S. A. (1951). Electromagnetic radiation fundamentals and their applications in food technology. *Adv. Food Res.* **3,** 119–196.

Pullen, M. M., and Genigeorgis, C. A. (1977). A study of coagulase-positive staphylococci in salami before fermentation. *J. Food Prot.* **40,** 704–708.

Quinn, D. J., Anderson, A. W., and Dyer, J. F. (1967). The inactivation of infection and intoxication microorganisms by irradiation in seafood. *Microbiol. Probl. Food Preserv. Irradiat., Panel Proc., Vienna, 1966* STI/PUB/168, pp. 1–13. IAEA, Vienna.

Raevuori, M., and Genigeorgis, C. (1975). Effect of pH and sodium chloride on growth of *Bacillus cereus* in laboratory media and certain foods. *Appl. Microbiol.* **29,** 68–73.

Raevuori, M., Nurmi, E., and Hill, P. (1978). The effect of nitrate, nitrite and a starter culture on the growth of *Yersinia enterocolitica* in dry sausage. *Proc. Annu. Meet. Eur. Meat Res. Workers, 24th, Kulmbach, W. Ger.*

Rahn, O., and Conn, J. E. (1944). Effect of increase in acidity on antiseptic efficiency. *Ind. Eng. Chem.* **36,** 185–187.

Rajagopalan, S., and Shiffman, M. A. (1974). "Guide to Simple Sanitary Measures for the Control of Enteric Diseases." World Health Organ., Geneva.

Rand, M. C., Greenberg, A. E., and Taras, M. J. (1976). "Standard Methods for the Examination of Water and Waste Water," 14th Ed. Am. Public Health Assoc., Washington, D.C.

Ranson, S. L., Walker, D. A., and Clark, I. D. (1957). Inhibition of succinic oxidase by high CO_2 concentrations. *Proc. Biochem. Soc.* **66,** 57P.

Rauscher, H., Mayr, G., and Kaemmerer, H. (1957). Ethylene oxide for cold sterilisation. *Food Manuf.* **32,** 169–172.

Ray, B., and Speck, M. L. (1972). Metabolic process during the repair of freeze-injury in *Escherichia coli*. *Appl. Microbiol.* **24,** 585–590.

Ray, B., and Speck, M. L. (1973a). Discrepancies in the enumeration of *Escherichia coli*. *Appl. Microbiol.* **25,** 494–498.

Ray, B., and Speck, M. L. (1973b). Freeze-injury in bacteria. *Crit. Rev. Clin. Lab. Sci.* **4,** 161–213.

Ray, B., Jezeski, J. J., and Busta, F. F. (1971). Repair of injury in freeze-dried *Salmonella anatum*. *Appl. Microbiol.* **22,** 401–407.

Ray, B., Jezeski, J. J., and Busta, F. F. (1972). Isolation of salmonellae from naturally contaminated dried milk products. III. Influence of pre-enrichment conditions. *J. Milk Food Technol.* **35,** 607–614.

Reddish, G. F. (1957). "Antiseptics, Disinfectants, Fungicides, and Chemical and Physical Sterilization," 2nd Ed. Lea & Febiger, Philadelphia, Pennsylvania.

Reddy, S. G., Chen, M. L., and Patel, P. J. (1975). Influence of lactic cultures on the biochemical, bacterial and organoleptic changes in beef. *J. Food Sci.* **40,** 314–318.

Rehm, H. J. (1960). A study on the antimicrobial activity of mixtures of agents used for the preservation of liquids. *Ann. Inst. Pasteur, Lille* **11,** 217–226.

Rehm, H. J., and Wittmann, H. (1962). Beitrag zur Kenntnis der antimikrobiellen Wirkung der Schwefligen Saüre I. Ubersicht über einflussnehmende Faktoren auf die antimikrobielle Wirkung der Schwefligen Saüre. *Z. Lebensm.-Unters. -Forsch.* **118,** 413–429.

Reich, G. T. (1945). Carbon dioxide in the food industry, Part III. Uses of solid CO_2. *Food Ind.* **17,** 1170–1173.

Reichenback, H. (1911). Die Absterbeordnung der Bakterien und ihre Bedeutung für Theorie und Praxis der Desinfektion. *Z. Hyg. Infektionskr.* **69,** 171–222.

Reiss, J. (1976). Prevention of the formation of mycotoxins in whole wheat bread by citric acid and lactic acid (mycotoxins in foodstuffs IX). *Experientia* 32, 168–169.

Reiter, B., Pickering, A., and Oram, J. D. (1964). An inhibitory system—lactoperoxidase/thiocyanate/peroxide in raw milk. *In* "Microbial Inhibitors in Food" (N. Molin and A. Erichsen, eds.), p. 297-305. Almqvist & Wiksell, Stockholm.

Reuter, G. (1970). Untersuchungen zur Mikroflora von vorverpackten, aufgeschnittenen Brüh - und Kochwürsten. *Arch. Lebensmittelhyg.* **21,** 257-264.

Reuter, G. (1972). Ripening of German salami type dry sausages with lactobacilli and micrococci starter cultures. *Proc. Annu. Meet. Eur. Meat Res. Workers, 18th, Guelph, Can.* pp. 666-674. (In Ger.)

Riemann, H. (1966). Botulism, its dangers and prevention in the meat industry. *Proc. Meat Ind. Res. Conf., Am. Meat Inst. Found., Chicago, Ill.* pp. 27-41.

Riemann, H. (1967a). The effect of the number of spores on growth and toxin formation by *Cl. botulinum* type E in inhibitory environments. *In* "Botulism 1966" (M. Ingram and T. A. Roberts, eds.), pp. 150-157. Chapman & Hall, London.

Riemann, H. (1967b). Microbiological methods used for the evaluation of packaging and processing techniques of fresh and cured meats. *Proc. Annu. Reciprocal Meat Conf., 20th, Natl. Livestock Meat Board, Lincoln, Neb.* pp. 150-156.

Riemann, H. (1969). Food processing and preservation effects *In* "Food-Borne Infections and Intoxications" (H. Riemann, ed.), pp. 489-541. Academic Press, New York.

Riemann, H., Lee, W. H., and Genigeorgis, C. (1972). Control of *Clostridium botulinum* and *Staphylococcus aureus* in semipreserved meat products. *J. Milk Food Technol.* **35,** 514-523.

Riester, D. W., Warner, R. C., and Hoffman, H. T., Jr. (1958). Packaging requirements for pressure propelled foods. *Food Technol.* **12,** 310-324.

Riha, W. E., Jr., and Solberg, M. (1975a). *Clostridium perfringens* inhibition by sodium nitrite as a function of pH, inoculum size and heat. *J. Food Sci.* **40,** 439-442.

Riha, W. E., Jr., and Solberg, M. (1975b). *Clostridium perfringens* growth in a nitrite containing defined medium sterilized by heat or filtration. *J. Food Sci.* **40,** 443-445.

Roberts, A. C., and McWeeny, D. J. (1972). The uses of sulphur dioxide in the food industry. A review. *J. Food Technol.* **7,** 221-238.

Roberts, T. A. (1968). Heat and radiation resistance and activation of spores of *Clostridium welchii*. *J. Appl. Bacteriol.* **31,** 133-144.

Roberts, T. A. (1970). Symposium on bacterial spores. Recovering spores damaged by heat, ionizing radiations or ethylene oxide. *J. Appl. Bacteriol.* **33,** 74-94.

Roberts, T. A., and Garcia, C. E. (1973). A note on the resistance of *Bacillus* spp., faecal streptococci and *Salmonella typhimurium* to an inhibitor of *Clostridium* spp. formed by heating sodium nitrite. *J. Food Technol.* **8,** 463-466.

Roberts, T. A., and Hitchins, A. D. (1969). Resistance of spores. *In* "The Bacterial Spore" (G. W. Gould and A. Hurst, eds.), pp. 611-670. Academic Press, New York.

Roberts, T. A., and Hobbs, G. (1968). Symposium on Growth of Microorganisms at Extremes of Temperature: Low temperature growth characteristics of clostridia. *J. Appl. Bacteriol.* **31,** 75-88.

Roberts, T. A., and Ingram, M. (1966). The effect of sodium chloride, potassium nitrate and sodium nitrite on the recovery of heated bacterial spores. *J. Food Technol.* **1,** 147-163.

Roberts, T. A., and Ingram, M. (1973). Inhibition of growth of *Cl. botulinum* at different pH values by sodium chloride and sodium nitrite. *J. Food Technol.* **8,** 467-475.

Roberts, T. A., and Smart, J. L. (1976). The occurrence and growth of *Clostridium* spp. in vacuum-packed bacon with particular reference to *Cl. perfringens* (*welchii*) and *Cl. botulinum*. *J. Food Technol.* **11,** 229-244.

Roberts, T. A., and Smart, J. L. (1977). The occurrence of clostridia, particularly *Clostridium botulinum,* in bacon and pork. *In* "Spore Research 1976" (A. N. Barker, J. Wolf, D. J. Ellar, G. J. Dring, and G. W. Gould, eds.), pp. 911-915. Academic Press, New York.

Roberts, T. A., Jarvis, B., and Rhodes, A. C. (1976). Inhibition of *Clostridium botulinum* by curing salts in pasteurized pork slurry. *J. Food Technol.* **11,** 25–40.

Robinson, R. A., and Stokes, R. H. (1959). "Electrolyte Solutions." 2nd Ed., Academic Press, New York.

Rödel, W., Herzog, H., and Leistner, L. (1973). Wasseraktivitäts-Toleranz von lebensmittelhygienisch wichtigen Keimarten der Gattung Vibrio. *Fleischwirtschaft* **53,** 1301–1303.

Ronsivalli, L. J., Bernsteins, J. B., and Tinker, B. L. (1966). Method for determining the bacterial permeability of plastic films. *Food Technol.* **20,** 1074–1075.

Rose, A. H. (1968). Physiology of microorganisms at low temperatures. *J. Appl. Bacteriol.* **31,** 1–11.

Rose, A. H. (1976). Osmotic stress and microbial survival. *Symp. Soc. Gen. Microbiol.* **26,** 155–182.

Roth, L. A., and Clark, D. S. (1972). Studies on the bacterial flora of vacuum-packaged fresh beef. *Can. J. Microbiol.* **18,** 1761–1766.

Rowland, A. J., and Alder, V. G. (1972). Transmission of infection through towels. *Community Med.* **128,** 71.

Rubin, H. E. (1978). Toxicological model for a two-acid system. *Appl. Environ. Microbiol.* **36,** 623–624.

Ruenger, E. L., Reineccius, G. A., and Thompson, D. R. (1978). Flavor compounds related to the warmed-over flavor of turkey. *J. Food Sci.* **43,** 1198–1200.

Rushing, N. B., Veldhuis, M. K., and Senn, V. J. (1956). Growth rates of *Lactobacillus* and *Leuconostoc* species in orange juice as affected by pH and juice concentration. *Appl. Microbiol.* **4,** 97–100.

Sair, L., and Komarik, S. L. (1968). Curing emulsified meat products. U.S. Patent 3,391,006.

Saitanu, K., and Lund, E. (1975). Inactivation of enterovirus by glutaraldehyde. *Appl. Microbiol.* **29,** 571–574.

Sakaguchi, G. (1969). Botulism type E. *In* "Food-Borne Infections and Intoxications" (H. Riemann, ed.), pp. 329–358. Academic Press, New York.

Sakazaki, R. (1973). Control of contamination with *Vibrio parahaemolyticus* in seafoods and isolation and identification of the vibrio. *In* "The Microbiological Safety of Food" (B. C. Hobbs and J. H. B. Christian, eds.), pp. 375–385. Academic Press, New York.

Sato, K., and Hegarty, G. R. (1971). Warmed-over flavor in cooked meats. *J. Food Sci.* **36,** 1098–1102.

Sato, M., and Takahashi, H. (1968). Cold shock of bacteria. I. General features of cold shock in *Escherichia coli*. *J. Gen. Appl. Microbiol.* **14,** 417–428.

Sato, M., and Takahashi, H. (1969). Cold shock of bacteria. II. Magnesium mediated recovery from cold shock and existence of two critical temperature zones in various bacteria. *J. Gen. Appl. Microbiol.* **15,** 217–229.

Schack, W. R., and Taylor, R. E. (1963). Extending shelf life of cured meats. Can. Patent 674,171.

Scharding, J. H., Kelley, D. C., Cook, J. E., and Kropf, D. H. (1973). Experimental survival studies of *Sporothrix schenckii* in a meat product. *J. Milk Food Technol.* 36, 311–314.

Scheusner, D. L., Hood, L. L., and Harmon, L. G. (1973). Effect of temperature and pH on growth and enterotoxin production by *Staphylococcus aureus*. *J. Milk Food Technol.* **36,** 249–252.

Schiemann, D. A. (1976). Inhibitory substances in the milk supply of Southern Ontario. *J. Milk Food Technol.* **39,** 490–492.

Schmidt, C. F. (1957). Thermal resistance of microorganisms. *In* "Antiseptics, Disinfectants, Fungicides and Sterilization" 2nd Ed. (G. F. Reddish, ed.), pp. 720–759. Lea & Febiger, Philadelphia, Pennsylvania.

Schmidt, C. F., Lechowich, R. V., and Folinazzo, J. F. (1961). Growth and toxin production by type E *Clostridium botulinum* below 40°F. *J. Food Sci.* **26,** 626–630.

Schmidt-Lorenz, W. (1973). Untersuchungen zur Prüfung der Bakteriendichtigkeit von Heiszsiegelnähten halbstarrer Leichtbehälter aus Aluminium-Kunststoff-Verbunden. *Verpack.-Rundsch., Tech.-Wiss. Beil.* **24**(8), 59–60.

Schulz, J. H., and Ritz, H. (1910). Die thermoresistenz junger und alter Coli-bacillen. *Zentralbl. Bakteriol., Parasitenkd. Infektionskr., Abt. 2* **54,** 283–288.

Schultze, W. D., and Olson, J. C., Jr. (1960). Studies on psychrophilic bacteria. I. Distribution in stored commercial dairy products. *J. Dairy Sci.* **43,** 346–350.

Schwartz, A. (1964). Mikrobielle Korrosion von Kunststoffen und ihren Bestandteilen. *Zentralbl. Bakteriol., Parasitenkd., Infektionskr. Hyg., Abt. 1: Orig.* **191,** 201–203.

Scott, C. R., and Smith, H. O. (1971). Cottage cheese shelf-life and special gas atmospheres. *J. Food Sci.* **36,** 78–80.

Scott, D. B. M., and Lesher, E. C. (1963). Effect of ozone on survival and permeability of *Escherichia coli*. *J. Bacteriol.* **85,** 567–576.

Scott, W. J. (1938). The growth of microorganisms on ox muscle III. The influence of 10 per cent carbon dioxide on rates of growth at −1°C. *Aust. Counc. Sci. Ind. Res. J.* **11,** 266–277.

Scott, W. J. (1953). Water relations of *Staphylococcus aureus* at 30°C. *Aust. J. Biol. Sci.* **6,** 549–564.

Scott, W. J. (1957). Water relations of food spoilage microorganisms. *Adv. Food Res.* **7,** 83–127.

Scott, W. J. (1958). The effect of residual water on the survival of dried bacteria during storage. *J. Gen. Microbiol.* **19,** 624–633.

Scott, W. J. (1962). Available water and microbial growth. *Proc. Low Temp. Symp., 1961* pp. 89–105. Campbell Soup Co., Camden, New Jersey.

Seabaugh, V. M., Bayard, S. P., Osterberg, R. E., Porter, W. K., McCaulley, D. F., Hoheisel, C. A., Hehir, R. M., and Bierbower, G. W. (1977). Detergent toxicity survey. *Am. J. Public Health* **67,** 367–369.

Sears, D. F. and Eisenberg, R. M. (1961). A model representing a physiological rate of CO_2 at the cell membrane. *J. General Physiol.* **44,** 869–887.

Sebald, M., and Ionesco, H. (1972). Germination lzP-dependante des spores de *Clostridium botulinum* type E. C. R. Acad. Sci., Ser. D275, 2175–2177.

Senhaji, A. F. (1977). The protective effect of fat on the heat resistance of bacteria. (II). *J. Food Technol.* **12,** 217–230.

Senhaji, A. F., and Loncin, M. (1977). The protective effect of fat on the heat resistance of bacteria (I.) *J. Food Technol.* **12,** 203–216.

Shank, J. L., and Lundquist, B. R. (1963). The effect of packaging conditions on the bacteriology, color and flavor of table-ready meats. *Food Technol.* **17,** 1163–1166.

Shank, J. L., Silliker, J. H., and Harper, R. H. (1962). The effect of nitric oxide on bacteria. *Appl. Microbiol.* **10,** 185–189.

Shapton, D. A., Lovelock, D. W., and Laurita-Longo, R. (1971). The evaluation of sterilization and pasteurization processes from temperature measurements in degrees Celsius (°C). *J. Appl. Bacteriol.* **34,** 491–500.

Sharf, J. F. (1960). The evaluation of microbiological standards for bottled carbonated beverages in the U.S.A. *Ann. Inst. Pasteur, Lille* **11,** 117–132.

Shaw, B. G. (1972). The effect of temperature and relative humidity on the microbiological quality of carcass meat. *In* "Meat Chilling — Why and How." *Symp. ARC Meat Res. Inst., 2nd,* pp. 7.1–7.10 *Langford, Bristol, U.K.*

Shaw, F. B. (1977). Toxicological considerations in the selection of flexible packaging materials for foodstuffs. *J. Food Prot.* **40,** 65–68.

Shaw, M. K. (1968). Formation of filaments and synthesis of macromolecules at temperatures below the minimum for growth of *Escherichia coli*. *J. Bacteriol.* **95,** 221–230.

Shaw, M. K., and Nicol, D. J. (1969). Effect of the gaseous environment on the growth of some food-poisoning and food spoilage organisms. *Proc. Eur. Meet. Meat Res. Workers, 15th, Helsinki* pp. 226–230.

Shehata, T. E., and Marr, A. G. (1975). Effect of temperature on the size of *Escherichia coli* cells. *J. Bacteriol.* **124,** 857-862.

Sherman, J. M., and Albus, W. R. (1923). Physiological youth in bacteria. *J. Bacteriol.* **8,** 127-139.

Sherman, J. M., Stark, C. N., and Stark, P. (1929). An unappreciated but important factor in pasteurization of milk. *J. Dairy Sci.* **12,** 385-393.

Sheu, C. W., Salomon, D., Simmons, J. L., Streevalsan, T., and Freese, E. (1975). Inhibitory effects of lipophilic acids and related compounds on bacteria and mammalian cells. *Antimicrob. Agents Chemother.* **7,** 349-363.

Silliker, J. H., Greenberg, R. A., and Schack, W. R. (1958). Effect of individual curing ingredients on the shelf stability of canned comminuted meats. *Food Technol.* **12,** 551-554.

Silliker, J. H., Woodruff, R. E., Lugg, J. R., Wolfe, S. K., and Brown, W. D. (1977). Preservation of refrigerated meats with controlled atmospheres: treatment and post-treatment effects of carbon dioxide on pork and beef. *Meat Sci.* **1,** 195-204.

Simon, E. W., and Beevers, H. (1951). The effect of pH on the biological activities of weak acids and bases. 1. The most usual relationship between pH and activity. *New Phytol.* **51,** 163-190.

Simon, S., Ellis, D. E., MacDonald, B. D., Miller, D. G., Waldman, R. C., and Westerberg, D. O. (1972). Influence of nitrite on quality of packaged frankfurters. *Proc. Annu. Meet. Eur. Meat Res. Workers, 18th, Guelph, Can.*

Sinskey, T. J., and Silverman, G. J. (1970). Characterization of injury incurred by *Escherichia coli* upon freeze-drying. *J. Bacteriol.* **101,** 429-437.

Sirviö, P., Nurmi, E., Puolanne, E., and Niinivaara, F. P. (1977). Der Einfluss von Starterkulturen und verschiedenen Zusatzstoffen auf das Wachstum von *Salmonella senftenberg* in Rohwurst. *Fleischwirtschaft* **57,** 1007-1008, 1011-1012.

Skinner, F. A., and Hugo, W. B., eds. (1976). "Inhibition and Inactivation of Vegetative Microbes," Society of Applied Bacteriology Symposia Series, No. 5. Academic Press, New York.

Smith, C. A. (1973). Environmental sanitation and management; sanitation-entomology. *In* "Food Safety through the Hazard Analysis and Critical Control Point System," (J. J. Slavics, ed.), pp. 163-184. Pillsbury, Minneapolis, Minnesota.

Smith, H. W. (1970). The transfer of antibiotic resistance between strains of enterobacteria in chickens, calves and pigs. *J. Med. Microbiol.* **3,** 165-180.

Smith, H. W. (1971). Incidence of R plus *Escherichia coli* in coastal bathing waters of Britain. *Nature (London)* **234,** 155-156.

Smith, H. W. (1974). Veterinary and food aspects of drug resistance. *J. Sci. Food Agri.* **25**(2), 228.

Smith, H. W. (1975). Survival of orally administered *E. coli* K12 in alimentary tract of men. *Nature (London)* **255,** 500-502.

Smith, J. L., Palumbo, S. A., Kissinger, J. C., and Huhtanen, C. N. (1975a). Survival of *Salmonella dublin* and *Salmonella typhimurium* in Lebanon bologna. *J. Milk Food Technol.* **38,** 150-154.

Smith, J. L., Huhtanen, C. N., Kissinger, J. C., and Palumbo, S. A. (1975b). Survival of salmonellae during pepperoni manufacture. *Appl. Microbiol.* **30,** 759-763.

Smith, W. H. (1959). The application of high concentration of carbon dioxide in transport and storage of some fruits. *Proc. Int. Congr. Refrig., 10th, Copenhagen,* 193-197.

Smith, W. H. (1963). The use of carbon dioxide in the transport and storage of fruits and vegetables. *Adv. Food Res.* **12,** 95-146.

Smittle, R. B. (1977). Microbiology of mayonnaise and salad dressing. A review. *J. Food Prot.* **40,** 415-422.

Solberg, M. (1967). Microbial considerations of packaging. *Food Technol.* **21,** 758-759.

Solberg, P. (1970). Experiments with twice-a-week collection of farm tank milk in Norway. *Proc. Int. Dairy Congr. 18th,* **1E,** 503.

Sommer, E. W. (1930). Heat resistance of the spores of *Clostridium botulinum*. *J. Infect. Dis.* **46,** 85–114.
Sommer, N. (1973). The effect of ionising radiation on fungi. *In* "Manual on Radiation Sterilization of Medical and Biological Materials," Tech. Rep. Ser., No. 149, STI/DOC/10/149, pp. 73–80. IAEA, Vienna.
Sommer, R., Rommel, M., and Levetzow, R. (1965). Die Uberlebensdauer von Toxoplasmazysten in Fleisch und Fleischzubereitungen. *Fleischwirtschaft* **45,** 454–457.
Speck, M. L., Ray, B., and Read, R. B., Jr. (1975). Repair and enumeration of injured coliforms by a plating procedure. *Appl. Microbiol.* **29,** 549–550.
Spitz, H. D., and Weinberger, J. (1971). Determination of ethylene oxide, ethylene chlorohydrin and ethylene glycol by gas chromatography. *J. Pharm. Sci.* **60,** 271–274.
Splittstoesser, D. F., Wettergreen, W. P., and Pederson, C. S. (1961). Control of microorganisms during preparation of vegetables for freezing. I. Green beans. *Food Technol.* **15,** 329–331.
Spooner, D. F., and Sykes, G. (1972). Laboratory assessment of antibacterial activity. *Methods Microbiol.* **7B,** 211–276.
Stark, C. N., and Stark, P. (1929). The relative thermal death rates of young and mature bacterial cells. *J. Bacteriol.* **18,** 333–337.
Stewart, G. T. (1973). Allergy to penicillin and related antibiotics: antigenic and immunochemical mechanism. *Annu. Rev. Pharmacol.* **13,** 309–324.
Stokstad, E. L. R., and Jukes, T. H. (1950). Further observations on the "Animal Protein Factor." *Proc. Soc. Exp. Biol. Med.* **73,** 523–528.
Straka, R. P., and Stokes, J. L. (1959). Metabolic injury to bacteria at low temperatures. *J. Bacteriol.* **78,** 181–185.
Strange, R. E., and Dark, F. A. (1962). Effect of chilling on *Aerobacter aerogenes* in aqueous suspension. *J. Gen. Microbiol.* **29,** 719–730.
Strange, R. E., and Ness, A. G. (1963). Effect of chilling on bacteria in aqueous suspension. *Nature (London)* **197,** 819.
Strong, D. H., and Canada, J. C. (1964). Survival of *Clostridium perfringens* in frozen chicken gravy. *J. Food Sci.* **29,** 479–482.
Stumbo, C. R. (1949). Thermobacteriology as applied to food processing. *Adv. Food Res.* **2,** 47–115.
Stumbo, C. R. (ed.) (1973) "Thermobacteriology in Food Processing" 2nd Ed. Academic Press, New York.
Stumbo, C. R., Gross, C. E., and Vinton, C. (1945). Bacteriological studies relating to thermal processing of canned meats. III. Influence of meat-curing agents upon growth of a putrefactive anaerobic bacterium in heat-processed meat. *Food Res.* **10,** 293–302.
Subramanian, C. S., and Marth, E. H. (1968). Multiplication of *Salmonella typhimurium* in skimmilk with and without added hydrochloric, lactic and citric acids. *J. Milk Food Technol.* **31,** 323–326.
Sutherland, J. P., Patterson, J. T., and Murray, J. G. (1975a). Changes in the microbiology of vacuum-packaged beef. *J. Appl. Bacteriol.* **39,** 227–237.
Sutherland, J. P., Patterson, J. T., Gibbs, P. A., and Murray, J. G. (1975b). Some metabolic and biochemical characteristics of representative microbial isolates from vacuum-packaged beef. *J. Appl. Bacteriol.* **39,** 239–249.
Swartling, P., and Lindgren, B. (1960). On the influence of certain bacteria on the growth of clostridia. *Milk Dairy Res. Rep.* No. 62, 1–18.
Sykes, G. (1958). "Disinfection and Sterilization." Spon, London.
Sykes, G. (1974). "Disinfection and Sterilisation," 2nd Ed. Spon, London.
Symposium on Hygiene of the Hand (1974). *Congr. Cent. RAI, Amsterdam, Voeding Huishoudg. Zkh. Inst.* **41,** No. 492, 166.

Tabatabai, L. B., and Walker, H. W. (1970). Oxidation-reduction potential and growth of *Clostridium perfringens* and *Pseudomonas fluorescens*. *Appl. Microbiol.* **20,** 441–446.

Tai, P. C., and Jackson, H. (1969). Mesophilic mutants of an obligate psychrophile, *Micrococcus cryophilus*. *Can. J. Microbiol.* **15,** 1145–1150.

Takács, J., and Kovacs, S. (1969). Demonstration of antibiotic residues in the meat of slaughtered animals. *Acta Vet. Acad. Sci. Hung.* **19,** 11–19.

Tamminga, S. K., Beumer, R. R., van Kooij, J. G., and Kampelmacher, E. H. (1975). Microbiological possibilities to demonstrate that strawberries have been irradiated. *Eur. J. Appl. Microbiol.* **1,** 79–93.

Tändler, K., and Heinz, G. (1970a). Frischfleischreifung im Vakuumbeutel. *Fleisch* **50,** 1185–1190.

Tändler, K., and Heinz, G. (1970b). Frischfleischreifung im Vakuumbeutel. *Fleisch* **50,** 1370–1376.

Tanner, F. W., and Evans, F. L. (1933). Effect of meat curing solutions on anaerobic bacteria. I. Sodium chloride. *Zentralbl. Bakteriol., Parasitenkd. Infektionskr.* **88,** 45–54.

Tarr, H. L. A. (1941a). The action of nitrites on bacteria. *J. Fish. Res. Board Can.* **5,** 265–275.

Tarr, H. L. A. (1941b). Bacteriostatic action of nitrates. *Nature (London)* **147,** 417–418.

Tarr, H. L. A. (1942). The action of nitrites on bacteria. Further experiments. *J. Fish. Res. Board Can.* **6,** 74–89.

Tarr, H. L. A. (1944). Action of nitrates and nitrites on bacteria. *J. Fish. Res. Board Can.* **6,** 233–242.

Task Force (1977). "The Staphylococcal Enterotoxin Problem in Fermented Sausage." Rep. to U.S. Dep. Agric., Washington, D.C.

Tatini, S. R. (1973). Influence of food environments on growth of *Staphylococcus aureus* and production of various enterotoxins. *J. Milk Food Technol.* **36,** 559–563.

Tatini, S. R., Lee, R. Y., McCall, W. A., and Hill, W. M. (1976). Growth of *Staphylococcus aureus* and production of enterotoxins in pepperoni. *J. Food Sci.* **41,** 223–225.

Taylor, A. A. (1971). The gaseous environment of packaged fresh meat. *Proc. Eur. Meet. Meat Res. Workers, 17th* pp. 662–667.

Taylor, J. B., and Rowlinson, J. S. (1955). Thermodynamic properties of aqueous solutions of glucose. *Trans. Faraday Soc.* **51,** 1183–1192.

Thatcher, F. S. (1963). Some public health aspects of the microbiology of irradiated foods. *Int. J. Appl. Radiat. Isot.* **14,** 51–58.

Thatcher, F. S., and Clark, D. S., eds. (1968). "Microorganisms in Foods: Their Significance and Methods of Enumeration." Univ. of Toronto Press, Toronto.

Thatcher, F. S., Robinson, J., and Erdman, I. (1962). The "vacuum pack" method of packaging foods in relation to the formation of the botulinum and staphylococcal toxins. *J. Appl. Bacteriol.* **25,** 120–124.

Theophilus, D. R., and Hammer, B. W. (1938). Influence of growth temperature on the thermal resistance of some bacteria from evaporated milk. *Iowa Agric. Exp. Stn., Res. Bull.* No. 244.

Thomas, G. (1977). Effects of near ultraviolet light on microorganisms. *Photochem. Photobiol.* **26,** 669–673.

Thomas, G. A. (1969). "The Selection of Detergents and Disinfectants for Use in Food Plant Cleaning," Tech. Circ. No. 433. Br. Food Manuf. Ind. Res. Assoc., London.

Thomas, S., and Russell, A. D. (1974). Temperature-induced changes in the sporocidal activity and chemical properties of glutaraldehyde. *Appl. Microbiol.* **28,** 331–335.

Thomson, J. E. (1970). Microbial counts and rancidity of fresh fryer chickens as affected by packaging materials, storage atmosphere, and temperature. *Poult. Sci.* **49,** 1104–1109.

Thomson, W. K., and Thacker, C. L. (1973). Effect of temperature on *Vibrio parahaemolyticus* in oysters at refrigerator and deep freeze temperatures. *Can. Inst. Food Sci. Technol. J.* **6,** 156–158.

Tilgner, D. J. (1977). The phenomena of quality in the smoke curing process. *Pure Appl. Chem.* **49,** 1629-1638.

Tiwari, N. P., and Maxcy, R. B. (1972). Moraxella-Acinetobacter as contaminants of beef and occurrence in radurized product. *J. Food Sci.* **37,** 901-903.

Toepfer, C. T., and Kanz, E., (1976). Wechselbeziehungen zwischen Kunststoffen and Bakterien. *Zentralbl. Bakteriol., Parasitenkd., Infektionskr. Hyg., Abt. 1: Orig., Reihe B* **163,** 540-555.

Tomkins, R. G. (1932). The inhibition of the growth of meat-attacking fungi by carbon dioxide. *J. Soc. Chem. Ind.* **51,** 261T-264T.

Tomlins, R. I., and Ordal, Z. J. (1976). Thermal injury and inactivation in vegetative bacteria. *In* "Inhibition and Inactivation of Vegetative Microbes" (F. A. Skinner and W. B. Hugo, eds.), Society of Applied Bacteriology Symposia Series, No. 5, pp. 153-190. Academic Press, New York.

Tompkin, R. B., Christiansen, L. N., and Shaparis, A. B. (1978a). Antibotulinal role of isoascorbate in cured meat. *J. Food Sci.* **43,** 1368-1370.

Tompkin, R. B., Christiansen, L. N., and Shaparis, A. B. (1978b). The effect of iron on botulinal inhibition in perishable canned cured meat. *J. Food Technol.* **13,** 521-527.

Tompkin, R. B., Christiansen, L. N., and Shaparis, A. B. (1978c). Effect of prior refrigeration on botulinal outgrowth in perishable canned cured meat when temperature abused. *Appl. Environ. Microbiol.* **35,** 863-866.

Tompkin, R. B., Christiansen, L. N., and Shaparis, A. B. (1978d). Enhancing nitrite inhibition of *Clostridium botulinum* with isoascorbate in perishable canned cured meat. *Appl. Environ. Microbiol.* **35,** 59-61.

Tompkin, R. B., Christiansen, L. N., and Shaparis, A. B. (1978e). Causes of variation in botulinal inhibition of perishable canned cured meat. *Appl. Environ. Microbiol.* **35,** 886-889.

Tompkin, R. B., Christiansen, L. N., and Shaparis, A. B. (1979a). Iron and the antibotulinal efficacy of nitrite. *Appl. Environ. Microbiol.* **37,** 351-353.

Tompkin, R. B., Christiansen, L. N., and Shaparis, A. B. (1979b). Isoascorbate levels and botulinal inhibition in perishable canned cured meat. *J. Food Sci.* **44,** 1147-1149.

Tonge, R. J. (1964). Mechanism of inhibition of *Bacillus* spores by unsaturated fatty acids. *In* "Microbial Inhibitors in Food" (N. Molin and A. Erichsen, eds.), pp. 35-45. Almqvist & Wiksell, Stockholm.

Torricelli, A. (1959). Sterilization of empty containers for food industry. *In* "Ozone Chemistry and Technology" (H. A. Leedy, ed.), Advances in Chemistry Series, No. 21, pp. 375-380. Am. Chem. Soc., Washington, D. C.

Toth, L. Z. J. (1959). The sterilizing effect of ethylene oxide vapor on different microorganisms. *Arch. Mikrobiol.* **32,** 409-410.

Townsend, C. T., Esty, J. R., and Baselt, F. C. (1938). Heat resistance studies on spores of putrefactive anaerobes in relation to determination of safe processes for canned foods. *Food Res.* **3,** 323-346.

Townsend, C. T., Yee, L., and Mercer, W. A. (1954). Inhibition of the growth of *Clostridium botulinum* by acidification. *Food Res.* **19,** 536-542.

Traci, P. A., and Duncan, C. L. (1974). Cold shock lethality and injury in *Clostridium perfringens*. *Appl. Microbiol.* **28,** 815-821.

Trakulchang, S. P., and Kraft, A. A. (1977). Survival of *Clostridium perfringens* in refrigerated and frozen meat and poultry items. *J. Food Sci.* **42,** 518-521.

Trenk, H. L., Butz, M. E., and Chu, F. S. (1971). Production of ochratoxins in different cereal products by *Aspergillus ochraceus*. *Appl. Microbiol.* **21,** 1032-1035.

Troller, J. A., and Christian, J. H. B. (1978). "Water Activity and Food." Academic Press, New York.

Troller, J. A., and Stinson, J. V. (1975). Influence of water activity on growth and enterotoxin formation by *Staphylococcus aureus* in foods. *J. Food Sci.* **40,** 802–804.

U.S. Department of Agriculture (1975). "Regulations Governing the Inspection of Eggs and Egg Products," 7CFR Part 59. Agric. Mark. Serv., USDA, Washington, D.C.

U.S. Department of Agriculture (1977). "List of Chemical Compounds Authorized for Use under U.S.D.A. Meat, Poultry Rabbit and Egg Products Inspection Programs," MPI 8. U.S. Gov. Print. Off., Washington, D.C.

U.S. Department Health, Education and Welfare (1966). "Fabrication of Single Service Containers and Closures for Milk and Milk Products—Guide for Sanitation Standards." Food Drug Adm., Washington, D.C.

U.S. Department of Health, Education and Welfare (1969). "Bacteriological Analytical Manual," 2nd Ed, Sect. 42.03. pH ranges. Food Drug Adm., Washington, D.C.

U.S. Department of Health, Education and Welfare (1972). "Report to the Commissioner of the Food and Drug Administration by the FDA Task Force on the Use of Antibiotics in Animal Feeds." Food Drug Adm., Rockville, Maryland.

U.S. Department of Health, Education and Welfare, (1974). "Antibiotic Residues in Milk, Dairy Products and Animal Tissues. Methods, Reports and Protocols." Natl. Cent. Antibiot. Insulin Anal., Food Drug Adm., Washington, D.C.

U.S. Department of Health, Education and Welfare (1975). "Foodborne and Waterborne Disease Outbreaks," Annual Summary, 1974. Cent. Dis. Control, Atlanta, Georgia.

U.S. Department of Health, Education and Welfare (19761). "Food Service Sanitation Manual," HEW Publ. No. (FDA) 78-2081. U.S. Gov. Print. Off., Washington, D.C.

U.S. Department of Health, Education and Welfare (1976b). "Foodborne and Waterborne Disease Outbreaks," Annual Summary, 1975. HEW Publ. No. (CDC) 76-8185. Atlanta, Georgia.

U.S. Department of Health, Education and Welfare (1977). "Santitizing Solutions," 21 CFR 178.1010. U.S. Gov. Print. Off., Washington, D.C.

U.S. Department of Health, Education and Welfare (1978a). "Grade A Pasteurized Milk Ordinance." Recommendations of the Public Health Serv/Food Drug Adm., Washington, D.C.

U.S. Department of Health, Education and Welfare (1978b). Deaths from asphyxia among fishermen. *Cent. Dis. Control Morbid. Mortal. Weekly Rep.* **27,** 309–310, 315.

U.S. Department of Health, Education and Welfare (1979). Part 114—Acidified Foods, 114.90 Methodology. In: Title 21—Acidified Foods and Low-Acid Canned Foods in Hermetically Sealed Containers, Good Manufacturing Practice. U.S. Food and Drug Administration. *Fed. Regist.* **44,** (53), 16230–16238.

Upadhyay, J., and Stokes, J. L. (1962). Anaerobic growth of psychrophilic bacteria. *J. Bacteriol.* **83,** 270–275.

Valley, G. (1928). The effect of carbon dioxide on bacteria. *Q. Rev. Biol.* **3,** 209–223.

Valley, G., and Rettger, L. F. (1927). The influence of carbon dioxide on bacteria. *J. Bacteriol.* **14,** 101–137.

Van den Berg, L., and Yang, S. M. (1969). Effect of relative humidity on production of extracellular proteolytic enzymes by *Botrytis cinerea* and *Sclerotinia sclerotiorum Can. J. Bot.* **47,** 1007–1010.

Vanichseni, S., Haughey, D. P., and Nottingham, P. M. (1972). Water- and air-thawing of frozen lamb shoulders. *J. Food Technol.* **7,** 259–270.

van Schothorst, M. (1976). Resuscitation of injured bacteria in foods. *In* "Inhibition and Inactivation of Vegetative Microbes" (F. A. Skinner and W. B. Hugo, eds.), Society of Applied Bacteriology Symposia Series, No. 5, pp. 317–328. Academic Press, New York.

van Schothorst, M., and Peelen-Knol, G. (1970). Detection and identification of some antibiotics in slaughter animals. *Neth. J. Vet. Sci.* **3,** 85–93.

van Schothorst, M., and van Leusden, F. M. (1972). Studies on the isolation of injured salmonellae from foods. *Zentralbl. Bakteriol., Parasitenkd., Infektionskr. Hyg., Abt. 1: Orig., Reihe A.* **221,** 19–29.

van Schothorst, M., and van Leusden, F. M. (1975). Further studies on the isolation of injured salmonellae from foods. *Zentralbl. Bakteriol., Parasitenkd., Infektionskr. Hyg., Abt. 1: Orig., Reihe A* **230,** 186–191.

van Schothorst, M., Uijttenboogaart, T. G., and van Leusden, F. M. (1973). Über die Leistungsfähigkeit des Nierentestes bei der Bestimmung von Antibiotikarückständen in Fleisch. *Fleischwirtschaft* **5,** 703–705.

Vas, K (1970). Problems in thermal processing. *J. Appl. Bacteriol.* **33,** 157–166.

Vaughn, R. H., Ng, H., Stewart, G. F., Nagel, C. W., and Simpson, K. L. (1960). Antibiotics in poultry meat preservation. Development *in vitro* of bacterial resistance to chlortetracycline. *Appl. Microbiol.* **8,**, 27–30.

Vinton, C., Martin, S., Jr., and Gross, C. E. (1947). Bacteriological studies relating to thermal processing of canned meats. VIII. Thermal resistance of spores normally present in meat. *Food Res.* **12,** 184–187.

von Bockelmann, I. (1975). Mikrobiologische Eigenschaften von Verpackungsmaterial für Lebensmittel. *Alimenta* **14,** 145–149.

von Bockelmann, I., and von Bockelmann, B. (1974). Die Mikroflora in Papier. *Alimenta* **13,** 181–184.

Voss, E., and Moltzen, B. (1973). Untersuchungen über die Oberflächenkeimzahlen extrudierter Kunststoffe für die Lebensmittelverpackung. *Milchwissenschaft* **28,** 479–486.

Walker, H. W., and Ayres, J. C. (1958). Antibiotic residuals and microbial resistance in poultry treated with tetracyclines. *Food Res.* **23,** 525–531.

Walter, W. G., and Schillinger, J. E. (1975). Bacterial survival in laundered fabrics. *Appl. Microbiol.* **29,** 368–373.

Warseck, M., Ray, B., and Speck, M. L. (1973). Repair and enumeration of injured coliforms in frozen foods. *Appl. Microbiol.* **26,** 919–924.

Warth, A. D. (1977). Mechanism of resistance of *Saccharomyces bailii* to benzoic, sorbic and other weak acids used as food preservatives. *J. Appl. Bacteriol.* **43,** 215–230.

Washam, C. J., Olson, H. C., and Vedamuthu, E. R. (1977). Heat resistant psychrotrophic bacteria isolated from pasteurized milk. *J. Food Prot.* **40,** 101–108.

Wasserman, A. E. (1973). Nitrite and the flavour of cured meat. II. *Proc. Int. Symp. Nitrite Meat Prod., Zeist, Neth.* (B. Krol and B. J. Timbergen, eds.), pp. 173–178. Pudoc, Wageningen, Netherlands.

Wasserman, A. E., and Kimoto, W. (1976). Consumer evaluation of the flavour of bacon cured with and without sodium nitrite. *Proc. Int. Symp. Nitrite Meat Prod., 2nd, Zeist, Neth.* (B. J. Timbergen and B. Krol, eds.), pp. 73–74. Pudoc, Wageningen, Netherlands.

Wasserman, A. E., and Talley, F. (1972).The effect of sodium nitrite on the flavor of frankfurters. *J. Food Sci.* **37,** 536–538.

Watanabe, T. (1963). Infective heredity of multiple drug resistance in bacteria. *Bacteriol. Rev.* **27,** 87–115.

Watkins, J. H., and Winslow, C. E. A. (1932). Factors determining the rate of mortality of bacteria exposed to alkalinity and heat. *J. Bacteriol.* **24,** 243–265.

Watrous, G. H., Jr. (1975). Food soils, water hardness, and alkaline cleaner formulations. *J. Milk Food Technol.* **38,** 163–165.

Wei, L. S., Siregar, J. A., Steinberg, M. P., and Nelson, A. I. (1967). Overcoming the bacteriostatic activity of onion in making standard plate counts. *J. Food Sci.* **32,** 346–349.

Weil, R. (1899). Zur Biologie der Milzbrandbazillen. *Arch. Hyg.* **35,** 355–408.

Wewalka, G., Rotter, M., Koller, W., and Stanek, G. (1977). Wirkungsvergleich von 14 Verfahren zur Hygienischen Hände Desinfektion. *Zentralbl. Bakteriol., Parasitenkd., Infektionskr. Hyg., Abt. 1: Orig., Reihe B* **165,** 242-249.

Wheater, D. M., Hirsch, A. and Mattick, A. T. R. (1952). Possible identity of "lactobacillin" with hydrogen peroxide produced by lactobacilli. *Nature (London)* **170,** 623-624.

Whelton, R., Phaff, H. J., Mrak, E. M., and Fisher, C. D. (1946). Control of microbiological food spoilage by fumigation with epoxides. I and II. *Food Ind.* **18,** 23-25, 174-176, 318-319.

Whitbourne, J., and West, K. (1975). Ethylene oxide application. *Dev. Ind. Microbiol.* **16,** 57-66.

Whitten, L. K. (1971). The effect of freezing on the viability of *Taenia ovis* cysts. *N. Z. Vet. J.* **19,** 223.

Wicher, K., Reisman, R. E., and Arbersman, C. E. (1969). Allergic reaction to penicillin present in milk. *J. Am. Med. Assoc.* **208,** 143-145.

Wickerham, L. J. (1957). Presence of nitrite-assimilating species of *Debaromyces* in lunch meats. *J. Bacteriol.* **74,** 832-833.

Wiedemann, B., and Schmidt, G. (1972). Die Bedeutung der Lipopolysaccharid-Struktur von *E. coli* für die Ubertragung von R-Faktoren. *Zentralbl. Bakteriol., Parasitenkd. Infektionskr. Hyg., Abt. 1: Orig., Reihe A* **219,** 180-186.

Wierbicki, E., and Heiligman, F. (1973). Shelf stable cured ham with low nitrite-nitrate additions preserved by radappertization. *Proc. Int. Symp. Nitrite Meat Prod., Zeist, Neth.* (B. Krol and B. J. Tinbergen, eds.), pp. 189-212. Pudoc, Wageningen, Netherlands.

Wierbicki, E., Heiligman, F., and Wasserman, A. E. (1976). Irradiation as a conceivable way of reducing nitrites and nitrates in cured meats. *Proc. Int. Symp. Nitrite Meat Prod., 2nd Zeist, Neth.* (B. J. Tinbergen and B. Krol, eds.), pp. 75-81. Pudoc, Wageningen, Netherlands.

Williams, F. T. (1936). Attempts to increase the heat resistance of spores. *J. Bacteriol.* **32,** 589-597.

Williams, O. B. (1929). The heat resistance of bacterial spores. *J. Infect. Dis.* **44,** 421-465.

Wilson, B. R., Pearson, A. M., and Shorland, F. B. (1976). Effect of total lipids and phospholipids on warmed-over flavor in red and white muscle from several species as measured by thiobarbituric acid analysis. *J. Agric. Food Chem.* **24,** 7-11.

Wilson, J. M., and Davies, R. (1976). Minimal medium recovery of thermally injured *Salmonella senftenberg* 4969. *J. Appl. Bacteriol.* **40,** 365-374.

Wimpenny, J. W. T. (1969). Oxygen and microbial metabolism. *Process Biochem.* **4**(June), 19-22.

Winarno, F. G., and Stumbo, C. R. (1971). Mode of action of ethylene oxide on spores of *Clostridium botulinum* 62A. *J. Food Sci.* **36,** 892-895.

Windmueller, H. G., Ackerman, C. J., Bakerman, H., and Mickelsen, O. (1959). Reaction of ethylene oxide with nicotinamide and nicotinic acid. *J. Biol. Chem.* **234,** 889-894.

Windsor, M. L., and Thoma, T. (1974). Chemical preservation of industrial fish: New preservative mixtures. *J. Sci. Food Agric.* **25,** 993-1005.

Wirth, F., and Leistner, L. (1970). Redoxpotentiale in Fleischkonserven. *Fleischwirtschaft* **50,** 491-492.

Witheridge, W. N., and Yaglou, C. P. (1939). Ozone in ventilation—its possibilities and limitations. *Trans. Am. Soc. Heat. Vent. Eng.* **45,** 509.

Witter, L. D., Campbell, M. F., and Azuma, Y. (1966). Formation of bacterial pigments at low temperature by psychrophilic pseudomonads. *Dev. Ind. Microbiol.* **7,** 231-239.

Woolford, G., Cassens, R. G., Greaser, M. L., and Sebranek, J. G. (1976). The fate of nitrite: reaction with protein. *J. Food Sci.* **41,** 585-588.

World Health Organization (WHO) (1963). "The Public Health Aspects of the Use of Antibiotics in Food and Feed." Tech. Rep. Ser., No. 260. WHO, Geneva.

World Health Organization (WHO) (1969). "Specifications for the Identity and Purity of Food Additives and their Toxicological Evaluation: Some Antibiotics," Tech. Rep. Ser., No. 430. WHO, Geneva.

World Health Organization (WHO) (1971). "International Standards for Drinking Water," 3rd Ed. WHO, Geneva.

World Health Organization (1973). The public health aspects of antibiotics in feedstuffs. *In* "Report of a Working Group," WHO REg. Off., Copenhagen.

World Health Organization (WHO) (1976). "Microbiological Aspects of Food Hygiene," Tech. Rep. Ser., No. 598. WHO, Geneva.

World Health Organization (WHO) (1977). "Environmental Health Criteria. V. Nitrates, Nitrites and N-Nitroso Compounds." WHO, Geneva.

Xezones, H., Segmiller, J. L., and Hutchings, I. J. (1965). Processing requirements for a heat-tolerant anaerobe. *Food Technol.* **19,** 1001-1002.

Yackel, W. C., Nelson, A. I., Wei, L. S., and Steinberg, M. P. (1971). Effect of controlled atmosphere on growth of mold on synthetic media and fruit. *Appl. Microbiol.* **22,** 513-516.

Yanai, S., Kimura, S., Kojima, H., and Sato, H. (1969). Studies of food packaging. II. Microorganisms of plastics packaging materials and packaged noodles. *Shokuryo Kenkyujo Kenkyu Hokoku* **24,** 53.

Yong, F. M., and Wood, B. J. B. (1976). Microbial succession in experimental soy sauce fermentations. *J. Food Technol.* **11,** 525-536.

York, G. K., II, and Vaughn, R. H. (1954). Use of sorbic acid enrichment media for species of *Clostridium. J. Bacteriol.* **68,** 739-744.

Zaika, L. L., Zell, T. E., Smith, J. L., Palumbo, S. A., and Kissinger, J. C. (1976). The role of nitrite and nitrate in Lebanon bologna, a fermented sausage. *J. Food Sci.* **41,** 1457-1460.

Zottola, E. A. (1973). Microbiology—Sanitation. *In* "Food Safety through the Hazard Analysis and Critical Control Point System," pp. 57-72. Pillsbury, Minneapolis, Minnesota.

Appendix I

The International Commission on Microbiological Specifications for Foods: Its Purposes and Accomplishments

HISTORY AND PURPOSE

The International Commission on Microbiological Specifications for Foods (ICMSF) was formed in 1962 by the parent body, the International Association of Microbiological Societies (IAMS), in response to the need for internationally acceptable and authoritative decisions on microbiological limits for foods commensurate with public health safety, and particularly for foods moving in international commerce. Its overall purpose is to appraise public health aspects of the microbiological content of foods. Through the IAMS, the ICMSF is linked to the International Union of Biological Societies (IUBS) and to the World Health Organization (WHO) and, hence, is a body of the United Nations.

The founding terms of reference are as follows: (a) to assemble, correlate, and evaluate evidence about the microbiological quality of foods; (b) to consider whether microbiological criteria are necessary for any particular food; (c) where necessary, to propose such criteria; and (d) to suggest appropriate methods of sampling and examination. More descriptively, the ICMSF seeks to provide comparable standards of judgment in different countries, to foster safe movement of foods in international commerce, and to dissipate difficulties caused by disparate microbiological criteria and methods of analysis. Fulfillment of such objectives would be of great value to the food industry, to the expansion of international trade in foods, to national control agencies, to the international agencies more concerned with the humanitarian aspects of food distribution, and, eventually, to the health of the consuming public.

The ICMSF is essentially a scientific advisory body that provides basic information through extensive study and makes recommendations based on such information. The results of studies are published either as books or papers and are thus available to interested individuals, governments, and national and international organizations to use as desired. The group provides the facts without prejudices and thereby fills a useful role as an authoritative base. Primarily through cross-membership, close liaison is enjoyed with other organizations involved in international standards, such as the Codex Alimentarius Commission, the International Standards Organization, the International Dairy Federation, and the Association of Official Analytical Chemists.

At meetings, the ICMSF functions as a work party, not as a forum for the reading of papers. Much of the work is done by subcommittees during the interval between meetings, often with assistance of nonmember consultants. The general meetings are largely directed toward assessing the work of the subcommittees, debating to achieve a consensus, editing of draft submissions, and planning. Meetings have been held in Montreal, Canada (1962, the founding meeting); Cambridge, England (1965); Moscow, USSR (1966); London, England (1967); Dubrovnik, Yugoslavia (1969); Mexico City, Mexico (1970); Opatija, Yugoslavia (1971); Langford, England (1972); Ottawa, Canada (1973); Caracas, Venezuela (1974); Alexandria, Egypt (1976) and Cairo, Egypt (1977 and 1978).

MEMBERSHIP AND SUBCOMMISSIONS

The membership consists of 21 food microbiologists from 14 countries (pp. 306–7), whose combined professional interests include research, public health, official food control, education, and industrial research and development. They are drawn from government laboratories in public health, agriculture, and food technology, from universities, and from the food industry. In addition, the ICMSF engages consultants from time to time to assist with specific aspects of its studies. All members and consultants are chosen on the basis of their expertise in areas of food microbiology, not as national delegates; all work voluntarily without fees or honoraria.

To promote similar activities among food microbiologists on a regional scale, subcommissions have been created in various areas of the world. To date, three have been established (see memberships, pp. 307–8): one in the Balkan-Danubian region (the Balkan-Danubian Subcommission, BDS), composed of eight members; one in Latin America (the Latin American Subcommission, LAS), composed of five full-members and five observer-members; and one in the Middle East-North African region (the Middle East-North African Subcommission, MENAS), composed of eight members. Each is an autonomous body which conducts studies on problems of specific concern to its region.

ACCOMPLISHMENTS

Since all studies made by the Commission are published, a list of its publications is a record of its accomplishments:

Books

1. *Microorganisms in Foods 1: Their Significance and Methods of Enumeration*. 1978. 2nd Edition. Univ. of Toronto Press, Toronto, Canada. 434 pages. Spanish translation available for 1st edition, 1973; Editorial Ascribia, Zaragoza, Spain.

2. *Microbiological Specifications and Testing Methods for Irradiated Foods*. 1970. Compiled and edited in cooperation with the Food and Agriculture Organization and the International Atomic Energy Agency. Published as Tech. Ser. 104 by the International Atomic Energy Agency, Vienna, Austria. 122 pages. Available in English, French, German, and Russian.

3. *Microorganisms in Foods 2: Sampling for Microbiological Analysis: Principles and Specific Applications*. 1974. University of Toronto Press, Toronto, Canada. 213 pages. Spanish translation in progress; Editorial Ascribia, Zaragoza, Spain.

Microorganisms in Foods 1: Their Significance and Methods of Enumeration has been widely acclaimed. Over 4000 copies of three printings in English have been sold. It has proved invaluable to food microbiologists in government control agencies, industry, and in teaching and research institutions. Volume 2, *Sampling for Microbiological Analysis: Principles and Specific Applications,* has also received excellent reviews, and is proving to be of the greatest value to all agencies and food companies involved in assessing the microbiological quality of foods.

Articles

1. The microbiology of specific frozen foods in relation to public health: report of an international committee. 1963. Author, F. S. Thatcher. *J. Appl. Bacteriol.* **26,** 266–285.
2. Il comitato internazionale per la definizione delle caratteristiche microbiologiche degli alimenti ed il suo contributo per la tutela igienica dei medesimi. 1969. Author, C. Cominazzini. *Cronache di Chimica* **25,** 16–22.
3. Comisión internacional de Especificaciones Microbiológicas de los Alimentos. 1971. Authors, S. Mendoza and F. Quevedo. *Bol. Inst. Bacteriol. de Chile* **13,** 45–48.

4. The International Committee on Microbiological Specifications for Foods: Its purposes and accomplishments. 1971. Author, F. S. Thatcher. *J. Ass. Offic. Anal. Chem.* **54,** 836–841.
5. ICMSF methods studies: I. Comparison of analytical schemes for detection of *Salmonella* in dried foods. 1973. Authors, J. H. Silliker and D. A. Gabis. *Can. J. Microbiol.* **19,** 475–479.
6. ICMSF methods studies: II. Comparison of analytical schemes for detection of *Salmonella* in high moisture foods. 1974. Authors, D. A. Gabis and J. H. Silliker. *Can. J. Microbiol.* **20,** 663–669.
7. ICMSF methods studies: III. An appraisal of 16 contemporary methods for the detection of *Salmonella* in meringue powder. 1974. Authors, E. Idziak and I. E. Erdman. *Can. J. Microbiol.* **19,** 475–479.
8. ICMSF methods studies: IV. International collaborative assay for the detection of *Salmonella* in raw meats. 1973. Author, I. E. Erdman. *Can. J. Microbiol.* **20,** 715–720.
9. ICMSF methods studies: V. The influence of selective enrichment media and incubation temperatures on the detection of salmonellae in frozen meats. 1974. Authors, J. H. Silliker and D. A. Gabis. *Can. J. Microbiol.* **40,** 813–816.
10. ICMSF methods studies: VI. The influence of selective enrichment media and incubation temperatures on the detection of salmonellae in dried foods and feeds. 1974. Authors, D. A. Gabis and J. H. Silliker. *Can. J. Microbiol.* **20,** 1509–1511.
11. ICMSF methods studies: VII. Indicator tests as substitutes for direct testing of dried foods and feeds for *Salmonella*. 1976. Authors, J. H. Silliker and D. A. Gabis. *Can. J. Microbiol.* **22,** 971–974.
12. ICMSF methods studies: VIII. Comparative study for the enumeration of *Clostridium perfringens* in foods. 1977. Authors, A. H. W. Hauschild, R. J. Gilbert, S. M. Harmon, M. F. O'Keeffe, and R. Vahlefeld. *Can. J. Microbiol.* **23,** 884–892.
13. ICMSF methods studies: IX. The influence of selective enrichment broths, differential plating media, and incubation temperatures on the detection of *Salmonella* in dried foods and feed ingredients. 1977. Authors, D. A. Gabis and J. H. Silliker. *Can. J. Microbiol.* **23,** 1225–1231.
14. ICMSF methods studies: X. An international comparative study of four media for the enumeration of *Staphylococcus aureus* in foods. 1978.

Authors, M. K. Rayman, J. J. Devoyod, U. Purvis, D. Kusch, J. Lanier, R. J. Gilbert, D. D. Till, and G. A. Jarvis. *Can. J. Microbiol.* **24,** 274–281.

15. The International Commission on Microbiological Specifications for Foods. 1978. Author, D. S. Clark. *Food Technology* **32,** 51–54, 57.
16. ICMSF methods studies: XI. Collaborative/comparative studies on the determination of coliforms using the most probable number procedure. 1979. Authors, J. H. Silliker, D. A. Gabis, and A. May. *J. Food Protection* **42,** 638–644.
17. ICMSF methods studies: XII. Comparative study for the enumeration of *Clostridium perfringens* in feces. 1979. Authors, A. H. W. Hauschild, P. Desmarchelier, R. J. Gilbert, S. M. Harmon, and R. Vahlefeld. *Can. J. Microbiol* **25,** 953–963.
18. ICMSF methods studies: XIII. An international comparative study of the MPN procedure and the Anderson—Baird-Parker direct plating method for the enumeration of *Escherichia coli* biotype I in raw meats. (1979). Authors, M. K. Rayman, G. A. Jarvis, C. M. Davidson, S. Long, J. M. Allen, T. Tong, P. Dodsworth, S. McLaughlin, S. Greenberg, B. G. Shaw, H. J. Beckers, S. Qvist, P. M. Nottingham, and B. J. Stewart. *Can. J. Microbiol.* **25,** 1321–1327.

Most of the articles deal with the results of the ICMSF methods-testing program which is described in the next section. All studies have proved to be of substantial value, but perhaps the most significant are those described in ICMSF Methods Studies I and II, which report on compositing food samples for *Salmonella* analysis. Combining multiples of standard 25-gm sample units into one composite for analysis gives the same assurance of detection as separate examination of each 25-gm sample unit. This finding alone has made the testing program worthwhile, because statistically valid quality control sampling programs to demonstrate *Salmonella*-negative lots are now economically feasible. The cost of *Salmonella* testing can be reduced to a fraction of what it was with no loss of accuracy.

Methods-Testing Program

The overall objective is to determine by detailed comparative analysis, involving laboratories in various countries, the best methods for the enumeration and identification of indicator and food-poisoning bacteria. In *Microorganisms in Foods 1: Their Significance and Methods of Enumeration,* the ICMSF describes several of the best-known methods for some of the microbial categories, because it could not distinguish which one, if any, is superior. International comparative testing is seen as the only way to determine the most accurate methods.

The studies are being coordinated in Silliker Laboratories, Chicago Heights,

Illinois, in the United States (Drs. J. H. Silliker and D. A. Gabis, coordinators), and at the Health Protection Branch of Health and Welfare Canada, in Ottawa, Canada (Drs. H. Pivnick and K. Rayman, coordinators). All projects are planned by a subcommittee of ICMSF members and consultants and are approved by the ICMSF in plenary session. Studies completed to date or in progress include those on methods for *Salmonella,* coliforms, *Staphylococcus, Clostridium perfringens,* and *Escherichia coli.*

FINANCING

The ICMSF raises funds for its activities from government agencies in several countries, from the WHO, from the IUBS, and from the food industry. Assistance from government agencies has come in the form of grants for specific projects: the United States Department of Agriculture has given two grants in support of the methods-testing program; the United States Department of Health, Education and Welfare sponsored PL 480 grants to support two general meetings on sampling of foods and to support three others on the preparation of this book; the government of Kuwait also contributed financially to the preparation of this book; and Health and Welfare Canada supported one general meeting on the preparation of the 2nd edition of Volume 1. WHO has contributed annually in support of the methods-testing program and general expenditures, and the IUBS has granted funds in support of administration costs and to help meet expenses for the publication of this book. Over 60 food companies and agencies in eight countries contribute to the ICMSF Sustaining Fund, mostly on a yearly basis (see Appendix II).

Appendix II

Contributors to the Sustaining Fund of ICMSF

Amatil Ltd., Box 145, GPO Sydney NSW 2001, Australia

American Can Co., America Lane, Greenwich, CT, 06830, USA

Arnotts Biscuits Pty Ltd., 170 Kent St., Sydney NSW 2000, Australia

Atlantic Sugar, Ltd., P.O. Box 7, Montreal, Quebec, Canada, H3C 1C5

Bacon and Meat Manufacturers Association, 1-2 Castle Lane, London, England, SWIE 6DU

Beatrice Foods Co., 1526 South State St., Chicago, IL, 60605, USA

Beecham Group Ltd., Beecham House, Great West Rd., Brentford, Middlesex, England

Brooke-Bond Oxo Ltd., Trojan Way, Purley Way, Croydon, England, CR9 9EH

Brown and Polson Ltd., Clay Gate House, Littleworth Rd., Esher, Surrey, England

Burns Foods Ltd., P.O. Box 1300, Calgary, Alta., Canada, T2P 2L4

Cadbury Schweppes Foods Ltd., Bournville, Birmingham, England

Cadbury Schweppes Ltd., P.O. Box 88, St. Kilda West, Victoria 3182, Australia

Campbell Institute for Research and Technology, Campbell Place, Camden, N.J. 08101, USA

Campbell Soup Co. Ltd., 60 Birmingham St., Toronto, Ont., Canada, M8V 2B8

Canada Packers Ltd., 2211 St. Clair Ave. W., Toronto 9, Ont., Canada, M69 1K4

Carlo Erba Institute for Therapeutic Research, Milan 20159, Italy

Central Alberta Dairy Pool, 5302 Gaetz Ave., Red Deer, Alta., Canada

Centro Studi sull' Alimentazione, Gino Alfonso Sada, P.za Diaz 7-20123, Milan, Italy

Christie Brown and Co., Ltd., 2150 Lakeshore Blvd. W., Toronto, Ont., Canada, M8V 1A3

Coca-Cola Co., 310 North Ave. N.W., Atlanta, GA, 30301, USA

CPC International Inc., International Plaza, Englewood Cliffs, NJ, 07632, USA

CSR Ltd., Box 1630 GPO., Sydney NSW 2001, Australia

Del Monte Corporation, 205 North Wiget Lane, Walnut Creek, CA, 94598, USA

Difco Laboratories, Detroit, MI, 48232, USA

Distillers Co. Ltd., 21 St. James Square, London, SW 1, England

Export Packers Company Ltd., 250 Summerlea Rd., Bramalea, Ontario L6T 3V6, Canada

Findus Ltd., Bjuv, Sweden

Geinoca Food Services Ltd., 6205 Airport Road, Bldg. B, Suite 205, Mississauga, Ont., Canada L4V 1E3

General Foods Corporation, Technical Center, White Plains, NY, 10625, USA

General Foods Canada Ltd., Box 4019, Terminal A, Toronto, Ont., Canada

Gerber Products Co., 445 State St., Fremont, MI, 49412, USA

H. J. Heinz Co. Ltd., Hayes Park, Hayes, Middlesex, England

Home Juice Co. Ltd., 175 Fenmar Dr., Weston, Ont., Canada

Horne & Pitfield Foods Ltd., 14550 112th Ave., P. O. Box 2266, Edmonton 15, Alta., Canada, T5J 2P6

Infant Formula Council, 64 Perimeter Center East, Atlanta, GA, 30346, USA

International Union of Biological Societies, 51 Bd. de Montmorency, 75016, Paris, France

Indulac, Caracas, Venezuela

ITT Continental Baking Co., P.O. Box 731, Rye, NY, 10580, USA

J. Sainsbury Ltd., Stamford House, Stamford St., London, England, SE1 9LC

John Labatt Ltd., 150 Simcoe St., P.O. Box 5050, London, Ont., Canada, N6A 4M3

Joseph Rank Ltd., Millcrat House, Eastcheap, London, EC 3, England

Kellogg/Salada Canada Ltd., 6700 Finch Ave. W., Rexdale, Ont., Canada M9W 5P2

Kraft Foods Ltd., Box 1673N, G.P.O., Melbourne, 3001, Australia

Kuwait Ministry of Health, Kuwait, Kuwait

Langnese-Iglo GmbH, Hauptverwaltung, Postfach 10 40 29, 2000 Hamburg 1, West Germany

Maizena Gesellschaft mbH, Postfach 560, Knorrstr. 1, 71 Heilbronn/Neckar, Germany

Maple Leaf Mills Ltd., P.O. Box 710, Station K, Toronto, Ont., Canada, M4P 2X5

Marks and Spencer Ltd., Michael House, Baker St., London, W1, England

Mars Ltd., Dundee Rd., Trading Estate, Slough, Bucks., England

McCormick and Co. Inc., Baltimore, MD, 21202, USA

Milk Marketing Board, Thames Ditton, Surrey, England, KT7 OEL

National Fisheries Institute Inc. 1101 Connecticut Ave., NW, Washington, DC, 20036, USA

New Zealand Food Manufacturers' Federation, Industry House, 38–44 Courtenay Place, Wellington, NZ

Oscar Mayer Co., Caracas, Venezuela

Plasmon SPA, Corso, Garibaldi 97.99, 20121 Milano, Italy

RHM Research Ltd., Lincoln Road, High Wycombe, Bucks., England, H12 3QN

R.J.R Foods Inc., 4th and Main Sts., Winston Salem, NC, 27102, USA

Reckitt and Colman Ltd., Carrow, Norwich, England, NR1 2DD

Ross Laboratories, 615 Cleveland Ave., Columbus, OH, 43216, USA

Royal Co., Caracas, Venezuela

Silliker Laboratories, 1304 Halsted St., Chicago Heights, IL 60411, USA

Spillers Ltd., Old Charge House, Cannon St., London E.C.4, England

Standard Brands Ltd., 1 Dundas St. W., Toronto, Ont., Canada M5B 2H1

Swift Canadian Co. Ltd., 2 Eva Rd., Etobicoke, Ont., Canada M9C 4V5

Tate and Lyle Refineries Ltd., 21 Mincing Lane, London, England

Terre de Crodo, via Cristoforo Gluck 35, Milano 20125, Italy

Tesco Stores Ltd., Tesco House, Delamere Rd., Cheshunt, Waltham Cross, Herts., England

The Borden Co. Ltd., 1275 Lawrence Ave. E., Don Mills (Toronto), Ont., Canada M3A 1C5

The J. Lyons Group of Companies, Cadby Hall, London, England W14 09A

The Nestle Company, Societe d'Assistance Technique pour Produits Nestle S.A., Case Postale 88, CH-1814 La Tour-de-Peilz, Switzerland

The Pillsbury Co., 608 Avenue S., Minneapolis, MN, 55402, USA

The Quaker Oats Co., 617 West Main St., Barrington, IL, 60010, USA

The Quaker Oats Co. Canada Ltd., Quaker Park, Peterborough, Ont., Canada K9J 7B2

Thomas J. Lipton, Inc., 800 Sylvan Ave., Englewood Cliffs, NJ, 07632, USA

Thomas J. Lipton, Ltd., 2180 Yonge St., Toronto, Ontario, Canada M4S2C4

Unigate (Australia) Pty Ltd., P.O. Box 13, Dandenong, Victoria 3175, NZ

Unilever Ltd., Unilever House, Blackfriars, London, E.C. 4, England

Watties Industries Ltd., P.O. Box 439, Hastings, New Zealand

World Health Organization, Geneva, Switzerland.

Appendix III

Members and Consultants of ICMSF and Its Subcommissions

MEMBERS OF THE ICMSF

Dr. H. Lundbeck (Chairman) Director, The National Bacteriological Laboratory, S-105 21, Stockholm, Sweden

Dr. D. S. Clark (Secretary-Treasurer), Director, Bureau of Microbial Hazards, Food Directorate, Health Protection Branch, Health and Welfare Canada, Tunney's Pasture, Ottawa, Ont., Canada, K1A 0L2

Dr. A. C. Baird-Parker, Head, Microbiological Research, Unilever Research, Colworth Laboratory, Unilever Limited, Colworth House, Sharnbrook, Bedford, England, MK44 1LQ

Dr. Frank L. Bryan, Chief, Foodborne Disease Training, Instructional Services Division Bureau of Training, Center for Disease Control, Public Health Service, Dept. of Health, Education and Welfare, Atlanta, GA 30333, USA

Dr. J. H. B. Christian, Associate Chief, Division of Food Research, C.S.I.R.O., P.O. Box 52, North Ryde, N.S.W. 2113, Australia

Professor C. Cominazzini, Professor in Charge of Hygiene, Faculty of Medicine, State University of Turin, Via Monte Nero 46 - 28100, Novara, Italy

Professor Otto Emberger, Chief, Department of Microbiology and Associate Professor of Hygiene, Faculty of Medical Hygiene, Charles' University of Prague, Srobárova 48, Praha 10, Vinohrady, Czechoslovakia

Dr. Betty C. Hobbs, Microbiology Department, Christian Medical College and Brown Memorial Hospital, Ludhiana, Punjab, India. Also, 1000, High Road, Whetstone, London N20 OQG, England (Formerly Director, Food Hygiene Reference Laboratory, Central Public Health Laboratory, Colindale Avenue, London NW9 5HT, England)

Dr. Keith H. Lewis, Professor of Environmental Health, School of Public Health, University of Texas, Health Sciences Center at Houston, P.O. Box 20186, Houston, TX 77025, USA

Dr. G. Mocquot, Chargé de mission à l'I.N.R.A., Technologie laitière, CNRZ, 78.350 Jouy-en-Josas, France

Dr. N. P. Nefedjeva, Chief, Laboratory of Food Microbiology, Institute of Nutrition, AMS USSR, Ustinsky pr. 2/14, Moscow G-240, USSR

Dr. C. F. Niven, Jr., Director of Research, Del Monte Research Center, 205 North Wiget Lane, Walnut Creek, CAL. 94598, U.S.A.

Dr. P. M. Nottingham, Head, Science Division, The Meat Industry Research Institute of

New Zealand (Inc.), P.O. Box 617, Hamilton, New Zealand

Dr. J. C. Olson, Jr., Consultant in Food Microbiology, 4982 Sentinel Dr., #204, Bethesda, MD 20016, U.S.A.
Formerly Deputy Assistant to the Director, Bureau of Foods, Food and Drug Administration, US Dept. of Health, Education and Welfare, Washington, DC

Dr. H. Pivnick, Bureau of Microbial Hazards, Food Directorate, Health Protection Branch, Health and Welfare Canada, Tunney's Pasture, Ottawa, Ont. Canada K1A 0L2

Dr. Fernando Quevedo, Head, Food Microbiology and Hygiene Unit, Pan American Zoonoses Centre, Casilla 3092, 1000 Buenos Aires, Argentina

Dr. T. A. Roberts, Head of Food Quality and Control Division, Agricultural Research Council, Meat Research Institute, Langford, Bristol, BS18 7DY, England

Dr. J. H. Silliker, President, Silliker Laboratories, 1139 Dominguez St., Suite I, Carson, CA 90746, U.S.A.

Mr. Bent Simonsen, Chief Microbiologist, Danish Meat Products Laboratory, Ministry of Agriculture, Howitzvej 13, DK-2000 Copenhagen F, Denmark

Professor H. J. Sinell, Director, Institute of Food Hygiene, Free University of Berlin, Koserstr. 20, 1000 Berlin 33, Germany.

Dr. M. van Schothorst, Head of the Microbiology Section, Central Control Laboratory, Nestle Products Technical Assistance Co., Ltd. Case Postale 88 CH-1814 la tour-de-Peitz Switzerland

BALKAN AND DANUBIAN SUBCOMMISSION

Professor Dr. J. Takács (Chairman), Director, Institute of Food Hygiene, University of Veterinary Sciences, 1400 Budapest, P.O. Box 2, Hungary (Deceased, 1979).

Dr. Milica Kalember-Radosavljević (Secretary), Food Bacteriologist, Military Medical Academy, Institute of Hygiene, 2 Pasterova Avenue, Belgrade, Yugoslavia

Dr. Vladimir Bartl, Head, Hygiene Laboratories, Hygiene Station for Middle Czech Region, Safarikova 14, 120 00 Praha 2, Czechoslovakia

Dr. Deac Cornel, Institutul de Igiena, Si Sănătate Publică, R3400 Cluj, Napoca, R. S., România

Professor Dr. O. Prandl, Director, Institute of Meat Hygiene and Veterinary Food Technology, Vienna III/40 linke Bahngasse 11, Austria

Dr. S. Tzannetis, Faculty of Medicine, Dept. of Microbiology, National University of Athens, Athens 609, P.O. Box 1540, Greece

Dr. Fuat Yanc, Sehir Hifzissihha Müessesesi, Sarachanebasi, Istanbul, Turkey

Professor Dr. C. Zachariev, Director, N.P.O. Veterinarno Delo, U.L. Rabotničeska Klasa No. 1 Sofia, Bulgaria

LATIN-AMERICAN SUBCOMMISSION

Professor Josefina Gomez-Ruiz (Chairwoman), Central University of Venezuela, Apartado 50259, Caracas 105, Venezuela

Dra. Silvia Mendoza G. (Secretary-Treasurer), Division of Biological Sciences, Department of Bioengineering, Simon Bolivar University, Apartado 80659, Caracas 108, Venezuela

Professor Dra. Nenufar Sosa de Caruso, Director, Dairy Institute, Veterinary Faculty, University of Uruguay, Casilla de Correo 753, Montevideo, Uruguay

Dr. Fernando Quevedo, Head, Food Microbiology and Hygiene Unit, Pan American Zoonoses Centre, Casilla 3092, 1000 Buenos Aires, Argentina

Dr. Sebastiao Timo Iaria, Instituto de Ciencias Biomedicas, Universidade de Sao Paulo, Av. Dr. Arnaldo 715-C.P. 8099, Sao Paulo, E. Sao Paulo, Brazil

Dra. Ethel G. V. Amato de Lagarde, Head of Division Bacteriologia Sanitaria, Instituto Nacional de Microbiologia "Carlos G. Malbrán", Avda Velez Sársfield 563, Buenos Aires, Argentina

Dra. Elvira Regús de Valera, Calle, José Contreras No. 98 (Altos), Zona 7 St. Domingo, República Dominicana

Dr. Mauro Faber de Freitas Leitao, Head, Department of Food Microbiology, Instituto de Tecnologia de Alimentos, Caixa Postal 139, 13.100 Campinas Sao Paulo, Brazil

Dr. Hernán Puerta Cardona, Chairman, Food Hygiene Section, Escuela Nacional de Salud Pública, Universidad de Antioquia, Apartado Aéreo 51922, Medellin, Colombia

Dra. María Alina Ratto, Head, Centro Latinoamericano de Ensenanza e Investigación de Bacteriología Alimentaria (CLEIBA), Universidad Nacional Mayor de San Marcos de Lima - PERU, Apartado 5653

MIDDLE EAST-NORTH AFRICAN SUBCOMMISSION

Professor Refat Hablas (Chairman), Bacteriological Department, Faculty of Medicine, Al-Azhar University, El Houssein Hospital, Eldarrasa, Cairo, Egypt

Dr. Hassan Sidahmed (Secretary), Head, Department of Bacteriology, P.O. Box 287, National Health Laboratory, Khartoum, Sudan

Professor A. Alaoui, Director of Institute Pasteur Maroc, Professor of Microbiology, Casa School of Medecine, Casablanca, Morocco

Dr. Abdul-Kareem Nasir Al-Dulaimi, Head, Department of Food and Water Hygiene, Central Public Health Laboratories, Andulis Square, Alwia, Baghdad, Iraq

Mr. I. Kashoulis, Analyst, Government Laboratory, Ministry of Health, Nicosia, Cyprus

Professor El-Sayed El-Mossalami, Head, Meat Hygiene Department, Faculty of Veterinary Medicine, Cairo University, Giza, Egypt

Mr. Yacoub Khalid Motawa, Head of Food Control Laboratory, Microbiological Section, Preventative Health, Public Health Laboratory, Ministry of Health, Kuwait

Dr. Neji Othman, Chief, Laboratory of Food Microbiology, Ministry of Health, National Institute of Nutrition, Tunis, Tunisia

CONSULTANTS FOR THIS VOLUME

Mr. J. P. Accolas, INRA Laboratoire de microbiologie laitière et de génie alimentaire, 78.350 Jouy-en-Josas, France

Dr. E. M. Barnes, Agriculture Research Council, Food Research Institute, Colney Lane, Norwich, England NR4 7UA

Dr. H. J. Beckers, National Institute of Public Health, P.O. Box 1, 3720 BA Bilthoven, The Netherlands

Dr. C. F. Böhme, Kalle, Niederlassung der Hoechst AG, Rheinqaustrasse 190, 6200 Wiesbaden-Biebrich, West Germany

Dr. A. W. M. Brooymans, Quaker Europe, Avenue Henri Matisse 16, Bte 6, B-1140 Brussels, Belgium

Mr. M. H. Brown, Unilever Research, Colworth Laboratory, Colworth House, Sharnbrook, Bedford, England MK44 1LQ

Dr. J. Bruijn, Sugar Milling Research Institute, University of Natal, King George V Avenue, Durban 4001, South Africa

Dr. D. A. Corlett, Jr., Del Monte Research Center, 205 North Wiget Lane, Walnut Creek, CA 94598, U.S.A.

Dr. C. M. Davidson, Silliker Laboratories of Canada, 2222 South Sheridan Way, Mississauga, Ont., Canada L5J 2M4

Dr. A. B. Dickinson, Henshelwood Terrace, Jesmond, Newcastle-upon-Tyne, England

Dr. J. C. de Man, Nestlé Products Technical Assistance Co. Ltd. (nestec), Case Postale 88, CH-1814 La Tour-de-Peilz, Switzerland

Mr. R. Paul Elliott, Consultant in Food Microbiology, 1095 Lariat Lane, Pebble Beach, CA 93953, USA

Mr. R. M. Friesen, Griffith Laboratories Limited, 757 Pharmacy Ave., Scarborough, Ont., Canada M1L 3J8

Dr. J. M. Goepfert, Research Centre, Canada Packers Ltd., 2211 St. Clair Ave. W., Toronto, Ontario, Canada, M69 1K4

Mr. M. Greenall, Overseas Egg Company, London, England

Dr. R. A. Greenberg, Research and Development Laboratory, 1919 Swift Dr., Oak Brook, IL 60521.

Dr. A. Hurst, Bureau of Microbial Hazards, Food Directorate, Health Protection Branch, Health and Welfare Canada, Tunney's Pasture, Ottawa, Ont., Canada K1A 0L2

Professor M. Ingram (Deceased), Formerly Director, Meat Research Institute, Langford, Bristol BS18 7DY England

Dr. M. Jemmali, INRA Laboratoire de technologie alimentaire, Service des mycotoxines, 16 Rue Nicolas Fortin, 75013 Paris, France

Dr. J. J. Jezeski, Research and Development H. B. Fuller Co., Monarch Chemicals Division, 3900 Jackson St. Northeast, Minneapolis, MN 55421 USA

Professor P. G. Keeney, Department of Food Science, College of Agriculture, Penn. State University, University Park, PA 16801 USA

Mr. A. H. Klopp, Griffith Laboratories Limited, 757 Pharmacy Avenue, Scarborough, Ont., Canada M1L 3J8

Dr. C. Knowles, Egg Marketing Board, London, England

Dr. O. E. Kolari, Armour Food Company Research Center, 15101 N. Scottsdale Road, Scottsdale, AZ 85260 USA

Mr. W. J. Kooiman, Unilever Research, Duiven, Helhoek 30, Groessen, Postbus 2, Zevenaar, The Netherlands

Professor J. Liston, Director, Institute for Food Science and Technology, College of Fisheries, University of Washington, Seattle, WA 98105 USA

Ms Helen Lynch, Corn Products Company, Federal Street, Yonkers, NY 10702 USA

Dr. J. H. McCoy, Public Health Laboratory, Hull, U.K.

Miss L. McMaster, Department of Microbiology and Plant Pathology, University of Natal, Pietermaritzburg, South Africa

Dr. A. J. Mercuri, US Dept. of Agriculture, Science and Education Administration, Richard B. Russell Agricultural Research Center, P.O. Box 5677, Athens, GA 30604, USA

Dr. E. H. Meursing, Cacaofabriek de Zaan b.v. Koog aan de Zaan, Holland

Dr. M. J. M. Michels, Unilever Research, P.O. Box 7, Zevenaar, The Netherlands

Dr. R. W. A. W. Mulder, Spelderholt Institute for Poultry Research, 7361 DA Beekbergen, The Netherlands

Dr. G. Tuynenburg Muys, Unilever Research Vlaardingen/Duiven Olivier vanNoortlaan 120 Vlaardingen, postbus 114 Nederlandse Unilever Bedrijven B.V. stat. zetel Rotterdam, h.reg. Rotterdam nr 53802, The Netherlands

Dr. A. Neitzert, Kalle, Niederlassung der Hoechst AG, Rheingaustrasse 190, 6200 Wiesbaden-Biebrich, West Germany

Dr. M. D. Northolt, National Institute of Public Health, P.O. Box 1, 3720 BA Bilthoven, The Netherlands

Dr. I. S. H. W. Notermans, National Institute of Public Health, P.O. Box 1, 3720 BA Bilthoven, The Netherlands

Dr. S. A. Palumbo, Agricultural Research Service, Eastern Regional Research Center, 600 East Mermaid Lane, Philadelphia, PA 19118 USA

Dr. J. T. Patterson, Ministry of Agriculture, Agricultural and Food Bacteriology Research Division, Newforge Lane, Belfast, BT9 5PX Northern Ireland

Dr. A. B. Ravno, Huletts Sugar Limited, Private Bag X04, Mount Edgecombe 4300, Natal, South Africa

Mr. R. A. Seward, Department of Food Science and Technology, University of Wisconsin, Madison, WI 53706 USA

Mr. B. H. Siebers, The Nestle Company, Inc., Fulton, NY 13069, USA

Dr. R. D. Skole, Research and Development Department, Amstar Corporation, 49 South Second Street, Brooklyn, NY 11211, USA

Dr. Richard B. Smittle, Silliker Laboratories of New Jersey, Inc. 2353 Beryllium Road Scotch Plains, NJ 07076, USA

Dr. R. H. Tilbury, Group Research and Development, Tate and Lyle, Ltd., Philip Lyle Memorial Research Laboratory, University of Reading, Whiteknights, Reading, Berks., England

Dr. R. B. Tompkin, Swift Research and Development Laboratory, 1919 Swift Drive, Oak Brook, IL 60521 USA

Dr. R. O. Wagenaar, Technical Center, General Mills, Inc. 9000 Plymouth Ave. N. Minneapolis MN 55427, USA

Index

D

E

I

J

K

L

N